Cybersecurity

Audun Jøsang

Cybersecurity

Technology and Governance

Audun Jøsang
Oslo, Norway

ISBN 978-3-031-68482-1   ISBN 978-3-031-68483-8   (eBook)
https://doi.org/10.1007/978-3-031-68483-8

© The Editor(s) (if applicable) and The Author(s), under exclusive license to Springer Nature Switzerland AG 2025

Commercial ML (Machine Learning) based on the material in the book is not allowed without explicit permission from the author and the publisher.

This work is subject to copyright. All rights are solely and exclusively licensed by the Publisher, whether the whole or part of the material is concerned, specifically the rights of reprinting, reuse of illustrations, recitation, broadcasting, reproduction on microfilms or in any other physical way, and transmission or information storage and retrieval, electronic adaptation, computer software, or by similar or dissimilar methodology now known or hereafter developed.
The use of general descriptive names, registered names, trademarks, service marks, etc. in this publication does not imply, even in the absence of a specific statement, that such names are exempt from the relevant protective laws and regulations and therefore free for general use.
The publisher, the authors and the editors are safe to assume that the advice and information in this book are believed to be true and accurate at the date of publication. Neither the publisher nor the authors or the editors give a warranty, expressed or implied, with respect to the material contained herein or for any errors or omissions that may have been made. The publisher remains neutral with regard to jurisdictional claims in published maps and institutional affiliations.

This Springer imprint is published by the registered company Springer Nature Switzerland AG
The registered company address is: Gewerbestrasse 11, 6330 Cham, Switzerland

If disposing of this product, please recycle the paper.

Preface

This book gives a complete introduction to cybersecurity, by covering both technological and governance aspects. The book's primary target audience is bachelor's and master's students in study programs in the areas of computer science or business management, who are taking modules in cybersecurity and/or security governance and risk management. Because digitization in all industry sectors makes cybersecurity relevant in many other subject areas, the book is similarly relevant for students in other fields of study. The book is also a valuable reference for employees in private business or in the public sector who want to update their knowledge about cybersecurity both from a technological and governance perspective.

Cybersecurity is a very broad field of study. The introduction that this book provides is broader than it is deep. Each chapter could easily have been expanded to become a separate book in its own right. The book is precisely intended to give the reader a broad introduction to the entire field of cybersecurity.

Cybersecurity: Technology and Governance is easy to read with many color figures. There are also exercises and study cases at the end of each chapter, with additional material on the book's website.

I would like to thank colleagues in the Department of Informatics and the Digital Security Group at the University of Oslo for the stimulating environment and the fruitful discussions we have. This has contributed to the maturation and articulation of the cybersecurity concepts that this book describes. I would also like to thank collaborators in private companies, government agencies, universities, and research institutes, who have directly or indirectly contributed to this book by bringing knowledge and experience to teaching, guidance, and research in cybersecurity at the University of Oslo. I would also like to thank the publisher Springer Nature for its professional guidance and thorough proofreading checks. Finally, I would like to thank my family for always supporting me in all my endeavors.

You will find presentations and a series of exercises for each chapter on the webpage: https://www.mn.uio.no/ifi/english/research/groups/sec/instructional.

I hope you enjoy reading the book.

Oslo, Norway
July 2024

Audun Jøsang

Contents

1 Basic Concepts of Cybersecurity 1
   1.1 Cybersecurity Terminology 1
   1.2 What Is Security? 2
   1.3 What Is Cybersecurity? 3
   1.4 Threats, Vulnerabilities, Incidents, and Impacts 5
   1.5 Information Security Controls 7
   1.6 People – Product – Partner – Process 10
   1.7 Sources of Requirements for Cybersecurity 12
   1.8 Cyber Risk 13
   1.9 CIA Security Goals 14
      1.9.1 Confidentiality 14
      1.9.2 Data Integrity 15
      1.9.3 System Integrity 16
      1.9.4 Availability 17
   1.10 Other Security Goals 17
      1.10.1 Authentication 17
      1.10.2 Accountability and Non-repudiation 20
      1.10.3 Reliability 21
   1.11 Access Authorization and Access Control 21
   1.12 Tasks 23

2 Attack Vectors and Malware 25
   2.1 Attack Vectors 25
      2.1.1 Phishing 27
      2.1.2 ID Spoofing with Cracked or Stolen Credentials 27
      2.1.3 Fake or Malicious Websites 28
      2.1.4 Physical Attacks on IT Infrastructures 29
      2.1.5 Direct Attacks Against Vulnerable Systems and Applications 29
      2.1.6 Supply Chain Attacks 30
      2.1.7 Malicious Peripherals 31

		2.1.8 Hacking Unpatched Vulnerable IoT Devices.	32
		2.1.9 Deepfake Attacks	33
		2.1.10 Insider Attacks	34
		2.1.11 DDoS Attacks.	34
		2.1.12 Tracking	35
	2.2	Malware	35
		2.2.1 Computer Virus	36
		2.2.2 Ransomware.	37
		2.2.3 Spyware	37
		2.2.4 Bot Malware.	38
		2.2.5 Exploit	39
		2.2.6 Macro-viruses	39
		2.2.7 Trojan.	39
		2.2.8 Computer Worm.	40
		2.2.9 Rootkit	40
		2.2.10 Back Door	40
		2.2.11 Malicious JavaScript	41
		2.2.12 Logic Bomb.	41
	2.3	Tasks.	41
3	**System Security**		43
	3.1	System Architecture.	43
	3.2	The Importance of System Security.	45
	3.3	Vulnerabilities Management for Systems and Software	46
	3.4	Privilege Levels for Processes in the Microprocessor	48
	3.5	Buffers Overflow Vulnerabilities, Exploits and Countermeasures.	50
	3.6	Virtualization	53
		3.6.1 Virtual Machines	54
		3.6.2 Virtualization Architectures.	55
		3.6.3 Security Aspects of Virtualization	56
	3.7	Trusted Computing	57
	3.8	Secure Boot	57
	3.9	Intel Management Engine	59
	3.10	Side Channels and Covert Channels	60
	3.11	Tasks.	61
4	**Cryptography**		63
	4.1	What Is Cryptography?	63
	4.2	Cryptographic Features	64
	4.3	Symmetric Algorithms.	65
	4.4	Block Cipher Modes of Operation.	66
		4.4.1 ECB: Electronic Codebook	67
		4.4.2 CTR: Counter Mode	68
		4.4.3 CBC: Cipher Block Chaining	69
	4.5	Hash Functions and MAC	69

4.6	Diffie-Hellman Key Exchange		71
	4.6.1	Traditional Diffie-Hellman	71
	4.6.2	ECDH: Elliptical Curve Diffie-Hellman	73
4.7	Asymmetric Algorithms		74
	4.7.1	The RSA Algorithm	74
	4.7.2	Hybrid Encryption	75
	4.7.3	Digital Signature	77
	4.7.4	Authenticated Encryption with Forward Secrecy	80
4.8	The History of Cryptography		81
	4.8.1	Classical Ciphers	81
	4.8.2	Ciphers in the Middle Ages	82
4.9	Ciphers Leading Up to World War I		83
	4.9.1	Kerckhoffs's Principle	83
	4.9.2	One-Time Pad	84
4.10	Ciphers Around World War II		85
	4.10.1	World War II Rotor Ciphers	85
	4.10.2	Shannon's Theory of Cryptography	86
4.11	Ciphers Up to the Year 2000		88
4.12	Ciphers After the Year 2000		89
	4.12.1	Block Cipher: Replace the Old, Tired Horse with a Young, Strong One	89
	4.12.2	Hash Functions: New Sponges Needed	90
	4.12.3	Postquantum Crypto: A Lifeboat in Case the Ship Sinks	90
4.13	Cryptography and Energy Consumption		93
	4.13.1	TLS Everywhere	93
	4.13.2	Cryptocurrency	94
4.14	Tasks		96
5	**Key Management and PKI**		**99**
5.1	Key Management		99
	5.1.1	Key Types	100
	5.1.2	Crypto Periods	101
	5.1.3	Key Sizes	102
	5.1.4	Key Lifecycle	104
5.2	PKI		105
	5.2.1	The Challenge of Key Distribution	105
	5.2.2	X.509 Certificates and PKI Components	107
	5.2.3	Generating X.509 Certificates	109
	5.2.4	Validation of X.509 Certificates	110
	5.2.5	Trust Models for PKI	111
5.3	Challenges and Solutions for PKI		112
	5.3.1	PKI Trust Scope Limited to Authenticity	112
	5.3.2	Certificate Revocation	113
	5.3.3	CA Authorization and Certificate Transparency	114
5.4	Use of Blockchains		114
5.5	Tasks		115

6 Network Security 117
6.1 Computer Networks and the Internet 117
6.1.1 The Domain Name System 118
6.1.2 The Internet Stack 121
6.2 Communication Security 126
6.2.1 TLS: Transport Layer Security 127
6.2.2 IPSec: Internet Protocol Security 130
6.2.3 VPN: Virtual Private Networks 132
6.2.4 Steganography 133
6.3 Computer Network Security 134
6.4 Firewalls 135
6.4.1 Stateless Packet Filter 136
6.4.2 Stateful Packet Filter 136
6.4.3 Application Firewall 137
6.5 Intrusion Detection 137
6.5.1 Signature-Based Detection 138
6.5.2 Anomaly-Based Detection 138
6.6 Network Architecture for Security 138
6.7 TLS Inspection 139
6.8 Tasks 142

7 Wireless Security 143
7.1 Radio Communications 143
7.1.1 Radio Signals 144
7.1.2 The Radio Spectrum 145
7.2 Security in Wi-Fi 146
7.2.1 Basic Wi-Fi Concepts 147
7.2.2 Development of Security in Wi-Fi 147
7.2.3 Secure Network Access with WPA 148
7.2.4 SAE: Simultaneous Authentication of Equals 150
7.3 Bluetooth Security 150
7.3.1 Bluetooth Technologies 151
7.3.2 Pairing and Connection 153
7.3.3 SSP and Authentication Between Devices 154
7.3.4 Bluetooth Security Recommendations 154
7.4 Mobile Network Security 155
7.4.1 Mobile Network Technologies 155
7.4.2 2G Security Architecture 157
7.4.3 IMSI Catchers 158
7.4.4 Security in 4G 160
7.4.5 Security in 5G 160
7.4.6 SIM, eSIM and iSIM 162
7.5 Tasks 163

8 User Authentication ... 165
- 8.1 What Is User Authentication? ... 165
- 8.2 Authentication Methods ... 166
- 8.3 Knowledge-Based Authenticators: Passwords and Learned Patterns ... 167
  - 8.3.1 Password Protection Against Cracking ... 167
  - 8.3.2 Strong Password Advice ... 170
  - 8.3.3 Password Managers ... 171
- 8.4 Ownership-Based Authenticators: Devices ... 173
  - 8.4.1 OTP Tokens ... 174
  - 8.4.2 Passkeys and Online Tokens ... 175
  - 8.4.3 Access and ID Cards ... 177
  - 8.4.4 Secondary Channels ... 178
- 8.5 Biometrics ... 179
  - 8.5.1 Requirements for Biometric Systems ... 179
  - 8.5.2 Mode of Operation and Components of Biometric Systems ... 180
  - 8.5.3 Quality Aspects of Biometric Systems ... 182
  - 8.5.4 Safety Aspects of Biometric Systems ... 184
- 8.6 Multi-factor Authentication ... 184
- 8.7 e-Authentication Frameworks ... 185
  - 8.7.1 USA: NIST SP 800-63-4 Digital Identity Guidelines ... 187
  - 8.7.2 EU: eIDAS ... 187
  - 8.7.3 India: e-Pramaan ... 188
  - 8.7.4 ISO/IEC 29115 Entity Authentication Assurance Framework ... 189
- 8.8 Tasks ... 189

9 IAM—Identity and Access Management ... 191
- 9.1 Definition of IAM ... 191
- 9.2 Identity Management ... 193
  - 9.2.1 Identity ... 194
  - 9.2.2 The Silo Model for Identity Management ... 195
  - 9.2.3 Federated Identity Management ... 196
  - 9.2.4 ID Federation Protocols ... 198
  - 9.2.5 OpenID Connect ... 199
  - 9.2.6 Indian eID and e-Pramaan Federation ... 200
  - 9.2.7 European eID and Identity Federation ... 201
  - 9.2.8 eID in the USA ... 202
  - 9.2.9 The Identity Federations Facebook, Twitter, Google etc. ... 202
  - 9.2.10 Categorization of Identity Federation ... 203
- 9.3 Access Control ... 206
  - 9.3.1 DAC: Name-Based Access Control ... 206
  - 9.3.2 MAC: Label Access Control ... 207
  - 9.3.3 RBAC: Role-Based Access Control ... 209

	9.4	OAuth and Distributed Access Management	210
		9.4.1 ABAC: Attribute-Based Access Control	211
		9.4.2 OAuth for Online Services	212
		9.4.3 OAuth for Access and Interaction Between Healthcare Institutions	213
	9.5	Tasks	214
10	**Information Privacy**		**215**
	10.1	What Is the Difference Between Privacy and Information Privacy?	216
	10.2	Privacy in the Digital Age	216
	10.3	Privacy-Invasive Technologies	218
		10.3.1 Tracking with Cookies	218
		10.3.2 Tracking with Email Addresses and Phone Numbers	220
		10.3.3 Cross-Platform Fingerprint Tracking	221
		10.3.4 Mobile App Tracking	221
	10.4	Anti-tracking	223
	10.5	GDPR: The EU General Data Protection Regulation	224
	10.6	Roles in GDPR	225
		10.6.1 The Data Subject and the Associated Personal Data	226
		10.6.2 The Data Controller	227
		10.6.3 Data Processor	227
		10.6.4 Data Protection Officer	227
		10.6.5 The Data Protection Authority and Penalties for Infringement of the GDPR	228
	10.7	Particularly Relevant Articles in GDPR	228
		10.7.1 Article 5: Principles Relating to Processing of Personal Data	229
		10.7.2 Article 6: Lawfulness of processing	230
		10.7.3 Article 25: Data Protection by Design and by Default	232
		10.7.4 Art. 32: Security of Processing	233
		10.7.5 Articles 45 and 46: Transfer of Personal Data to Countries Outside the EU/EEA	233
	10.8	Article 35: Data Protection Impact Assessment—DPIA	235
		10.8.1 The Process Around DPIA	235
		10.8.2 When Is It Necessary to Perform a DPIA?	236
		10.8.3 Threat Actors as an Element in Risk Assessment	237
		10.8.4 Who Should Perform DPIA?	238
		10.8.5 The Steps of the DPIA	238
	10.9	Notification of Personal Data Breaches	241
	10.10	Tasks	242

11 Secure by Design ... 243
- 11.1 Secure by Design ... 243
- 11.2 Privacy by Design ... 244
- 11.3 The Seven Phases of Secure by Design ... 244
  - 11.3.1 Training ... 245
  - 11.3.2 Cybersecurity and Privacy Requirements ... 246
  - 11.3.3 Secure Design ... 246
  - 11.3.4 Implementation and Secure Coding ... 247
  - 11.3.5 Software Security Testing ... 247
  - 11.3.6 Release ... 249
  - 11.3.7 Operation and Incident Management ... 250
- 11.4 Secure Software Development ... 251
  - 11.4.1 The Waterfall Method ... 251
  - 11.4.2 Agile Software Development ... 252
  - 11.4.3 Secure, Agile Software Development ... 253
  - 11.4.4 Security Champion ... 254
- 11.5 Identification of Threats During Software Development ... 255
  - 11.5.1 Threat Modelling ... 255
  - 11.5.2 User Stories and Use Cases ... 256
  - 11.5.3 Attacker Stories and Threat Scenarios ... 256
  - 11.5.4 STRIDE Threat Modelling for Software Development ... 257
- 11.6 Application Security ... 258
  - 11.6.1 Web Applications' Exposure to Threats ... 258
  - 11.6.2 OWASP: The Open Web Application Security Project ... 259
  - 11.6.3 OWASP Top 10 ... 259
  - 11.6.4 OWASP ASVS ... 261
- 11.7 Examples of Attacks Against Applications ... 261
  - 11.7.1 SQL Injection ... 261
  - 11.7.2 XSS: Cross-Site Scripting ... 263
  - 11.7.3 Security in the cloud ... 264
  - 11.7.4 Cloud services ... 264
  - 11.7.5 Cloud Security ... 266
  - 11.7.6 DevOps ... 268
  - 11.7.7 Cloud Security Alliance ... 269
- 11.8 Tasks ... 269

12 Physical Information Security ... 271
- 12.1 Physical Security Goals and Threats ... 271
- 12.2 Safety of Staff and Their Environment ... 273
  - 12.2.1 HSE: Health, Safety and Environment ... 273
  - 12.2.2 Weighing Priorities for Safety Against Security ... 273
- 12.3 Controlled Authorized Access ... 274
  - 12.3.1 Secure Entry Points ... 274
  - 12.3.2 Physical Security Monitoring ... 274
- 12.4 Shielding of Sites and Equipment ... 275
  - 12.4.1 Physical Perimeter Defences ... 275

		12.4.2	Working in Secure Areas	276
		12.4.3	Security of Assets Off-Site	277
		12.4.4	Protecting Against Physical and Environmental Threats	277
		12.4.5	Emission Security	277
	12.5	Continuity of Technical Support Services		278
		12.5.1	Uninterruptible Power Supply	279
		12.5.2	Water, Gas, Air Temperature and Humidity	280
	12.6	Tasks		281
13	**Security Culture**			283
	13.1	Definition of Security Culture		283
	13.2	Building the Security Culture		285
	13.3	The Insider Threat		286
		13.3.1	Personal Integrity	287
		13.3.2	Management's Responsibility for Handling the Insider Threat	289
	13.4	Social Engineering		290
	13.5	Techno-social Engineering		290
		13.5.1	Phishing Attacks	290
		13.5.2	Detection of Phishing Attacks	291
		13.5.3	When Realising that You Have Been Phished	292
	13.6	Physical Social Engineering		292
		13.6.1	Physical Social Engineering Attack Strategies	293
		13.6.2	Defense Against Physical Social Engineering	295
	13.7	Security Usability and Security Learning		297
	13.8	Tasks		298
14	**Cybersecurity Readiness, Security Testing and Audit**			301
	14.1	Background for Cyber Contingency Planning		301
	14.2	Contingency Planning Principles		303
	14.3	Technical Concepts of Cyber Contingency Planning		304
	14.4	Cyber Contingency Planning		306
	14.5	Cyber Incident Management		307
		14.5.1	Preparation	309
		14.5.2	Triage	310
		14.5.3	Response	312
		14.5.4	Post-Incident	313
	14.6	Digital Forensics		314
	14.7	Security Testing of Systems, Networks, and Enterprises		316
		14.7.1	Pentesting	316
		14.7.2	Red-Teaming and Blue-Teaming	317
		14.7.3	TIBER	318
	14.8	Security Audit		319
	14.9	Tasks		320

15 AI and Cybersecurity ... 321
- 15.1 Introduction to AI ... 321
  - 15.1.1 Artificial Neural Networks (ANN) ... 322
  - 15.1.2 Machine Learning Paradigms ... 323
  - 15.1.3 Training Methods and Model Tasks ... 324
- 15.2 Offensive Use of AI ... 326
  - 15.2.1 Deepfakes ... 326
  - 15.2.2 Malware Engineering ... 327
  - 15.2.3 Attack Automation ... 328
- 15.3 Defensive Use of AI ... 329
  - 15.3.1 Detection of Malicious Activities ... 329
  - 15.3.2 Malware Detection and Analysis ... 329
  - 15.3.3 AI Support for Cyber Threat Intelligence ... 330
  - 15.3.4 Automated Response and Mitigation ... 330
- 15.4 Vulnerabilities and Exploitation of AI Systems ... 331
  - 15.4.1 Poisoning Attacks ... 331
  - 15.4.2 Polluted Learning ... 332
  - 15.4.3 AI Supply Chain Risks ... 332
  - 15.4.4 Evasion Attacks ... 333
  - 15.4.5 Privacy Attacks ... 333
  - 15.4.6 Prompt Injection Attacks ... 334
- 15.5 Tasks ... 334

16 Cyber Operations ... 337
- 16.1 Advanced Cyber Threats ... 337
  - 16.1.1 APT: Advanced Persistent Threat ... 338
  - 16.1.2 Cyber Kill Chain: A Model for APT Attacks ... 339
- 16.2 CTI: Cyber Threat Intelligence ... 340
  - 16.2.1 Categories and Levels of Digital Threat Intelligence ... 341
  - 16.2.2 MITRE ATT&CK ... 343
  - 16.2.3 CTI Cycle ... 344
  - 16.2.4 Sharing CTI ... 346
  - 16.2.5 Representation and Use of CTI ... 348
- 16.3 Cyber Warfare ... 348
  - 16.3.1 Comparison of Weapons ... 350
  - 16.3.2 Cyber Deterrence and Cyber Privateering ... 351
  - 16.3.3 The Role of Big Tech in Cyber Warfare ... 353
- 16.4 Tasks ... 354

17 Cyber Organizational Structures and Regulation ... 355
- 17.1 The Importance of Cybersecurity Regulations ... 355
  - 17.1.1 Responsibility and Accountability ... 356
  - 17.1.2 Hierarchy of Regulation ... 357
- 17.2 Cyber-Organizational Structures ... 358

17.3 USA 358
    17.3.1 CFAA (Computer Fraud and Abuse Act) 360
    17.3.2 FISMA (Federal Information Security Management Act) 361
    17.3.3 CISA (Cybersecurity Information Sharing Act) 361
    17.3.4 HIPAA (Health Insurance Portability and Accountability Act) 362
17.4 Europe 362
    17.4.1 EU Regulation 363
    17.4.2 NIS2 Directive 364
    17.4.3 Cybersecurity Act 365
    17.4.4 Cyber Resilience Act (in Progress) 365
    17.4.5 Cyber Solidarity Act (in Progress) 365
17.5 Russia 366
    17.5.1 Federal Law on Personal Data 367
    17.5.2 Criminal Code of the Russian Federation 368
    17.5.3 Sovereign Internet Law 369
    17.5.4 FSB Law 369
17.6 China 369
    17.6.1 Cybersecurity Law (CSL) 371
    17.6.2 Data Security Law (DSL) 372
    17.6.3 Personal Information Protection Law (PIPL) 372
    17.6.4 National Intelligence Law 373
17.7 General Remarks on Cyber-Organizational Structures 374
17.8 Tasks 374

18 Governance and Information Security Management 377
18.1 Information Security Management Levels 377
    18.1.1 Information Security Governance 378
    18.1.2 Questions that the Board and Executive Should Ask Themselves 381
    18.1.3 Information Security Management 381
    18.1.4 Administration and Operation of Cybersecurity 383
18.2 NIST Cybersecurity Framework 383
    18.2.1 NIST CSF Core 385
    18.2.2 CSF Profiles 387
    18.2.3 CSF Tiers 387
18.3 The Cyber Defense Matrix 388
18.4 CIS Critical Security Controls 390
18.5 ISO/IEC 27000 Series of Security Standards 392
    18.5.1 The History of ISO/IEC 27001 and 27002 392
    18.5.2 ISO/IEC 27001 ISMS: Requirements 394
    18.5.3 ISMS Process Cycle 397
    18.5.4 ISO/IEC 27002 Information Security Controls 398
    18.5.5 27000 Family of Standards 399
18.6 Maturity in Information Security Management 400
18.7 Tasks 402

19	**Cyber Risk Management**		405
	19.1	Interpretation of Risk and Risk Management	405
		19.1.1 Definition of Information Security Risk	406
		19.1.2 Information Security Risk Models	407
		19.1.3 Definition of Information Security Risk Management	409
	19.2	Risk Management Process	410
		19.2.1 Context Establishment	411
		19.2.2 Risk Assessment	412
		19.2.3 Risk Treatment	412
		19.2.4 Risk Treatment Plan and Accepted Risk	415
	19.3	Risk Assessment Process	416
	19.4	Risk Identification	417
		19.4.1 Identification of Assets	418
		19.4.2 Threat Modeling: Identifying Threats	418
		19.4.3 Identification of Impacts	420
		19.4.4 Risk Description	420
	19.5	Risk Analysis	421
		19.5.1 Qualitative Risk Analysis	421
		19.5.2 Relative Risk Analysis	423
		19.5.3 Quantitative Risk Analysis	425
	19.6	Risk Evaluation and Reporting	427
	19.7	Tasks	428
References			429
Index			433

Chapter 1
Basic Concepts of Cybersecurity

Nothing is so practical as a good theory.

Kurt Lewin, German-American psychologist

In a nutshell, cybersecurity is the protection of information assets from harm. Information assets can be devices, applications, networks, data and even users. With the migration of most business processes to digital online platforms, all these information assets become exposed to potential cyber threats from anywhere. The numerous high-profile attacks being reported in the press clearly demonstrate how cyberthreats cause serious material impacts that organization struggle to prevent and deal with. For this reason, cybersecurity has become a critical concern for the global economy, for politics and national security, for organizations as well as for individuals. While cybersecurity has traditionally been a technological discipline only, mature cyber governance is now seen as a critical condition for having effective cybersecurity programs.

In this chapter, you will learn what cybersecurity is about, and also the meaning of other key concepts related to cybersecurity. There is some confusion about the meaning of some of these concepts, where this book takes a stand. When you have clear understanding of each concept and you are familiar with possible alternative interpretations, you will not get confused if somebody uses a term in different meaning. In addition, you will have the knowledge and confidence to argue for the interpretation that you find correct.

1.1 Cybersecurity Terminology

We have many names for the things we love, which is clearly the case for cybersecurity. Security terms with approximately the same meaning are for example information security, computer security, IT security, cybersecurity, and digital security. New words appear continuously, and their popularity changes over time. "Information security" was coined in the 1970s and has enjoyed steady popularity. Moreover, the

© The Author(s), under exclusive license to Springer Nature
Switzerland AG 2025
A. Jøsang, *Cybersecurity*, https://doi.org/10.1007/978-3-031-68483-8_1

term information security is enshrined in many international standards and frameworks. "Computer security" has been popular since the 1980s, mainly carrying the meaning of principles for protecting the computer platforms. "Cybersecurity" emerged and became popular from the 2010s to denote the security of computers and business processes that are connected to the Internet. "Digital security" is the most recent addition from around 2020, and reflects the security of everything digital, i.e. that has IT components. This book mainly uses the term cybersecurity, but we also use the term information security with the same meaning as cybersecurity.

For purists, information security is also about the protection of information on paper or conveyed through analogue sound waves, while cybersecurity can be interpreted as only dealing with the protection of digital assets. For example, if a group of colleagues discuss sensitive information while having lunch in a café so that other people nearby can hear what they are saying, then it is not strictly a breach of cybersecurity, but a breach of information security. However, it is of little use to focus on this difference, because in practice people use the terms in the same meaning.

1.2 What Is Security?

Generally speaking, security is the protection of assets from harm, where there are many different categories of assets that can each be harmed in different ways.

> Security is the protection of assets from harm.

Key asset categories can be, for example, property, infrastructure, democracy, societal order, life and health, environment, information, and personal data. For each of these asset categories, there are well-established disciplines for security:

- **Physical security** is the protection of physical assets and infrastructure from theft, penetration, disruption and destruction.
- **Safety** is the protection of life and health.
- **Environmental safety** is the protection of the environment by preventing pollution and invasion of alien species into nature.
- **Civic security** is the preservation of law and order, also called legal security.
- **Societal security** is the protection of critical infrastructure and basic functions such as energy supply, communication and transport.
- **National security** is the maintenance of our national sovereignty, territorial integrity, and the existing form of government in a country.
- **Cybersecurity** (and information security) is the protection of information assets from harm.
- **Data privacy** is the protection of personal data through requirements for collecting, storing, deleting, sharing and processing information about persons (see Chap. 10 about privacy).

1.3 What Is Cybersecurity?

The asset/security categories above often have interdependencies. For example, cybersecurity is of course essential for societal security because critical infrastructure is controlled by IT systems. Figure 1.1 illustrates that there are different threat sources that each can breach different security goals.

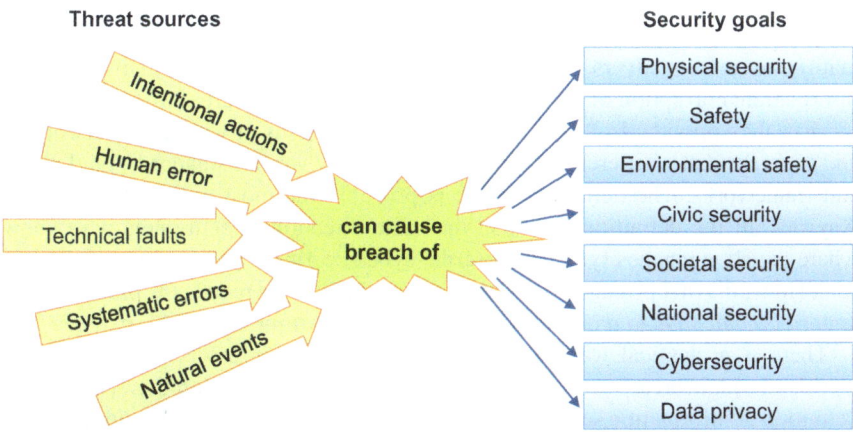

Fig. 1.1 Threat sources for security breaches

In the field of cybersecurity, the main focus is typically on intentional adversarial actions as causes of incidents. At the same time, it is important to bear in mind that information assets can also be harmed by human error, technical faults, systematic errors and natural events, i.e. all hazards. Systematic errors mean that security breaches occur due to inadequate specifications of systems and processes. Put another way: Security breaches occur precisely because systems are correctly built according to incorrect specifications. Technical errors are relatively easy to correct by making sure that the system complies with specifications. Systematic errors can be difficult to correct because the specifications must first be corrected so that systems can be changed, or new systems must be built. If the fault already exists in billions of systems, these must first be phased out, which can take many years. Weaknesses in mobile networks are an example of this, as described in Sect. 7.4.

1.3 What Is Cybersecurity?

"Cybersecurity (and information security) is the protection of information assets from harm. An informational asset can be the information itself, but also resources related to the processing of information.

> Information assets are information and resources used for the processing of information.

Examples of information assets are data, devices, applications, networks, and even their users, as explained in Sect. 18.3.

> Cybersecurity is the protection of information assets from harm.

In a more verbose form, we can define cybersecurity as the protection of information assets from adversarial attacks that may result in unauthorized disclosure of information, corruption of data, software and hardware, as well as disruption of the services they provide. In fact, there is no exact consensus of what cybersecurity is, and there are many different definitions proposed by various standards, guidelines, and frameworks. The definition above is probably the most concise of them all. Anyway, all the definitions point in the same direction. Practitioners who know where the shoe pinches typically describe cybersecurity in practical terms such as Rick Howard's first principle of cybersecurity which is *"to reduce the probability of material impact due to a cyber incident over the next three years."* [1, p. 39].

There is a classical way to describe how information assets can be harmed, namely through breaches of confidentiality, integrity, or availability. The classic definition of information security is in fact the protection of *Confidentiality, Integrity and Availability* of information assets. This is often called the CIA triad, typically illustrated as in Fig. 1.2. Note that different information assets have different security needs, where some assets need protection of confidentiality, while other assets need protection for ensuring availability and integrity.

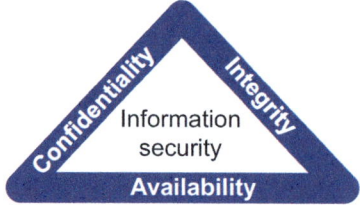

Fig. 1.2 The CIA triad of information security

This and similar articulations of information security can be found in several standards and frameworks, including the international standard ISO/IEC 27000 [2]. This standard defines information security as follows:

> *Information security is the preservation of confidentiality, integrity and availability of information.*
> *In addition, other properties, such as authenticity, accountability, non-repudiation, and reliability can also be involved.* (ISO/IEC 27000:2018)

The first sentence of the above definition is intuitively easy to understand as the basis for the CIA triad, but it only mentions "information", not "information assets" which would have been better. The second sentence brings in relatively advanced concepts that can make it difficult to understand exactly what is meant by information security. To clear up the confusion, the second sentence in the definition of ISO/IEC 27000 can be seen as an attempt to say that information security is not only to protect the information (data) itself, but also resources related to the processing of information, which are *information assets*. The first sentence in the definition above is so enshrined in the 30-year history of the ISO/IEC security standards that it would be very difficult to adjust to cover more than just information. In today's world where we are surrounded by information technology it is obvious that the term "information" in the definition above must be interpreted to also include "information assets" which constitute the fabric for storing, communicating, and processing information, but this is not obvious for people who encounter the definition for the first time. A standard definition should be expressed in plain language without the need for subtle interpretations.

The terms in the second sentence of the definition from ISO/IEC 27000 are in fact *security controls* that support the general CIA security goals of confidentiality, integrity, and availability. It is instructive to distinguish between the abstract CIA security goals and more specific security controls that support CIA. Examples of how different security controls support CIA are mentioned in Sect. 1.9, which describes confidentiality, integrity, and availability more in detail. The concepts of authenticity, accountability, non-repudiation, and reliability from the second sentence of the definition from ISO/IEC 27000 are explained later in the present chapter.

1.4 Threats, Vulnerabilities, Incidents, and Impacts

It is important to have a clear understanding of security concepts to be able to talk precisely about cybersecurity. Central concepts are threats, vulnerabilities, assets, incidents, risks and controls. It is also essential to understand the connection between these concepts, which are described below.

- **Threat actor:** A threat actor is an active entity that can trigger or execute a threat scenario. Threat actors can be intelligent entities with adversarial intent, or forces of nature that are too strong or unpredictable for effective prevention.
- **Threat scenario:** A threat scenario is a sequence of steps that can be triggered or controlled by a threat actor and that can harm information assets. A threat scenario is realistic and can lead to an incident when vulnerabilities exist.
- **Vulnerability:** A vulnerability represents a weakness, flaw or defect that allows (a step in) a threat scenario to be executed, thus making it more likely that a threat actor will succeed in an attack that leads to an incident. A vulnerability can also be interpreted as the absence of security controls, i.e. that the vulnerability could otherwise be removed by implementing one or more security controls. The three main categories of vulnerabilities are:
    1. Technical vulnerabilities, which are weaknesses in systems, networks, software and hardware,

2. Process vulnerabilities, i.e. weaknesses in functions and activities, and weaknesses in their composition,
3. Human vulnerabilities, which are weaknesses in human consciousness, attitude, and behavior.

- The *vulnerability surface* are all the attack vectors a threat actor can leverage as part of an attack. Reducing the vulnerability surface will reduce the risk, which is of course desirable, but will usually have a cost and can affect usability, so that there is a trade-off. Sometimes a distinction is made between *weaknesses* and *vulnerabilities*, where a weakness represents an abstract type of vulnerability without reference to a real system, while a vulnerability is an instance of a weakness in a specific system. This distinction is made, for example, by the Common Vulnerability Scoring System (CVSS) described in the Sect. 3.3. Note that when vulnerabilities exist in the supply chains of an organization, these vulnerabilities cannot be directly controlled by the organization. The typical approach to achieve some control is to ensure that contracts require suppliers to undergo security audits and certify a level of maturity in security management, as described in Sect. 2.1.6.
- **Incident:** In general, a security incident is a breach of confidentiality, integrity or availability for information assets that results in a negative impact for the organization. During risk assessment, an attempt is made to estimate the likelihood of incidents occurring. A distinction can be made between incidents resulting from an intentional act as opposed to incidents resulting from other causes, such as human errors, technical faults, and natural disasters. Both types of incidents can result in cybersecurity breaches, but incidents that are intentionally caused by a threat actor are cyber incidents. The field of information security covers both types of incidents but focuses mostly on intentional incidents.
- **Impact:** An impact of a security incident is a form of loss that can have various aspects, such as lost profits/profits (for commercial companies), loss of service provision, loss of reputation, costs resulting from breach of legal compliance or costs of recovery after the incident. During risk assessment, an attempt is made to estimate the magnitude or severity of the potential impacts. Various forms of impact of a security incident also provide guidelines for how we describe the incident. For example, it can be called non-conformity, security threatening incident, security compromise or security breach.
- **Risk:** An information security risk is a relevant combination of a potential threat that can exploit an existing vulnerability and thereby harm information assets. Risk identification involves mapping such relevant combinations. To do a risk analysis is to combine the estimated likelihood of an incident and the estimated severity of the impacts to compute the level of the associated risk. Impact and risk are described in detail in Chap. 19

Figure 1.3 shows the dynamics between threat actor, threat scenario, vulnerability, incident and impact. It shows how the different vulnerabilities are exploited at different steps of the threat scenario, which in the end allows the threat actor to reach the attack target and cause negative impact for the organization. When an incident

1.5 Information Security Controls

occurs, it can have different impacts, which also depend on the organization's ability to respond and recover. Missing backup or lack of readiness in general are also vulnerabilities that typically make impacts worse than they otherwise would have been.

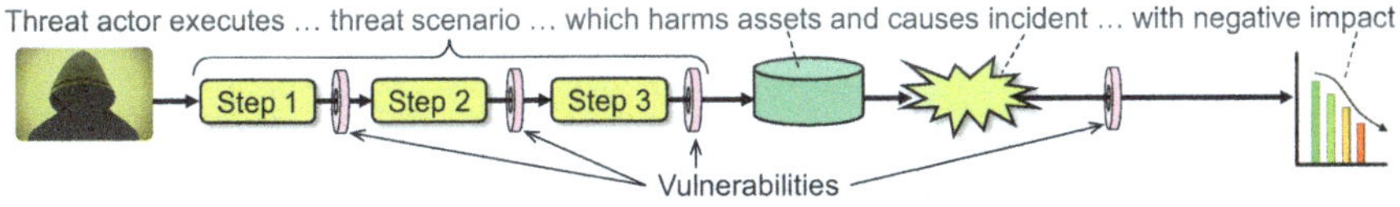

Fig. 1.3 The dynamics between threat, vulnerabilities, assets, incident, and impact

A security control is typically intended to prevent a threat scenario, or to reduce negative impacts. Implementing such a security control removes or reduces vulnerabilities that could be exploited as part of the threat scenario, or it strengthens cyber readiness to handle an incident. The purpose is to reduce the likelihood that the threat scenario can be completed and cause an incident, and to reduce the impacts in case the incident occurs anyway.

1.5 Information Security Controls

An information security control is a way of preventing or mitigating a threat scenario so that it cannot (easily) be executed to harm assets. There are also security controls for detecting and responding to incidents. Implementing security controls is thus a way of removing vulnerabilities, and for reducing the impact of incidents.

Figure 1.4 illustrates a modified version of the previous figure where security controls have replaced the vulnerabilities. In that way, the threat actor is unable to execute the threat scenario, and hence the assets stay protected.

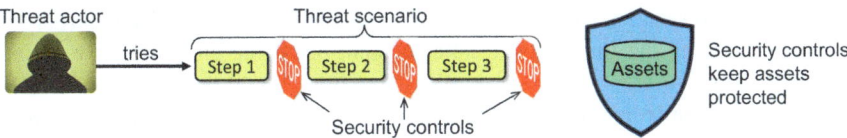

Fig. 1.4 Principle of stopping threats with security controls

The principle that security controls and mechanisms support the CIA security goals is illustrated in Fig. 1.5, which also shows an analogy to how the goal of civic security is supported by having police and security guards with appropriate equipment (controls).

To say that a security goal must be met says nothing about how it should be done in practice. To state a security goal is basically independent of implementation. It is up to those who work with information security in an organization to choose the most appropriate controls to achieve security goals for information assets. When the choice of controls is made, specific services and mechanisms that may be tied to individual products and suppliers are also selected.

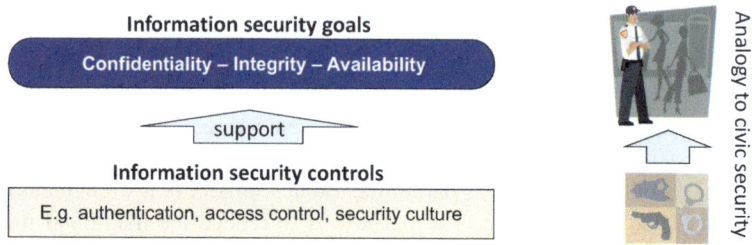

Fig. 1.5 Relationship between security controls and security objectives, with analogy to civic security

Security controls can be categorized in different ways. A logical categorisation is to separate between *preventive*, detective, and *corrective* controls, as well as *governance* for selecting and managing controls in a rational way.

- **Preventive controls** shall prevent cyber incidents so that they do not occur, or that they are less likely to occur. Examples include authentication and access control, which is intended to ensure that unauthorized persons do not gain access to information and resources. These types of controls are "likelihood-reducing controls" since they reduce the likelihood of security incidents.
- **Detective controls** should be able to reliably detect attacks and incidents. One example is IDS (Intrusion Detection System) which triggers alarms according to specific criteria. In some cases, such controls can detect attacks at such an early stage that no security breach has yet occurred, but usually detective controls are about reducing the time between the incident occurring and corrective action being taken.
- **Corrective controls** are used to recover from attacks and restore operation after an incident. An example is backups that can be retrieved and reinstalled if operational data is lost or corrupted. Another example is making contingency plans, and conducting cyber exercises to establish readiness for responding rapidly end effectively when cyber incidents occur. These types of controls are "impact-reducing controls" since they reduce the negative impacts of the security incidents that the preventive controls could not stop.
- **Governance** is a foundational layer necessary to organize the mentioned preventive, detective and corrective controls in a rational way. Such a governance structure is often called an ISMS (Information Security Management System). At the

1.5 Information Security Controls

highest management level, the organization should define its mission, vision, values, and risk tolerance regarding cybersecurity. The organization must ensure that cyber risk management is aligned with business objectives, and must define its approach to identifying, assessing, and mitigating cybersecurity risks. An ISMS is articulated in terms of policies and procedures that define e.g. strategy for risk management, requirements for security skills and training, requirements for incident response, and maturity objectives. The organization must also define roles and responsibilities for cybersecurity, as well as its strategy for collaborating with strategic cybersecurity partners.

To prevent, detect, correct, and govern are high-level security functions as illustrated in Fig. 1.6. This framework is quite intuitive because it reflects a logical approach for managing and reducing cybersecurity risk. This framework corresponds to that of NIST CSF (Cybersecurity Framework) [3] described in Sect. 18.2. Note that NIST CSF does not describe security controls under each function, but instead refers to other frameworks where specific security controls are described.

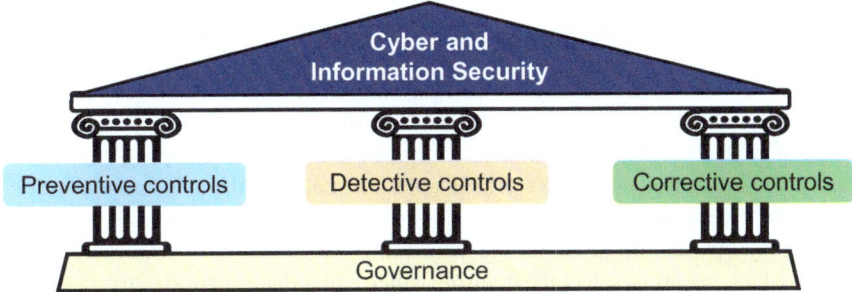

Fig. 1.6 Fundamental cyber and information security functions

Until around year 2000, the focus was almost exclusively on implementing preventive security controls. The thinking was that if we protect our systems and networks well enough, then security incidents will never happen. However, the security community slowly realised that this was a flawed approach, and that perfect security is unattainable in practice. When realising that organizations will experience cyber incidents no matter how strong defences they put in place, the focus was expanded to also cover detective and corrective security controls. Hence, the current thinking is that organizations must focus on all three functions, as well as cybersecurity governance, to demonstrate due care and achieve a reasonable level of security.

Another type of categorisation is to define domain categories such as organizational controls, people controls, physical controls, and technological controls, as shown in Fig. 1.7. This categorization is used in the standard *ISO/IEC 27002 Security Controls* [4]. People controls refer to personnel security. The accompanying standard *ISO/IEC 27001 Information Security Management System - Requirements* [5] describes governance requirements. This framework is described in Sect. 18.5.

Fig. 1.7 Information security control domains in ISO/IEC 27002, and governance requirements in ISO/IEC 27001

With reference to Fig. 1.7, most cybersecurity controls of ISO/IEC 27002 are described under the technological domain, where example controls are encryption (preventive), IDS (detective), and backup (corrective).

A third categorisation of security controls goes according to operational activities such as identity and access management, awareness and training, network security, software security, cloud security, secure configuration, incident management, etc. This type of operational categorization of security controls is used, for example, in CIS Critical Security Controls [6] and in NIST SP 800-53 Security and Privacy Controls for Information Systems and Organizations [7]. NIST CSF (Cyber Security Framework) refers to these and other guidelines for the description of specific security controls.

In the end, the different ways of categorizing security controls described above cover the same controls. The different standards and frameworks also provide mappings of controls so that practitioners easily can find the corresponding controls in each standard or framework. An organization must choose the standards or frameworks that seem most appropriate according to the country or industry sector in which they operate. In any case, the selection of security controls should be risk-based, which means to select the controls which can reduce cyber risk to an acceptable level in a cost-effective way. This type of assessment is described in the next two sections, and in more detail in Chap. 19.

1.6 People – Product – Partner – Process

ITIL (Information Technology Infrastructure Library) [8] describes practices for IT management according to "the four P's" which represent the dimensions of People, Product, Partner, and Process, as shown in Fig. 1.8. The aim of the ITIL framework is to manage IT to align with the needs of the organization.

Fig. 1.8 The four P's of ITIL

1.6 People – Product – Partner – Process

Information security management is in essence a specific case of IT management. Hence the four P's also apply to information security management. The four P's correspond roughly to the "People – Process – Technology" (PPT) dimensions of cyber governance described e.g. by NIST CSF [9] and the Cyber Defense Matrix [10], except that these do not explicitly mention "Partner" which is an essential dimension of modern IT and information security management. We therefore think that it is more helpful for cybersecurity practitioners to consider the "four P's" instead of "PPT". Alternatively, practitioners who follow the PPT approach must remember to also include the "Partner" dimension. Each of the "P" dimensions are briefly described below with reference to their significance for cybersecurity.

- **People:** People are the weakest link in information security. People can bypass security measures intentionally or by accident. It is therefore essential to focus skills training and on building a good culture in the organization, as described in Chap. 13. There also need to be roles and organizational structures for people to work effectively and efficiently with cybersecurity.
- **Product:** From an information security perspective, products are technical security controls, such as a firewall which can filter and block malicious traffic. In general, it is necessary to have technology/products that can prevent, detect, correct and recover from security incidents, as illustrated in Fig. 1.6.
- **Partner:** In the current landscape of IT outsourcing and online management of IT products by 3rd party providers, organizations are dependent on external partners to be able to operate functions and deliver services. Hence, organizations need to engage with partners when buying products and services from 3rd parties. It can be a waste of money to buy an expensive security product if the vendor is not able to provide adequate support for operating the product.
- **Process:** All use of technology (Product) consists of processes, so when an organization acquires or implements security technology, they must ensure that they have an adequate process for using the technology. If the process is ineffective or inefficient, the technology will have little value.

As an example of applying the four P's, think of an organization which considers buying a specialized firewall (Product) for filtering traffic in an industrial network. The organization must ask whether staff (People) have the skills to operate the firewall, whether the vendor (Partner) can provide continuous 24/7 support and regular updates of filtering rules, and finally how the operation (Process) of the firewall shall be handled and integrated in the organization's other security processes such as incident response (Sect. 14.5), CTI (Cyber Threat Intelligence)(Sect. 16.2) and cyber risk management (Chap. 19). The investment in the new firewall (Product) will be risky if the organization cannot adequately answer these questions.

1.7 Sources of Requirements for Cybersecurity

For an organization, it is important to identify requirements for cybersecurity, both for their own interest and in relation to others. It takes good security management to be able to choose appropriate security controls. It is the top-management's responsibility to facilitate that the organization knows how to identify relevant requirements for cybersecurity. In general, there are three types of sources of cybersecurity requirements:

1. **Requirement for adequate security according to standard good practice**

    For each type of system or application that is implemented and put into operation, it is important to know what represents standard good practice for incorporating security mechanisms and functions. For example, it is standard good practice that the interface between the Internet and the organization's computer network is protected with a firewall. It is also standard practice that access to online services operated by the organization is protected with user authentication and access control. An example of good practice cybersecurity is the set of control activities from IG1 (Implementation Group 1) of the CIS Controls [6] described in Sect. 18.4. Most organizations will benefit greatly from implementing these or the equivalent controls from ISO/IEC's security controls standard [4] described in Sect. 18.5.4. However, it is insufficient for a business to simply follow standard good practices. According to NIST's maturity scale "Tiers" [9] described in Sect. 18.6, an organization is considered to have low maturity in cybersecurity governance if it does not additionally carry out its own risk assessments to identify necessary security controls.

2. **Requirement to limit security risk to an acceptable level**

    Security controls for this purpose are identified as part of security risk management. For example, risk assessment may point to the necessity of installing an advanced firewall with the capacity to withstand DDoS attacks, because the probability and impacts of DDoS attacks are considered to be high. DDoS (Distributed Denial of Service, see Sect. 2.1.11) is an overload attack that means a website is bombarded with so much traffic that it is unable to serve legitimate users. If standard good practice requires a (simple) firewall, risk assessment may indicate the need for an advanced *next generation* firewall that can filter out DDoS traffic. The risk assessment may also point to the necessity of introducing MFA (Multi-Factor Authentication) for access to highly sensitive resources. If common practice requires (simple) user authentication, risk assessment may also require MFA. Risk management is described in Chap. 19.

3. **Legal, regulatory, and contractual cybersecurity requirements**

    All businesses are subject to cybersecurity laws and regulations that must be complied with, and for specific sectors and organizations, additional regulations often apply. It is each organization's duty to have an overview of which legal requirements for cybersecurity must be complied with. A trend with modern security regulations is to specify that businesses must identify security requirements based on risk assessments, which is precisely the source of security requirements just mentioned above. Laws and regulations for cybersecurity are described in Chap. 17.

An organization should be aware of all three sources of requirements for cybersecurity. In the end, it all comes down to the general requirement of reducing cyber risk in the organization. To simply implement security controls according to good practice will reduce cyber risk even without assessing that risk. *Compliance risk* is the potential negative impact resulting from non-compliance, and to implement legally required security controls will of course reduce that risk.

1.8 Cyber Risk

Cyber risk, or information security risk, is the potential that cyberthreats will exploit vulnerabilities around information assets and thereby cause a harmful incident. In other words, a cyber risk is a relevant combination of a cyberthreat, a vulnerability, and an asset. There are many different risks, and risk identification consists of mapping the most significant risks one can think of.

In general, the magnitude of a particular risk is considered proportional to asset value, threat strength, and vulnerability severity. In other words, the more valuable assets you have, the more threats you are exposed to, and the more vulnerabilities you have, then the greater the risk. This principle is illustrated in Fig. 1.9, where the size of the triangle represents magnitude of risk. The figure shows that when trying to reduce risk, in theory one has the options of reducing assets, reducing threats, or reducing vulnerabilities. In practice, it is difficult to reduce assets or threats, which leaves reducing vulnerabilities as the best option in most cases. A practical way to remove or mitigate vulnerabilities is to implement security controls, as explained in the previous section.

Fig. 1.9 The risk triangle

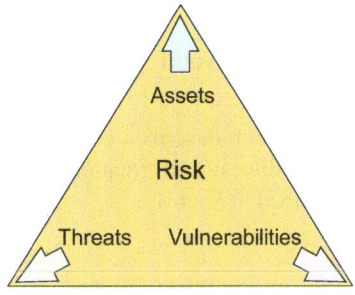

One could ask if it were possible to eliminate all security risk simply by implementing enough security controls? The answer is No, for several reasons:

1. New vulnerabilities in existing systems are constantly being discovered by threat actors,
2. New services are constantly being put online, thereby exposing them to cyber threats,

3. New attack tools, e.g. based on AI, are constantly being developed,
4. Threat actors are often very good innovators; they constantly invent new attack techniques,
5. Threat actors have little to fear, law enforcement is typically ineffective at stopping them,
6. Security controls cost money, and organizations have a limited budget.

The conclusion is that cybersecurity governance and risk management must always be a continuous process aimed at stopping threats and responding to incidents. An appropriate budget for cybersecurity activities must be based on business considerations within the organization. On an abstract level, on can say that organizations should aim at a balance between the magnitude of cyber risks and the budget for security controls, as illustrated in Fig. 1.10.

Fig. 1.10 Balance between cyber risks and security controls

The definition of information security (Sect. 1.3) sets as goal the preservation of confidentiality, integrity and availability of information assets. Another way of defining the goal of information security is to manage and reduce cyber risk to achieve an acceptable balance between risks and investments in security controls.

1.9 CIA Security Goals

This section describes in detail the security goals of confidentiality, integrity, and availability of information assets from the definition of information security described in Sect. 1.3.

1.9.1 Confidentiality

Confidentiality is perhaps the security goal people primarily think of in connection with information security. All organizations and private individuals need to protect information they possess against unauthorized access. The standard ISO/IEC 27000 provides a simple definition:

> *Confidentiality is the property that information is not made available or disclosed to unauthorized individuals, entities, or processes.* (ISO/IEC 27000)

1.9 CIA Security Goals

This definition is intuitively easy to understand, but it is important to note the term "unauthorized", because if it is unclear who is authorized or not, the term confidentiality suddenly becomes meaningless. There is some confusion here that needs to be steered clear of, which we will return to in Sect. 1.11 about authorization and access control.

General threats to confidentiality are *data theft* committed by external actors, or *data leakage* caused by internal actors, either accidentally or intentionally. If it is intentional, then it is called an *insider threat* which is described in Sect. 13.3.

Security controls that support confidentiality include the following:

- User authentication and access control to protect stored information
- Encryption of data for protection at rest or in transit
- Security mechanisms in operating systems to protect information during processing
- Security guards and physical access control

The list of controls above is not at all exhaustive. Many other controls can contribute directly or indirectly to maintaining confidentiality, such as security culture, security policies, procedures, and practice.

1.9.2 Data Integrity

The term "integrity" can have different meanings, with two types of integrity in particular being central to information security. We call these data integrity and system integrity. Data integrity is described here, while system integrity is described in the next section.

A good definition of data integrity can be found in the X.800 standard [11]:

> *Integrity is the property that data has not been altered or destroyed in an unauthorized manner.* (X.800)

The definition of integrity from X.800 is intuitively easy to understand and is in line with the definition of confidentiality from ISO/IEC 27000 in that it also uses the term "unauthorized". For the definition of integrity from X.800, it is also essential to know who is authorized or not, because if it is unclear, data integrity becomes meaningless.

General threats to data integrity are corrupted data, for example caused by external actors, or inside actors either accidentally or intentionally. Unauthorized deletion of data or unauthorized creation or alteration of data and files also constitute a breach of data integrity. It is interesting to note that data integrity in principle is equivalent to data authenticity, because if data is altered in an unauthorized way, the data is no longer authentic; Or vice versa, if data lacks authenticity, its integrity is uncertain.

Security controls that support data integrity can be:

- User authentication and access control to protect stored information
- Encryption or cryptographic checksum of data for protection at rest or in transit
- Data backups
- Security guards and physical access control

Security controls for data integrity are many of the same as for confidentiality. The list of controls above is also not exhaustive. Many other controls can also contribute directly or indirectly to maintaining data integrity.

1.9.3 System Integrity

A good definition of system integrity can be found in the ISO/IEC 27000 standard:

> *Integrity is the property of accuracy and completeness (of information assets).* (ISO/IEC 27000)

The definition of integrity from ISO/IEC 27000 is relatively abstract, allowing for interpretation of what is meant by accuracy and completeness. The definition does not refer to authorized entities, so that it is only implicit who decides what is correct and complete, but the natural thing is that the owner of the information asset in question is the one who decides. If the asset in question is information represented as data, then it simply means data integrity. If the asset in question is an IT resource like a computer, a network, or an application, then *system integrity* is a natural interpretation of the definition. System integrity means, for example, that systems, applications and network components have the correct configuration, correct software and updated patch status.

General threats to system integrity are misconfigured systems caused by external threat actors or by internal accidental or intentional and outdated software that may be caused by inadequate operation of systems and software. Security controls to maintain system integrity can be for example:

- User authentication and access control for all, and strong authentication for admin users
- Configuration management
- Change management

Many other controls can also contribute directly or indirectly to maintaining system integrity.

1.9.4 Availability

Availability of information resources is an essential security goal, because without availability, IT systems are of no use. A reasonable definition of availability can be found in ISO/IEC 27000:

> *Availability is the property of being accessible and usable on demand by an authorized entity.* (ISO/IEC 27000)

The term "authorized" is again important to note, because if it is unclear who is authorized, it is also unclear for whom the information asset is supposed to be available.

General threats to availability include denial of service (DoS/DDoS), obstruction of authorized access to resources and delay of time-critical functions.

Security controls that support availability include filtering of harmful traffic, redundancy of resources, load balancing of processing, backups, and incident response.

1.10 Other Security Goals

Since the ISO/IEC 27000 standard also mentions authenticity, accountability, non-repudiation, and reliability as elements of information security, these terms are briefly described in the following sections.

1.10.1 Authentication

Authentication is a very important feature that supports the CIA security goals of confidentiality, integrity, and availability. The concept of authentication is relatively general, with several different meanings. Figure 1.11 provides an overview of the main types of authentication.

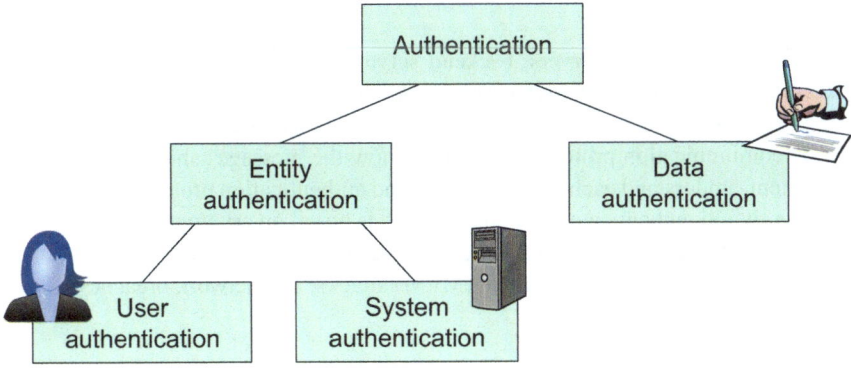

Fig. 1.11 The main types of authentication

The different authentication types are explained below.

- **User authentication**

> User authentication is to verify the correctness of a claimed user ID.

User authentication is the security mechanism we most often encounter and use when interacting with our digital environments. Typically, user authentication consists of two steps: (1) provide your user ID and (2) prove that you are the legitimate owner of that user ID by providing authenticators, such as passwords, hardware/software authentication tokens, or biometrics. Users are traditionally humans, but system processes can act as users to perform routine functions on behalf of a human or organization. In such applications, authentication and access control also apply to system processes.

General threats to authentication are identity theft, impersonation and false login. The term *spoofing* means that an attacker pretends to have somebody else's identity. Security controls to strengthen user authentication are, for example, multi-factor authentication (MFA) based on two (2FA) or more independent authenticators. Simply making sure to use strong passwords, to keep them safe and not get tricked to reveal them in a phishing attack will also go a long way to strengthen authentication.

Attacks against user authentication are a very common attack vector. Here, the principle that a chain is not stronger than its weakest link applies, i.e. if only one user in an organization has a weak password, attackers manage to gain unauthorized access. Put another way: There is a lot to be gained from making sure that everyone follows good routines around user authentication and the handling of authenticators.

User authentication is described in detail in Chap. 8 about user authentication and in Chap. 9 about identity and access management.

- **System authentication**

> System authentication is to verify the correctness of a claimed system identity.

Systems can be web servers, backend servers, database servers, routers, cloud servers, etc. As a rule, when such systems communicate with each other, authentication is integrated. The communication, i.e. the exchange of data and messages, follows a communication protocol that defines how the messages should look so that the systems understand each other. Security and authentication protocols define how systems should authenticate each other, which is based on cryptographic methods.

General threats to system authentication are fake nodes in networks, fraudulent data transactions, man-in-the-middle (MitM) attacks, and network breaches.

1.10 Other Security Goals

System authentication is based on cryptographic authentication protocols that ensure authentication and integrity of the exchanged data. Cryptography is described in Chap. 4 and communication and security protocols for authentication are described in Chap. 6.

- **Data authentication**

> Data authentication is to verify the identity of the origin or sender of data.

Data is not an entity. Data is passive, virtual and independent of any specific physical representation. Data authentication means that we want to know where data comes from or who generated it. Put another way, data authentication means to trace data to a sender or author entity which can be a human user or system entity. While user and system authentication typically only apply at login or for a specific session, data authenticity might be required for many years or for as long as possible.

General threats to data authentication are forgery of digital documents and forgery of messages sent through computer networks. One mechanism for data authentication is *MAC (Message Authentication Code)*, which is a cryptographic checksum on a message that can be verified by a specific recipient with a secret key. A cryptographic checksum is a bit sequence, typically of 256 bits, that is calculated over a document of arbitrary size. If a single bit is changed in the document, the checksum will no longer match. Thus, anyone who has the secret key can check the cryptographic checksum to verify that the document has not been altered. A stronger mechanism for data authentication, and which supports *non-repudiation*, is *digital signature* which like a MAC is also a type of cryptographic checksum on data. The difference is that a digital signature can be verified by third parties (i.e. by anyone), not just by a specific recipient who knows the secret key. The term non-repudiation means precisely that authenticity is undeniable because there exists a cryptographic proof of origin that any third party can verify. Fig. 1.12 illustrates this.

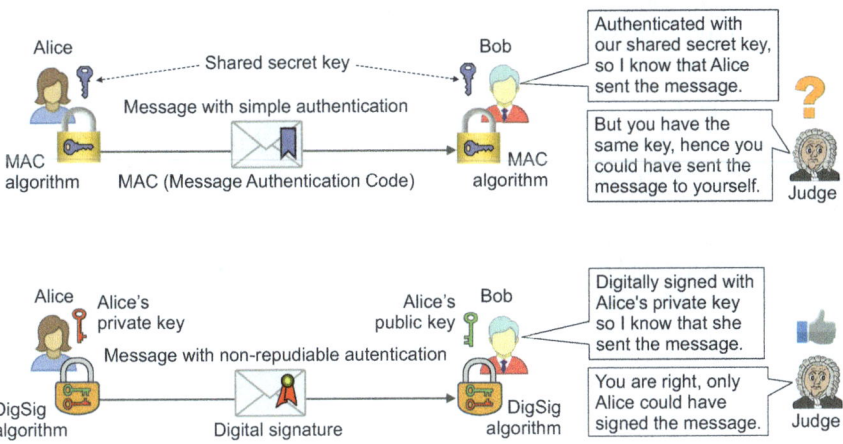

Fig. 1.12 Difference between simple authentication with MAC and non-repudiable authentication with digital signature

To clarify the difference between simple data authentication with MAC, on the one hand, and non-repudiable authentication with digital signature, on the other, suppose a situation where Alice has sent an order to stockbroker Bob to buy shares in a company. The next day, the share price drops sharply, which will cause huge losses for Alice. To try to get out of the situation, Alice claims she never sent the order and demands the transaction be voided, while Bob claims she sent the order. Then there is a court trial to determine who is right. Here, it is crucial whether the authenticity of the order is based on MAC or digital signature, as shown in Fig. 1.12. For MAC, the secret key shared between Alice and Bob is used to both generate and verify the MAC cryptographic checksum. For digital signature, Alice uses her private key to generate the signature, while Bob or other third parties use her public key to verify the signature.

Figure 1.12 shows that the MAC proves to Bob that Alice sent the message, but that the MAC does not allow Bob to prove this to the judge. A digital signature provides evidence both to Bob and the judge that Alice sent the message. The difference is that MAC provides authenticity, while digital signature provides non-repudiation, which is essentially a strong form of authenticity.

The use of digital signature relies on a *PKI (Public Key Infrastructure)*, and is a relatively complex mechanism compared to MAC, which only needs the exchange of a secret key between two parties (although that can be complicated enough). PKI is described in Sect. 5.2.

Unfortunately, neither data authentication with MAC nor with digital signature can provide authenticity assurance for eternity because the underlying cryptographic algorithms have limited validity, typically less than 30 years.

MAC and digital signature are described in Chap. 4 about cryptography.

1.10.2 Accountability and Non-repudiation

> Accountability means that activities in a system or computer network can be traced to someone who can be held accountable for the activities.

Accountability is a very important feature for detection, investigation and recovery after cyber incidents. Accountability is based on logging activities in systems and networks, and maps which user or other entity is behind each logged activity.

General threats to accountability are e.g. that it is not possible to identify who was behind an act, and that there is a lack of sufficient evidence to be able to report an incident and for presenting evidence in a potential court case.

Security mechanisms for accountability include authentication of all users and logging of system events. Logging can be done in different ways and can be combined with data authentication to ensure the integrity of data logs. Digital forensics (described in Sect. 14.6) can also contribute to accountability by revealing correlations that are not directly evident from logs through analysis of equipment, data and other evidence.

> Non-repudiation means that an entity cannot deny having performed an action.

Non-repudiation is based on a digital signature which has the effect that the signer of a message, transaction or document cannot deny having done just that. Undeniability is related to accountability because a digital signature is assumed to be a strong proof of who has sent a message, made a transaction, or authored a document. However, a digital signature is never absolutely non-repudiable. If you want to repudiate a digital signature on a document, you can simply claim that your private signature key was stolen and hence that the attacker made the signature. Digital forensics may be necessary to assess the veracity of such a claim.

1.10.3 Reliability

> Reliability of a system means that it does not contain significant flaws and that the system is able to maintain important functionality even when certain faults occur.

Reliability mainly means the ability to withstand unintentional events (i.e. faults), but also means that important functionality can be maintained even in the presence of some faults. Reliability is particularly important for the security goal availability.

General threats to reliability are low quality in the development, configuration or operation of systems.

Controls for good reliability are simply good practices for the secure development and operation of systems, which is called *secure-by-design*, as described in Chap. 11.

1.11 Access Authorization and Access Control

It is important to have a good understanding of *access authorization* because the consistency of the CIA security goals described in Sect. 1.9 depend on the correct interpretation of authorization.

> Access authorization is the act of specifying access rights for users, roles, and processes.

Access rights are typically represented as *access rules* or *access policy*. Access authorizations for a user are specified according to the user's role in an enterprise and are formalized as rules and configuration in access control systems.

Authorities in an organization are responsible for authorizing users, but the authorization function can be delegated in an authorization hierarchy such as: Manager → System Administrator → User.

> Access control is to check an entity's access rights to determine whether or not to grant access to resources when the entity makes an access request.

The user entity must be authenticated and logged in so that the system knows the identity and thus can find the applicable access rights to decide whether the user should be granted access to the requested resources.

The term *IAM (Identity and Access Management)* can be described in terms of a configuration phase and a usage phase, each consisting of steps, as illustrated in Fig. 1.13. User accounts and authenticators must be created and configured before it is possible to perform user authentication during the usage phase. Similarly, access authorizations must be specified and configured before the user can access resources to perform tasks in the usage phase. IAM links the steps for registering and authorizing users in the configuration phase, with user authentication and access control in the usage phase. User authentication is described in Chap. 8, and IAM is described in Chap. 9.

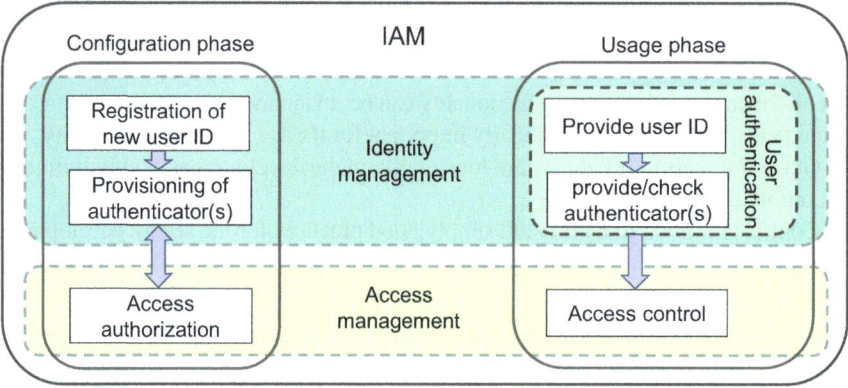

Fig. 1.13 Phases and steps of IAM (Identity and Access Management)

In the configuration phase, new users are registered with a new user ID and are given the necessary authenticator(s). Based on the user's role in the organization, access rights are specified during the authorization step. Typically, a new employee in the human resources department is authorized to read and edit all employee profiles that contain payroll information and other personal details. On the other hand, as a rule, a new employee in the sales department will not be authorized to read employee profiles about colleagues.

In the usage phase, user authentication takes place during login by having the user entering the user ID and authenticator(s) that are checked by the system. After login, the user can make a request to open documents or use services. Access control consists of the system checking the current authorization policy/rules to determine whether the user should be able to access requested documents and services.

Often, systems/applications will only make visible the resources for which the user is authorized, while other times all resources are visible, and the system denies access if the user requests something for which they are not authorized. Defining authorization policy is a form of configuration, while access control takes place during usage. Models for access authorization and access control are described in Sect. 9.3.

There is a clear difference between authorization and access control. Unfortunately, these terms are often mixed up in literature and textbooks by falsely saying that authorization is the same as access control. However, it becomes completely absurd if the step of access control in Fig. 1.13 is defined as authorization. As explained in Sect. 1.9 above, the CIA security goals of confidentiality, integrity and availability are defined on the basis of who is authorized or not. By defining authorization as equivalent to access control, any person who actually obtains access would by definition be authorized, whether that person is a genuine user or an intruder. That would make the CIA security goals meaningless, and would make absurd many laws and regulations including the US Computer Fraud and Abuse Act (CFAA) described in Sect. 17.3.1. This can be illustrated by the following scenario:

Let us assume that you fraudulently obtained another person's password, which you use to steal information from their account. Your criminal act is discovered, you are prosecuted for unauthorized access and retrieval of information, and you end up on trial. When the judge asks you to explain your action, you pull out a cybersecurity textbook (not this one!) that says, for example, *"After logging in, the system authorizes the user to access the computer resources."* You argue that you logged in, and hence that you were authorized to access the resources.

Based on explanations of authorization in such textbooks you are logically right in your line of argument. However, it would probably not be good enough to win the case. The judge could, for example, refer to the textbook you are now holding in your hand.

1.12 Tasks

1. Access authorization

    (a) X.800 is a standard for security services in OSI (Open Systems Interconnection). Search and find the X.800 standard, or visit https://www.itu.int/rec/T-REC-X.800-199103-I/en

    Read the definitions of confidentiality, integrity, and authorization in X.800. Are the definitions of confidentiality and integrity from X.800 meaningful in terms of how authorization is defined? Why or why not?

(b) How is authorization defined on Wikipedia? https://en.wikipedia.org/wiki/Authorization
(c) Explain whether the definitions of confidentiality, integrity, and availability (CIA) in standards X.800 and ISO/IEC 27000 make sense based on Wikipedia's definition of authorization.

2. Breach of the CIA security goals

   Describe examples of attacks that can cause security breaches for each CIA security goals, and possible controls that can prevent the attacks. The attacks should be described in a very abstract way, such as "a hacker steals a password and takes over another person's account".

   (a) Attacks causing breach of confidentiality
   (b) Attacks causing breach of integrity
   (c) Attacks causing breach of availability

3. Threats to IAM (identity and access management)

   A simple way to identify threats is to ask, *"What could go wrong?"* or *"How can this be attacked?"*. Look at Fig. 1.13 in Sect 1.11.

   (a) Mention relevant threats to (the steps in) the configuration phase of IAM (identity and access management).
   (b) Mention relevant threats to (the steps in) the usage phase of IAM (identity and access management).

Chapter 2
Attack Vectors and Malware

> *If you know the enemy and know yourself, you need not fear the result of a hundred battles. If you know yourself but not the enemy, for every victory gained you will also suffer a defeat.*
>
> Sun Tzu, ancient Chinese military strategist

Rapid technology innovation and frequent paradigm shifts in the IT industry have the side effect of bringing new security vulnerabilities and new attack opportunities that threat actors can exploit. Cybersecurity is therefore a continuous challenge of countering new and changing attack vectors and malware with new and increasingly effective security controls. An important emerging challenge is to protect against AI-driven attacks that can learn and adapt to circumvent current security controls, and that also can scale up the volume of attacks to overwhelm defenders. The security industry is already responding to this challenge by offering defensive AI solutions.

The learning objectives of this chapter are to give you an overview of the most common attack vectors and the most common forms of malware. Attack vectors are practical methods and channels used by threat actors to attack and compromise the assets of an organization. An attack vector is generally the same as a threat scenario, which is described in Sect. 1.4. Malware is data and program code that the threat actor transfers to and/or runs in the infrastructure of the company being attacked, and which can result in damage to the company's information assets.

2.1 Attack Vectors

An attack vector is a channel or interface by which an attack can be carried out. An attack typically uses one or more channels of communication, access, or control to send malicious content and/or carry out malicious activity in the organization's infrastructure. The purpose of an attack is typically to compromise systems and networks, or to steal or destroy data. Attack vectors exploit vulnerabilities including, for example, human lack of awareness through social engineering. Attack vectors are generally the same as threats. When undertaking threat modelling, which means mapping possible and relevant cyber threats to your own organization, classic attack vectors are the first thing to think about. Threat modelling is described in more detail in Chap. 11 which focuses on how to make systems secure by design, and in Chap. 19 about risk management.

The most common attack vector today is phishing, which consists of sending malware or links to malicious websites through deceptive emails and other messages such as SMS and chat. The most dangerous vulnerability to this type of attack is a lack of awareness among us users, so we are tricked into clicking on dangerous links, opening malicious documents, or installing malware.

Firewalls and antivirus filters can to some extent block against such attack vectors, but there are no methods that provide 100% protection. A given protection mechanism can quickly become outdated, as hackers continually improve and develop new attack methods. Figure 2.1 shows examples of typical attack vectors. Each attack vector can be utilized separately or in combination with others to further increase attack potency.

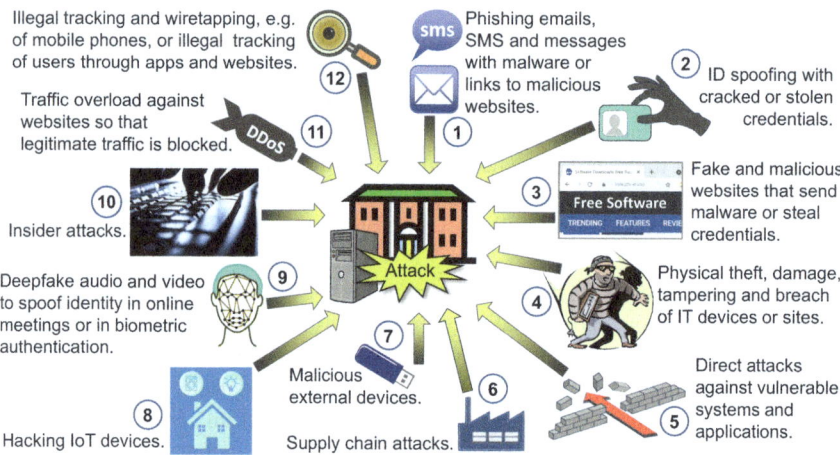

Fig. 2.1 Typical attack vectors. (Photo of insider attacks by Christoph Scholz, CC BY-SA)

The list of attack vectors in Fig. 2.1 is, of course, not exhaustive. A quite peculiar and uncommon attack vector is, for example, eavesdropping on emission from digital equipment, which is described in Sect. 3.10 and in Sect. 12.4.5.

Threat actors are highly innovative, and new attack vectors and methods are constantly being developed. This can be called threat innovation, where AI-enhanced attacks are a major trend, as described in Sect. 15.2. At the same time, classic attack methods are still very potent, and they are constantly being improved. Each numbered attack vector illustrated in Fig. 2.1 is described in the following sections.

2.1.1 Phishing

Phishing is a type of social engineering that is more thoroughly described in Sect. 13.4. For example, a phishing email is designed to trick the recipient into replying with sensitive information, visiting a fake website with an embedded link, or installing malware from an attachment. The name and email address of the sender are often spoofed (see next section), that is, misrepresented, which is technically easy to perform. If the victim clicks on a link to a fake website controlled by the attacker, the attacker can observe everything the victim does and e.g. perform a man-in-the-middle (MitM) attack against the legitimate website. MitM attacks mean that the fake website passes input from the user on to the real website, and the same in the opposite direction, which allows the attacker to observe and change data in the communication. Mass phishing is when the same phishing email is sent to a high number of recipients. Most often, the content of mass phishing is irrelevant to the recipient, so this type of phishing typically easy to spot. It is trickier to spot spear-phishing because it is specially crafted for a single targeted victim, thus increasing the likelihood that the recipient will fall for the phish. As of 2025, phishing is by far the most common attack vector for cyberattacks and cybercrime on the Internet. Other types of messages such as SMS or chat messages can also be used for this type of social engineering. For example, the phone number of the caller or of the sender of SMS can easily be spoofed, so that we must always consider whether SMS messages are genuine, or if the caller is who they claim to be. Lack of knowledge and awareness is the most dangerous vulnerability to this type of attack. The best protection, then, is to learn to recognize the signs of phishing, and to always be sceptical.

Quishing, or QR phishing, is an attack vector where threat actors use QR codes to redirect victims to malicious websites or prompt them to download malicious content. Quishing represents a cyberthreat in physical environments where victims can be tempted to follow a QR code on a poster or advertisement board. When received through the Internet, quishing can bypass conventional defenses like secure email gateways. Notably, QR codes in emails are interpreted by many secure email gateways as just images that are assumed to be irrelevant for security, thereby exposing users to potentially harmful QR codes.

2.1.2 ID Spoofing with Cracked or Stolen Credentials

"Identity is the new security perimeter" is a common phrase in the cybersecurity community. It simply means that in the omni-connected environment we are living, the protection of identities and credentials has become the most important security endeavour. Stealing identities and credentials is perhaps the most attractive tactic for threat actors. Stolen identities and credentials are being used for spoofing, i.e. to take on the identities of others. Spoofing of identities opens up for a myriad of

hard-to-detect attack and fraud opportunities. The challenge for threat actors is primarily to get hold of identities and credentials in the first place, and hence many different methods are being used for this purpose. Phishing, which is described in the previous section, is a common method whereby the victim e.g. is tricked to provide login details to a fake webpage. When threat actors successfully compromise and get access to a company network, a goal high on the priority list is to steal databases with user IDs and credentials. Such databases are being offered for sale or are being published on the *dark web*. Passwords are typically not stored in cleartext, but in hashed form, as described in Sect. 4.5. However, many cleartext passwords can be recovered through *password cracking* described in Sect. 8.3.1. When biometrics are being used as part of user authentication, an attack technique is to use forged biometrics as described in Sect. 8.5.3. Deepfake based on ML (Machine Learning) is a powerful technique for mimicking face and voice as described in Sects. 2.1.9 and 15.2.1. Artifacts for fingerprints can be manufactured based on a sample fingerprint of the victim. Additionally, it is relatively easy to get biometrics from a sleeping or drugged victim. Your mobile phone number represent a strong form of identity that can be stolen e.g. with SIM swapping attacks, as described in Sect. 7.4.6.

2.1.3 Fake or Malicious Websites

Fake websites can be almost indistinguishable from their corresponding genuine websites. Users may be misled into visiting a fake website, for example through phishing emails, through other fake or infected websites. A user who has landed on a fake website may be tricked into providing user ID and password or other sensitive information which leads to identity theft as described in the previous section. *Malwertising* (malicious advertising) is a stealthy way of turning otherwise genuine webpages into malicious webpages. Malwertising typically involves injecting malware-carrying advertisements into legitimate online advertising networks and webpages.

Drive-by attacks is when a victim lands on a webpage that sends malware to the client automatically and without user interaction. Criminal websites are owned and operated by attackers, while infected websites are legitimate websites where attackers have installed malicious elements. Criminal websites look like legitimate websites. Drive-by attacks take place through covert download of malware to the client. The special thing about drive-by attack is that the user does not need to click on anything, explicitly download anything, or open any attachment for the client to become infected. The user only needs to visit the website to be attacked.

Finally, installing malicious programs can happen through websites that the victim user explicitly visits to download software. The user may be tricked into installing malware in the belief that the software is legitimate. There are a number of different types of malware, as described in Sect. 2.2. Trojan (horse) is a type of malware that actually has a useful function, but at the same time has hidden harmful features.

2.1.4 Physical Attacks on IT Infrastructures

If threat actors are able to get physical access to IT infrastructure and devices, most technological security controls can be bypassed. Hence, it is essential to design physical barriers to prevent physical entry, damage or tampering with IT infrastructure and devices. Physical security is described in Chap. 12. Physical entry into buildings can e.g. performed by lockpicking, with brute force, or through social engineering (described in Sect. 13.6). However, IT equipment is located or carried out of physically protected sites. For example, people carry laptops, mobile phones and other devices e.g. in their pockets, in bags and in cars where they can be stolen. Breaking into devices and accessing data or executing functions can be done e.g. by forging biometric credentials, or by opening up the casing where hardware components get exposed and can be tampered with. The last layer of physical protection is therefore on the device itself, where robust authentication and tamper resistant/proof casing are possible techniques.

2.1.5 Direct Attacks Against Vulnerable Systems and Applications

Systems and applications with an online presence are exposed to threats from all over the Internet. Such systems and applications often have vulnerabilities, which are either unknown to system and software vendors, or which are known but that the business has not had time to patch. Two classic attacks against vulnerable applications are SQL-injection and XSS (Cross Site Scripting), which are described in Sect. 11.7. Attacks and exploits of such vulnerabilities are typically done with automated tools through the Internet.

Password spraying means that the attacker tries to log into a large number of accounts with default or typical passwords. An account is usually locked after three attempts with an incorrect password, but the attacker can bypass this obstacle with password spraying, because a potential password is tested against many different accounts instead of many times against the same account. When a new server is installed, the admin account often has a default password. If this password is not changed, the server is easy to compromise.

Automatic spread of malware can occur, for example, through a computer worm that automatically sends itself to other computers through computer networks and the Internet. A computer worm exploits unknown or unpatched vulnerabilities to infect a computer and uses it as a host to scan and infect other computers. This triggers a chain reaction that can produce exponential infection in a short period of time. Besides seizing significant bandwidth through the spreading, worms can have intended adverse effects.

2.1.6 Supply Chain Attacks

A *supply chain attack* means that the threat actor first attacks and compromises products and services of a supplier, which then makes it possible to attack the supplier's customers. Hence, this is a two-stage attack. A company which has implemented reasonable security controls in its own infrastructure may nevertheless be vulnerable if, for example, an attacker manages to install a hidden backdoor in software that the company buys from a supplier. A threat actor that successfully installs a backdoor in the software of a supplier could easily attack all the supplier's customers that buy the same software. Hence, supply chain attacks can be very potent. The principle of supply chain attacks is illustrated in Fig. 2.2.

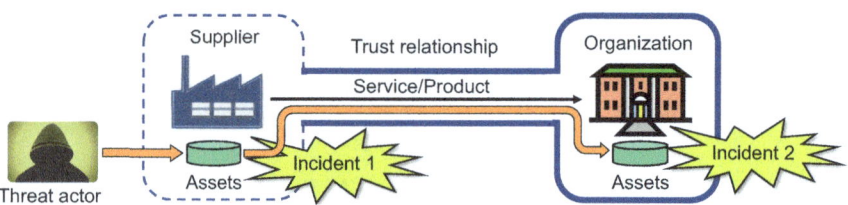

Fig. 2.2 Principle of supply chain attacks

As shown in the figure, two security incidents occur sequentially. The first incident is when the threat actor attacks the supplier where, for example, a hidden backdoor or ransomware is installed in the software that the supplier develops and sells. When attacking a supplier, the threat actor will usually focus on manipulating software developed by the supplier, but may also attempt to steal passwords, API keys, firewall rules, cryptographic keys, certificates, CRM data, and other confidential information. In addition, the threat actor may attempt to compromise hardware and processes used for generating digital signatures.

The second security incident occurs when the threat actor exploits compromised software or hardware by gaining unauthorized access to the customer's systems, or by conducting a ransomware attack.

From a security perspective, supply chains represent part of the attack surface of the organization. The reliance on long and complex supply chains has led to a dramatic enlargement of the attack surface of the typical organization in recent years. Any entity that contributes to a supply chain has the potential to impact security further up the supply chain. The longer and more complex your supply chain, the harder it is to get an overview of threats and vulnerabilities, and the harder it is to manage security incidents when they occur.

Cyber supply chain risk management (C-SCRM) covers managing vulnerabilities and security risks associated with the supply chain [12]. In Chap. 19, risk management is described such that the level of security control activities should be proportional to the level security risk. When the security risk of the customer organization is much higher than that of the supplier, a market failure occurs in that the supplier does not see the risk exposure of the customer, and thus does not have an incentive to dimension its security controls proportionally to the level of the security risk of customer organizations. This is a fundamental problem that can only be solved by organizations requiring suppliers to maintain an adequate level of security through agreements, and by auditing the suppliers' security programmes.

A variant of supply chain attacks is if the supplier itself is a threat actor to its customers. This can happen, for example, if the government of a state forces a national supplier to put hidden spyware or hidden backdoors into systems sold to other countries. The fear of this type of attack means that many countries are now taking precautions. For example, this may mean that host countries can require that suppliers of equipment for critical infrastructure must be located in a country with which the host country has political security cooperation. A typical example of this practice is that telecom operators in many Western countries are not allowed to buy equipment for the 5G network from the Chinese company Huawei. Presumably the same principle applies in, for example, China and Russia, that is, these countries are probably cautious about using ICT equipment manufactured in Western countries in their critical infrastructure.

A part of supply chain risk management, an organization should consider the requirements below included in contracts and SLAs (Service Level Agreements):

- Obligation to report and provide notification on security-related matters such as security incidents, security policies and changes in ownership structure,
- Audit reports on cybersecurity at subcontractors,
- Level of risk acceptance at subcontractors,
- Expected uptime of services, and subcontractor's risk acceptance for compliance with this,
- Change and exit clauses, e.g. based on deviations or changes in ownership structure.

Irrespective of formal contracts and service level agreements, the organization should map and monitor the subcontractor's ownership structure. A possible threat is that a subcontractor is acquired by a company located in a country with which the organizations host country has no political security cooperation. It is therefore plausible that the authorities in the supplier's country could ask or force the new owner's management to attack the organization via the delivery of products and services.

2.1.7 Malicious Peripherals

Malicious peripherals connected to the USB socket of a device can be used in attacks. For example, the device may contain malware that the user might install out of curiosity. Another possibility of attack is that a USB flash drive is configured as a Human Interface Device (HID) device, which tricks the computer into thinking it is a keyboard and sends a stream of keystrokes that constitute malicious commands. A *drop attack* is when the attacker leaves malicious USB flash drives in places like cafes, waiting for someone to find them and plug them into a computer.

When charging mobile devices through a USB socket in public places, a possible threat is that the USB socket attacks the device as described above. Therefore, USB condoms should always be used for charging through public charging socket. A USB condom blocks all data transmission and only allows electricity supply and charging.

2.1.8 Hacking Unpatched Vulnerable IoT Devices

IoT devices are physical devices with a built-in minicomputer that connects wirelessly to a network through Wi-Fi, mobile network (4G/5G) or Bluetooth, and has the ability to send and receive data. IoT stands for *Internet of Things*, which means they form networks to automate home and industrial tasks. They can transmit sensor data or receive commands to perform functions. IoT devices can be categorized into three main groups: for consumers, for business, and for industrial plants.

In a smart home, IoT devices can detect and react to a person's presence. When you drive home, your car communicates with the garage to open the gate. When you enter the house, the thermostat is adjusted to the desired temperature, and the lighting is set to intensity and color according to the time of day and the weather outside.

While regular computers with full-fledged operating systems like Windows or Linux get regular security updates and patching to fix software bugs and security vulnerabilities, IoT devices are usually never updated after they have been manufactured and installed. The absence of full-fledged operating systems in IoT devices makes it technically difficult or impossible to perform online software updates on them.

All software has vulnerabilities and bugs which regularly get discovered, and which can be exploited by threat actors. IoT devices that are never updated never get their vulnerabilities fixed, and therefore become sitting ducks waiting to be hacked. A smart home or an industrial facility with unpatchable and vulnerable IoT devices connected to the Internet can be an easy target for hackers.

Many countries are aware of this problem. As the first country in the world to regulate IoT security, the UK introduced the Product Security and Telecommunications Infrastructure (PSTI) regime in April 2024.[1] The act is designed to enforce smart device security for consumers by compelling device makers to ensure that they:

- do not ship products with default or easy-to-guess passwords
- provide a point of contact for consumers to report security vulnerabilities
- state the minimum period of time the device will receive security updates

---

[1] https://www.gov.uk/government/publications/the-uk-product-security-and-telecommunications-infrastructure-product-security-regime

The fines for non-compliance can be draconian, up to £10 m or 4% of global annual revenue, whichever is higher. The EU cyber resilience act (see Sect. 17.4.4) which is under preparation will have similar provisions as the UK PSTI act.

2.1.9 Deepfake Attacks

Deepfakes are AI-generated fake video, audio, images, or textual data (messages or documents) aimed at deceiving. Deepfake and other offensive uses of AI are described in Chap. 15. With deepfakes an attacker can appear with your face and voice in a video meeting with your colleagues who will think they are talking with you. The term "deep" refers to the use of "deep learning" technology to create such deceptions, as described in Chap. 15. For example, fake live video can be made by replacing the appearance (voice and face) of an attacker with the appearance of an employee in a company. In that way, the attacker can take part in online meetings by appearing as a colleague. Fake video can also be fully synthetic consisting of an avatar that looks and sounds like a real person. Both types of fake video are potent tools for social engineering as described in Sect. 13.4. For example, an attacker could spoof the identity of the company's director during an online meeting and instruct an employee with high authorizations to transfer a large sum of money to an account controlled by the attacker. Alternatively, the attacker could spoof the identity of an employee of a company during a video meeting, sending a chat message with a link to a malicious website. Because colleagues know and trust the employee, they will typically click on the link. There are already examples of such attacks, and businesses need to be increasingly aware of this attack vector. In February 2024, a company worker in Hong Kong was tricked to transfer US$25 million to fraudsters using deepfake technology to pose as senior officers in their company during a video call.[2]

Deepfakes can be used for business identity theft, which is a type of identity theft committed with the intent to misrepresent a business to defraud the businesses itself or to defraud partner companies. A 2021 report entitled from the US Department of Homeland Security entitled "Increasing Threats of Deepfake Identities" suggests methods for detecting deepfakes.[3] In addition a 2022 report entitled "Phase 2: Mitigation Measures" proposes measures to mitigate against deepfake deceptions [13].

As part of the reconnaissance and OSINT (open-source intelligence) phase of an attack, threat actors need sound and video recordings to produce deepfakes of a specific person. A possible security control could then be to never make recordings or not to publish such recordings online. However, for some people like politicians, appearing in the public is a necessary part of their job. Hence, other security controls must be designed to mitigate such threats.

---
[2] https://edition.cnn.com/2024/02/04/asia/deepfake-cfo-scam-hong-kong-intl-hnk/
[3] https://www.dhs.gov/sites/default/files/publications/increasing_threats_of_deepfake_identities_0.pdf

2.1.10 Insider Attacks

Insider attacks are attacks committed by an employee or other person who is authorized to access the company's systems and data. Insider attacks can be very damaging because the threat actor has knowledge and access authorizations to systems and assets, so typical security controls aimed at keeping external attackers out have little effect. In addition, insider attacks can be very difficult to detect because the threat actor often knows how not to leave traces. Insider attacks are described in more detail in Chap. 13 about cybersecurity culture.

2.1.11 DDoS Attacks

DDoS stands for "distributed denial of service", which is a rather vague description. A more precise description is to call it a "distributed overload attack" which means that these attacks aim to exhaust the target's resources like bandwidth, CPU & memory by overloading the target with superfluous requests. Mitigating these attacks involves diverting traffic, filtering malicious requests, and absorbing excess traffic. The main DDoS attack types are briefly described below.

- Volumetric attacks are aimed at exhausting the capacity of the victim server's network by sending massive amounts of Internet traffic towards it, with the effect that the server crashes or otherwise is unable to serve genuine requests.
- App layer attacks are aimed at exhausting the victim server's CPU resources. From the attacker's perspective, the effectiveness of this attack comes from the asymmetry between the amount of CPU resources it takes to send a fake request relative to the amount of resources it takes to process it at the server end. For example, when a website receives a simple request, it must typically make database queries or other API calls to produce the queried webpage. When a botnet of clients targets a single webpage, the effect can overwhelm the victim server, resulting in denial-of-service to legitimate traffic.
- Protocol attacks are aimed at exhausting the memory and CPU resources at the Internet layer of the victim server node (see Sect. 6.1). These attacks are typically done by sending large amounts of malformed packets in a way that forces the server to use up all its memory, or to spend CPU time. Example of protocol attacks are SYN floods, packet fragmentation, Ping of Death, and Smurf DDoS. These attacks can also harm intermediate communication equipment, such as firewalls and load balancers.

The distributed aspect of DDoS means is that the traffic against the target victim comes from thousands of different client hosts around the Internet which makes it difficult to block specific IP addresses to stop the traffic. The origin behind the DDoS traffic is typically a botnet, which means that the traffic originates from the computers of legitimate individuals and organizations, but that these computers have been infected with bot malware. A *botnet* is a large number of infected computers that are controlled by a *bot herder* which is a grouping of threat actors. Botnets can be used for various malicious purposes, where stealing computing resources for crypto currency mining and DDoS attacks are quite common. In fact, you do not need to create your own botnet to start a DDoS attack; in fact, botnets can be rented quite cheaply, typically for around $10 per hour. DDoS attacks are normally not considered to represent high risk, and DDoS gives little value for a threat actor other than that of expressing aggression against the target organization. However, DDoS attacks cause temporary disruptions which can be serious if continuous availability is critical, and unavailability can damage the reputation of the organization. Methods to mitigate DDoS attacks include redundant servers and load balancing.

2.1.12 Tracking

Espionage and tracking are very common threat activities in digital environments. Such attacks even represent a specific category of military cyber operations, as described in Sect. 16.3. Eavesdropping on mobile phones and tracking people can be done with IMSI catchers which are described in Sect. 7.4.3. Law enforcement agencies can legally use of IMSI catchers, but criminals also use them. Less spectacular, but maybe much more potent is the tracking of our online activities through mobile phone apps and cookies on webpages. Depending on the region where you live, data protection laws are supposed to protect us from abusive tracking on the Internet but it seems very difficult to effectively enforce such laws, as described in Sect. 10.3.

2.2 Malware

Malware is software with intentional malicious features. Software that causes accidental damage due to bugs is just bad software. Typically, malware is intentionally created to be used as part of an attack. Malware can also be legitimate software that is used as part of an attack. Therefore, it is not always clear what is malware, because it depends on which perspective we adopt. Software produced by legitimate companies may well be considered malware if it secretly acts against the interests of the user. As an example, in 2005 Sony distributed a rootkit stored on music CDs to prevent music copying. This rootkit created security vulnerabilities that were in turn exploited by other attackers. Sony engineers probably thought this was useful software, but to users it was obvious malware.

Fileless malware is an umbrella term for malware that is not stored on disk in the form of a file, but is only stored and run in system memory. Hence, fileless malware disappears when the system shuts down and reboots, which increases its ability to evade antivirus tools and leaves very little evidence for digital forensic investigators. Traditional viruses and Trojans are a form of malware stored on disk, while exploits and malicious JavaScript are fileless. Figure 2.3 lists the most common types of malware.

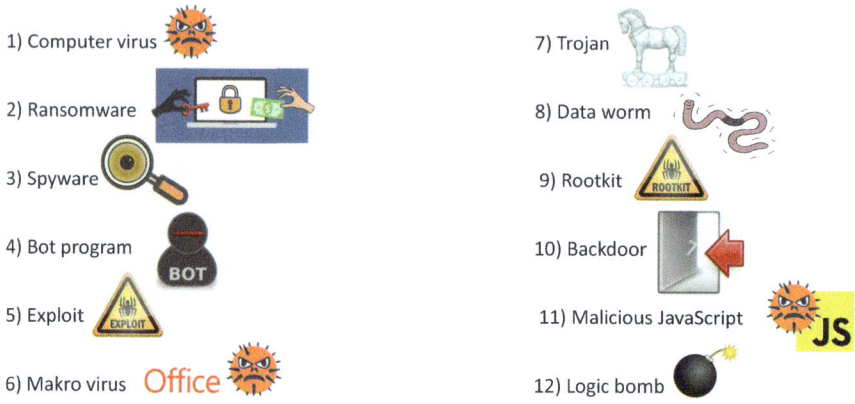

Fig. 2.3 Common types of malware

Each numbered malware type illustrated in Fig. 2.3 is described in the respective sections below.

2.2.1 Computer Virus

The term *computer virus* originally has the meaning of malicious program code that infects a legitimate program by injecting itself into the program to disturb its normal function. This is analogous to a biological virus which injects itself into ordinary cells of the body and disturbs the functioning of the cells and the body. The computer virus performs its harmful function only when the infected program is executed. That a virus has become part of another program makes it especially difficult to remove; even the best antivirus programs struggle to do this correctly. Such traditional computer viruses were widely used during the early age of computers in the 1980s and 1990s before the Internet, but are less common today, accounting for less than 10% of all malware.[4]

---

[4] https://www.csoonline.com/article/548570/security-your-quick-guide-to-malware-types.html

2.2 Malware

Note that the term "virus" is often used as a generic term for all types of malware, which can be a source of confusion.

2.2.2 Ransomware

Ransomware is a type of malware that, in the first stage, encrypts all or a selection of the data on a victim's systems, leaving data and applications unusable. In other words, this is a form of denial-of-service attack that primarily leads to a breach of availability. The attacker follows up on the initial encryption step and demands a ransom to hand over the decryption key. *Cryptovirus* is another name for ransomware, precisely because it encrypts and locks down data. An important security prevention against ransomware is to take regular backups of all important data, and to have good recovery routines. Backups should be stored offline from the network, if possible, to shield them from being attacked. Recovery routines must be tested regularly, for example once a year, to verify that they work, and that routines are rehearsed before an incident happens for real.

Doxing is the disclosure of personal information or sensitive information about a business without the consent of individuals or the business. Doxing is usually done by publishing on the web or on the *dark web*, and is carried out for various reasons, including blackmail, revenge and political motives. Threat actors can use ransomware to steal sensitive information and threaten doxing, i.e. blackmail the business by demanding a ransom for not publishing the information publicly.

In this type of blackmail, backup is useless. It can be very serious and challenging for businesses to deal with such forms of extortion if it involves sensitive personal data. The best advice is to provide good preventive security controls that can prevent data breaches and theft of data from happening at all.

2.2.3 Spyware

Spyware are programs that spy on the user by collecting different types of information. A *keylogger* is a type of spyware or a hardware device that logs keystrokes, allowing the attacker to steal passwords and other sensitive information when they are typed. Such programs are often installed through some form of social engineering combined with one of the attack vectors described in the previous section, and such spyware usually starts up every time the system boots.

The use of cryptography for data communication makes it practically impossible for spyware to read information intercepted as encrypted data. Instead, cryptographic protection is typically bypassed by installing spyware on endpoint devices to steal data before it is being encrypted.

Pegasus is a spyware used to steal data and remote-control mobile phones running iOS and Android. It is developed and sold by the Israeli cyber-arms company

NSO Group. The spyware is designed to be covertly and remotely installed, e.g. by simply sending an SMS to a user, or by tricking the user to click on a malicious link. As Apple and Google are continuously removing newly discovered vulnerabilities in iOS and Android that Pegasus may exploit, the NSO Group must continuously arm Pegasus with new exploits for unknown (to Apple and Google) vulnerabilities. While NSO Group markets Pegasus as a tool for fighting crime and terrorism, some governments use Pegasus to surveil journalists and political dissidents. The Israeli defense ministry must approve the sale of Pegasus licenses to foreign governments.

One type of spyware that is not usually considered outright criminal is that websites collect different types of information about users while browsing, which can be sold and used for marketing to users. GDPR, described in Chap. 10, regulates this type of data collection in the EU/EEA, but is unfortunately difficult to enforce.

2.2.4 Bot Malware

Bot malware performs automated tasks on command from a threat actor called *bot herder*. A computer with bot malware is called a *bot* (eng.: bot) which might as well be a private computer as a computer in a company. All types of computers with Internet connections can become infected and turn into a bot, including IoT devices. The name comes from "robot", i.e. something that works automatically. When not made for malicious purposes, bot software can be used for legitimate purposes, such as for collecting telemetry from many hosts or as a distributed search engines to perform indexing. Unfortunately, bots are most often used by threat actors. Bot malware is typically self-replicating, and is capable of receiving commands from a centralized or distributed bot herder. Attackers infect computers in large numbers to create a *Botnet* which is the collection of all infected bot computers controlled by a bot herder. The size of botnets ranges from a few hundred to several million infected computers. A botnet can carry out different types of attacks, such as DDoS attacks (see Sect. 6.4.3) or cryptocurrency mining (see Sect. 4.13.2). For cryptocurrency mining, electricity and computational resources are used in the infected bot computers. For DDoS attacks, the infected bott computers are used to send traffic towards attack targets on the Internet. Operators who control large botnets often operate botnet rentals as a way to generate income. This allows anyone who has the conscience to turn themselves into a criminal to carry out DDoS attacks cheaply and easily.

2.2 Malware

2.2.5 *Exploit*

Exploit is a small program, string of data, a file, or a sequence of commands that exploit one or more bugs or vulnerabilities in a software, hardware or other computer equipment to trigger corrupt or abnormal behavior. An exploit trigger typically a *buffer overflow*, which is described in Sect. 3.5. The attacker's intent in triggering such behavior is to gain control of a system, download backdoors, gain unauthorized access, or to carry out a denial-of-service attack. To find vulnerabilities that can be used to create exploits, fuzzing is often used,. Fuzzing is described in Sect. 11.3.5.

2.2.6 *Macro-viruses*

Malicious Office macros can reside in Office documents. Macros are small programs used to automate features in Office documents. Unfortunately, macros can be made to perform harmful functions, and attackers cand hide malicious macros in Office files sent as phishing email attachments. These files use names that are meant to trick people into opening them, and typically look like invoices, receipts, legal documents, etc. Previously, Office macros ran automatically, but no longer. An attacker now has to trick users into explicitly enabling macros to run.

2.2.7 *Trojan*

Trojans is malware masquerading as legitimate programs that actually (or apparently) have useful features, but at the same time have hidden harmful features. Trojans are popular among threat actors and have mostly taken over from traditional computer viruses. Trojans are typically downloaded from criminal or infected websites, or come as an attachment to phishing emails. The user is tricked into thinking that the program is useful, and explicitly chooses to install it. It is ironic that fake antivirus programs are a widespread type of Trojan. Trojans often come in the form of extension (aka. plugins) to browsers. Every browser lets you remove extensions, but it can be challenging to know if a specific extension is a Trojan.

The term "Trojan" comes from Greek mythology, where soldiers from Athens during their campaign against the city of Troy hid inside a large horse sculpture of planks placed outside the city. The inhabitants of Troy dragged the horse sculpture into the city, believing it to be a gift. At night, the Athenian soldiers would sneak out and open the city gate so that the Athenian army could storm in and take the city.

2.2.8 Computer Worm

Computer worms are self-contained programs that spread themselves to other computers within a computer network or all over the Internet in an automatic way, usually without users having to do anything. Computer worms accomplish this by connecting to other host nodes and exploiting vulnerabilities in applications exposed to the local network or the whole Internet. What makes computer worms so devastating is their ability to spread without user intervention. Viruses and Trojans at least require a user to launch a program. The infamous *SQL Slammer worm* exploited a buffer overflow vulnerability in the Microsoft SQL database server. In January 2003, the worm infected approximately 75,000 Microsoft SQL servers in about 10 minutes, demonstrating its exponential rate of spread.

2.2.9 Rootkit

A rootkit is an attack tool which enables attackers to hide their traces when they get unauthorized access to systems. A rootkit typically has mechanisms to hide itself as well as the activities of the attacker. This is for example done by hiding malicious processes from the system process list, by hiding back doors, or by hiding user accounts created by the attacker. The rootkit can also be used to hide other malware, such as spyware and keyboard loggers.

Rootkit infection often happens through some form of social engineering combined with an attack method such as. an exploit. Rootkit detection is challenging precisely because a rootkit is designed to hide itself and may even be able to subvert and blind anti-malware software.

The term "root" refers to the traditional Unix term which means privileges access so that an operator has full control of the system. The term "kit" refers to the set of components that the rootkit tool consists of. Hence, a rootkit is a tool that gives the attacker full control over the system. When a rootkit has been installed on a system, the legitimate operators can no longer trust what they see on the screen or that system commands do what they are supposed to. It is quite scary, it makes you blind. If a rootkit is suspected, the operators must install a new clean system.

2.2.10 Back Door

Back door is a hidden method of bypassing normal authentication and access control in a system. Backdoors can have legitimate uses, such as providing a way for the manufacturer of systems and software to perform maintenance and servicing. A backdoor can also be used by attackers to gain unauthorized access to systems. It can be a hidden part of a program, a separate program, code in the firmware to your hardware or part of the operating system. For example, a backdoor can be installed by a Trojan or by an exploit.

2.2.11 Malicious JavaScript

JavaScript (JS) is a programming language typically used to create advanced functionality on web pages. Most web pages contain JavaScript which automatically downloads and runs in the browser when a user visits the website. Typically, JavaScript code is perfectly legitimate, and is used for advanced functions on a web page, but it can also perform malicious functions, such as downloading an exploit from an attacker-controlled website. JavaScript can also be misused to turn the browser into cryptocurrency mining engine for the attacker or to steal sensitive information which is sent to the attacker. Malicious JavaScript can be injected into web pages through XSS attacks, described in Sect. 11.7.2. Because JavaScript executes automatically and without any interaction when the user visits a web page, JavaScript is typically used for drive-by attacks.

2.2.12 Logic Bomb

A logic bomb is malware that triggers a malicious function when specified conditions are met. A threat actor may, for example, be an employee of a company who for some reason wants to harm the company. A logic bomb can be programmed to trigger on a specific date, or to trigger if employee's salary is stopped, which will typically happen if they lose their job. The harmful function of a logic bomb can e.g. be to delete files.

Different types of malware, such as viruses and worms, often contain logic bombs that perform a specific function at a predefined time or when another condition is met. This technique can be used by a virus or worm to speed up and spread before it is noticed. A *time bomb* is when malware is programmed to attack the host system on a specific date.

2.3 Tasks

1. Attack vectors
    (a) What is the most common attack vector for cyberattacks today?
    (b) How can deepfake voice and video be used in cyberattacks?
    (c) How can USB devices be used to attack a computer?

2. Supply chain attacks
    (a) Explain what is meant by supply chain attacks.
    (b) Describe examples of supply chain attacks that have occurred in recent years.
    (c) What measures can a business take to reduce supply chain risk?

3. Botnet
    (a) What is a botnet?
    (b) What is a DDoS and how can a botnet be used to execute DDoS attacks?
    (c) Mirai is a malware that was used to "take down" the Internet in much of the United States through an attack on a DynDNS one day in October 2016. Briefly describe the Mirai malware and how this attack was carried out.

Chapter 3
System Security

> *Complexity is the enemy of security.*
>
> Bruce Schneier, American cryptographer

The IT infrastructure consists of systems and the connections between them. Systems are what we typically call computers, servers, hosts, nodes, smartphones, devices etc. The design architecture of systems consists of an IT stack, which roughly contain the hardware layer at the bottom, operating systems in between and applications as the top layer. There might be mezzanine layers in between. Anyway, for a system to be secure, it is of course necessary that every layer is secure, and most developers of systems make a good effort to ensure adequate security. Unfortunately, the totality of the IT stack in a system is extremely complex, with millions of lines of code at each layer. For this reason, every system always has security vulnerabilities that will be exploited as soon as attackers can find them. There is a continuous race between attackers searching for and exploiting vulnerabilities, and vendors and organizations removing and mitigating the same vulnerabilities.

The first learning objective of this chapter is to understand the overall architecture of computer systems. This forms the basis for the next learning objective, which is for you to understand the main security features and security mechanisms built into computers, both in software and in hardware. You will also learn about how the complexity of systems makes it impossible to build systems that are totally secure.

3.1 System Architecture

Simply put, our ICT infrastructure consists of systems and networks, where the systems contain operating systems and applications, while networks bind systems together.

A system has one or more microprocessors, also called CPUs (Central Processing Unit), that run/execute software that constitutes operating systems, applications, and other computational functions. There are thousands of different software modules in a server or a desktop computer. A software module resides on a storage disk when not in use and is copied into system memory when activated, turning it into an active running process. The system memory, which is called RAM (Random-Access Memory) can be accessed quickly from the microprocessor so that code (program instructions) and data can be retrieved very fast to execute the code and make calculations on data. The components of a system communicate over a common system bus, as shown in Fig. 3.1.

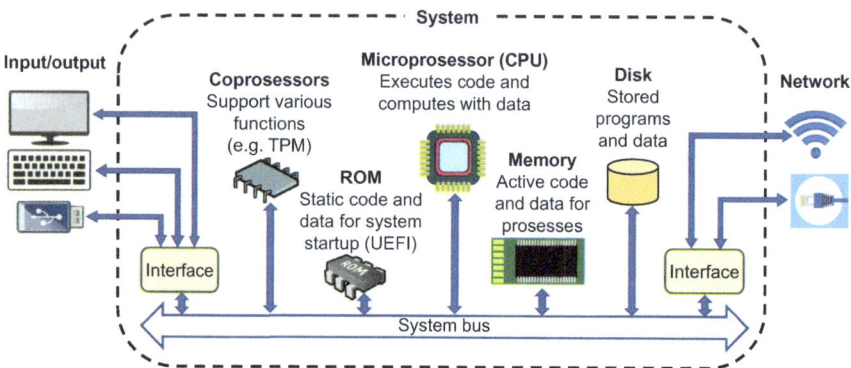

Fig. 3.1 Simplified system architecture with basic components

ROM (Read-Only Memory) is a type of memory that contains static code and data, that is, it (almost) never changes. Static code and data that cannot be changed is important for system security because it cannot (easily) be attacked and thus forms a so-called *trust anchor*. The term trust anchor in this context means that if we can have confidence that code and data in the ROM memory have integrity, this code can ensure that the system startup is correct, and perhaps even verify that other program modules are also correct.

ROM has become an overarching term that embraces many types of non-volatile memory, but which can nevertheless be updated through specific procedures. An example is *firmware* which stores low-level program code and data. It is usually possible to update firmware because it is necessary and desirable to be able to correct errors and change the configuration of systems. Firmware is usually stored in *flash memory*, which is a type of EEPROM (Electrically Erasable Programmable Read-Only Memory). Delivering systems with intentionally malicious firmware is a form of supply chain attack, which is described in Sect. 2.1.

Systems often have several *coprocessors* that supports various features, such as a graphics coprocessor. The Trusted Platform Module (TPM) is a coprocessor that supports specific security features and forms a form of trust anchor. The principle of having trust anchors based on hardware components is traditionally called Trusted Computing, which we return to in Sect. 3.7.

A system has several connectivity options to communicate with the outside world, such as keyboard, monitor, USB connectors, Wi-Fi and network cable. Attacks usually enter through one of the input channels, and will typically attempt to take control of (privileged) programs running in the microprocessor. In addition, information leakage may occur through *side-channel attacks*, which is described in Sect. 3.10.

Systems are complex products consisting of physical hardware components and many different software modules that are integrated into hierarchical levels of abstraction. In order to (try to) maintain overall security in such systems, there are several approaches that must be used in combination. Some of these approaches are described below.

3.2 The Importance of System Security

In the early days of the Internet, almost all data communication was unencrypted, and even passwords were sent unencrypted. This changed when HTTPS was introduced around the year 2000. HTTPS is the secure encrypted variant of HTTP (HyperText Transfer Protocol), which enables encryption of web traffic. With HTTPS and other cryptographic security protocols, the Internet can actually provide very strong communication security.

Unfortunately, we cannot say the same about system security and endpoint security. The term *"endpoint"* denotes host nodes with apps that exchange application data, such as between web servers and web clients (PCs and smartphones). Endpoint security is often inadequate because endpoints due to their complexity always have vulnerabilities, which means that there are a number of ways they can be attacked.

Attackers usually go after the weakest link, and then it is only natural that they go after the endpoints. This situation is nicely illustrated by the analogy in the quote below, which is also illustrated in Fig. 3.2.

> Using encryption on the Internet is equivalent to using an armored car to deliver credit card information from someone who lives in a cardboard box to someone who lives on a park bench. (Gene Spafford[1])

Fig. 3.2 Analogy: Strong transport security does not help if the endpoints are weak

The moral is that it is necessary to have both strong endpoint system security and strong communication security to achieve good overall security when using online services.

---

[1] Eugene Howard Spafford (born 1956) is an American professor of computer science at Purdue University and a leading expert in information security.

3.3 Vulnerabilities Management for Systems and Software

The complexity of computer systems means that there are many errors and vulnerabilities in the systems' software and hardware at any given time. Software manufacturers do their best to fix bugs and remove vulnerabilities as they are discovered. The vulnerability lifecycle includes several steps, from the time a vulnerability is discovered and reported to the time it is removed in enterprise software instances, as shown in Fig. 3.3.

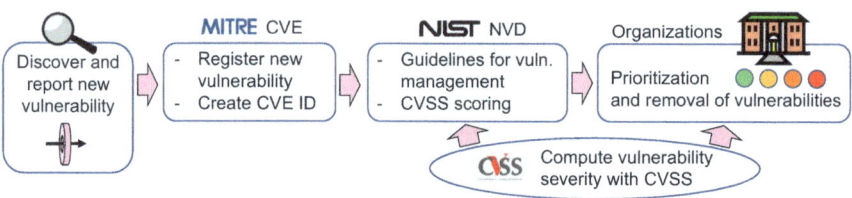

Fig. 3.3 Vulnerability lifecycle

MITRE Corporation manages CVE (Common Vulnerabilities and Exposures), a database of vulnerabilities in all types of software. When a new vulnerability is reported, MITRE assigns a CVE ID, and describes and catalogues the vulnerability. The next step is performed by NIST, also in the US, which manages the NVD (National Vulnerability Database). NIST imports data from CVE and adds guidelines for handling and removing the vulnerability in the software in question, typically with input advice from the software manufacturer. Note that China has its own vulnerability management scheme, as described in Sect. 17.6.

A distinction is made between the concepts *of weakness* and *vulnerability,* where weakness represents an abstract type of vulnerability without reference to any specific system, while a vulnerability is when a vulnerability actually exists in production software. In addition to CVE, MITRE therefore also manages CWE (Common Weakness Enumeration), which is a database of abstract weaknesses.

To help enterprises in prioritizing vulnerability management, it is necessary to assess the degree of severity, which is calculated using CVSS (Common Vulnerability Scoring System). CVSS, published by the Forum of Incident Response and Security Teams (FIRST), defines a method based on a set of different factors for calculating a vulnerability's severity score on a scale from 0 to 10, with the following ratings:

- Critical: 9.0–10
- High: 7.0–8.9
- Medium: 4.0–6.9
- Low: 0.1–3.9
- None: 0.0

An enterprise can use the calculation method from CVSS to calculate a score based on its own context, and prioritise removal of the vulnerability accordingly. In case timely removal of a critical vulnerability is difficult, the organization can instead opt

3.3 Vulnerabilities Management for Systems and Software

to mitigate the vulnerability by e.g. isolating the vulnerable system with strict firewall rules or by disconnecting it from the Internet. An alternative mitigation strategy is to strengthen the security monitoring of the vulnerable system so that a response can be triggered immediately upon a detected exploitation.

When systems are hosted by administrators in an organization, they usually want to have their own control over software updates because updates themselves can have unexpected negative consequences. For example, they can try out an update first on a system that is not in production to verify that the system is still stable after the update. Then, the update can be deployed to systems that are in production.

For consumer products such as PCs and mobile phones, software updating and security patching is typically an automated online process, but the user is usually prompted or asked to accept the update and to restart the system for the update to take effect. Most software vendors, and some hardware manufacturers, support automatic updating, Software and hardware vendors are constantly updating their products, whether these are microprocessor and hardware manufacturers or operating system and app manufacturers.

Unfortunately, it also happens that attackers or others discover a vulnerability that they do not report to the software manufacturer or to MITRE CVE, so that the vulnerability is kept out of the handling cycle in Fig. 3.3. It is then called a *zero-day vulnerability* because the manufacturer has had 0 days to find a solution and create a patch against the vulnerability. The vulnerability is therefore still present in all copies of the software, and can be freely exploited by attackers who have knowledge of the vulnerability. This is described in more detail in Sect. 16.3.1.

It must be mentioned that IoTdevices are a type of technology that is typically never updated, so vulnerabilities that are discovered are never removed. This is a serious issue that allows threat actors to attack through unpatched vulnerable IoT devices, as described in Sect. 2.1.8.

Modern systems and applications are composed of many ready-made software components. Instead of writing every line in a computer program for an application as was typically done in the 1980s, modern software development is done with tools that retrieve from software libraries ready-made software components for the specified functions that the application should have. It is difficult for organizations to know the quality and security of the code libraries they are using. Similarly with AI-based coding copilots, the quality and security of the produced code is no better than the corpus in which the coding copilot has been trained, as explained in Sect. 15.4.2.

Keeping track of vulnerabilities in an application requires that one has an overview of vulnerabilities in each software component that the application consists of. An approach to deal with this huge challenge is SBOM (Software Bill of Materials), which means that "digital bills of materials" are created of all components included in systems or software applications. This requires tools and infrastructure to automatically maintain an SBOM for every application.

With SBOM it is possible to automatically map all known vulnerabilities in a system or application. The SBOM initiative started in 2018 and is gradually becoming an important part of software security and supply chain risk management.

SBOM is a collaboration between different players, managed by CISA (Cybersecurity and Infrastructure Security Agency) in the US.

3.4 Privilege Levels for Processes in the Microprocessor

An operating system (OS) is a set of application modules, where the OS kernel is the one that has overall control over programs running in the microprocessor. When a program is running, we say that it is a process running in the microprocessor. Processes have different levels of privilege based on their role in the system, just as officers in the military have different levels of command based on their military rank.

When Intel launched its X86 microprocessor in 1985, it built in support for processes to have four different privilege levels, with level 0 being the highest privilege level and level 3 the lowest. The idea was that the OS kernel would run at privilege level 0, other OS features and drivers would run at level 1 and 2, while user applications would run at level 3.

It turned out that OS developers (e.g. Microsoft, Apple, RedHat) found it impractical to use levels 1 and 2, partly because it causes unnecessary delay of processing. Additional delay is due to the fact that processes at levels 1, 2, and 3 often have to call functions in the OS kernel when a feature requires privilege level 0, such as displaying graphics on screen or sending data to the network interface. Graphics and network drivers perform such functions all the time, so it soon became clear that drivers could not call the OS kernel all the time—creating excessive delays in processing. Therefore, the OS developers decided that processes for drivers and other OS functions must also run at privilege level 0. The result is that levels 1 and 2 are no longer in (normal) use in modern systems.

In contrast, processes for user applications have retained privilege level 3 for security reasons, although it also creates some delay. To maintain good security in systems, it is absolutely necessary to limit what user applications are allowed to do. Therefore, they must run at level 3, which is the lowest privilege level.

When we log into a system as a normal user, the user process will run with privilege level 3. When we log in as administrator (or "root" in the Linux world), the admin user process runs with privilege level 0, which allows the administrator to have unlimited ability to read and modify all data and programs in the system. Thus, an administrator could accidentally make mischief if not careful, so the best practice is to only log in as an administrator when really needed.

Virtualization is an advanced way to manage operating systems so that multiple operating systems can run in parallel on the same physical machine. Traditionally, the operating system performs functions directly on physical hardware, such as moving data between the disk and memory, retrieving programs from disk, and starting new processes. Virtualization means installing a type of software called *hypervisor*, which performs these functions on behalf of the operating system. This means that the hypervisor appears as physical hardware from the perspective of traditional operating systems such as Windows and Linux. To the OSs, the

3.4 Privilege Levels for Processes in the Microprocessor

hypervisor looks like hardware, but in reality it is software. The hypervisor is thus virtual hardware and the stack of hypervisor with OS and user applications on top is called a *virtual machine*. From the hypervisor's perspective, the operating systems running "on top of it" are applications. Thus, it is possible to simultaneously run many operating systems, as if they were applications, on the same physical machine. When operating systems are run on a virtual machine, they are called *guest operating systems*.

Machine virtualization has security implications. The hypervisor controls operating systems, and hence should logically be more privileged than operating systems. To support virtualization security, a new process mode was defined in microprocessors from manufacturers such as Intel and AMD around 2006. Unofficially, it is called privilege level-1, but it is also called *hypervisor mode*. Thus, the process for a hypervisor runs with privilege level-1, which is thus more privileged than level 0, which is the traditional privilege level for operating systems. The architecture with privilege levels is called *protection rings*, illustrated in Fig. 3.4. The term "ring" is metaphorical and means that processes within a specific ring are protected from processes in outer rings. In reality, the privilege level of a process is defined with a flag (i.e., a numerical value) in the code segment that describes the process in memory. The privilege level is determined when the application is compiled and/or installed on a system, and is enforced by the operating system when the application runs as a process.

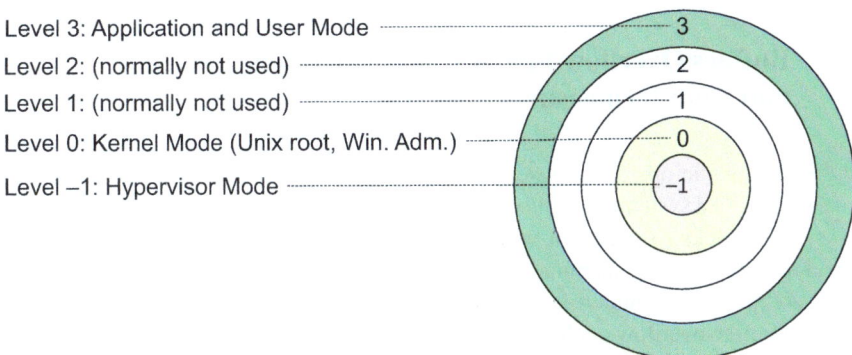

Fig. 3.4 Privilege levels in microprocessors, illustrated as protection rings

There are two important effects of a process running with a specific privilege level. The first effect is that certain instructions in the microprocessor are allowed to be executed only when the calling process has a sufficiently high level of privilege. If a process nevertheless tries to call a disallowed instruction, the operating system will stop the process, which means that the program crashes.

The second effect of a process running with a specific privilege level is that it limits which data segments and code segments a process can access. The principle is that a process can access data and code defined with the same or lower privilege level (i.e. higher numerical value). Figure 3.5 illustrates a process with privilege level 0, indicated by the dark blue box called "Process", where solid arrows point to code and data it can access, while dotted arrows with No-Entry traffic signs point to code and data it cannot access.

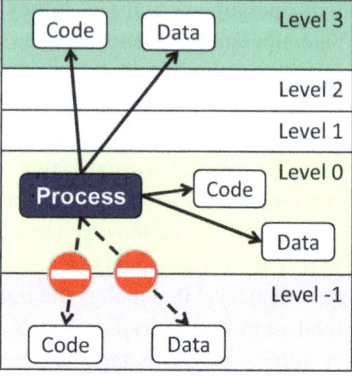

Fig. 3.5 Permitted and prohibited access to data and code based on privilege

Virtualization of systems has been used since the 1960s, but became popular from around the year 2000 when many new hypervisor products became available on the market. Virtualization allows hardware to be utilized very efficiently and is how all cloud data centres manage their computing resources for hire.

3.5 Buffers Overflow Vulnerabilities, Exploits and Countermeasures

Vulnerabilities in operating systems and software can be exploited by attackers to take control of a system. *An exploit* is a form of malware (described in Sect. 2.2) that uses smart tricks to exploit a vulnerability in this way. As attackers continue to develop new exploits, manufacturers of microprocessors, operating systems, and software must take countermeasures to stop them. Here we briefly explain the principle of buffer-overflow exploits and associated countermeasures.

Buffer overflow is a classic vulnerability that still occurs frequently in software. The drama takes place by corrupting the memory, and specifically by attackers being able to overwrite data on the stack, which they replace with their own data that, by using clever tricks, allows them to gain control of the system. Figure 3.6 provides a simple representation of the memory address space available for a single process. Note that this address block represents virtual memory, and that all processes have an equal address block of virtual memory space, but where each virtual address block in reality maps to a separate physical memory address block for each process, to not step on each other's toes. The OS keeps track of which physical memory block belongs to which process. The virtual address space has lowest address 0000 0000 and highest address FFFF FFFF (depending on the operating

3.5 Buffers Overflow Vulnerabilities, Exploits and Countermeasures

system). This virtual address space theoretically provides 4GB to each process, but systems are typically configured so that only 2GB or 3GB is available for each process. Memory addresses are specified in the hexadecimal number system, which has sixteen different digits as opposed to only ten different digits in the standard decimal number system. Digits in the hexadecimal number system consist of the numbers 0–9, and the "numbers" A–F with values 10–15. For example, the hexadecimal number 10 is equal to the decimal number 16, and the hexadecimal number F1 equals the decimal number $15 \times 16 + 1 = 241$.

Virtual memory addresses are translated by the operating system into physical memory addresses before they are read and written in the physical memory illustrated with the component "Memory" in Fig. 3.1. Although all processes have the same virtual address space, in reality they have different physical address ranges. This principle means that a process does not have access to physical addresses of other processes, only to its own physical address space, indirectly through the translation from virtual to physical memory addresses.

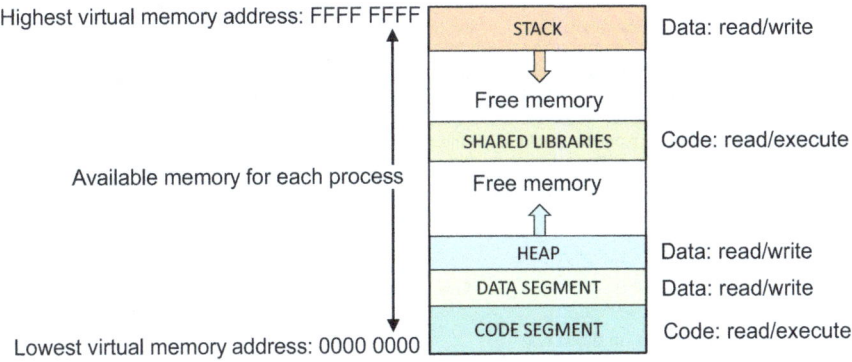

Fig. 3.6 The structure of the virtual memory address space of a process

The microprocessor retrieves code from the SHARED LIBRARIES and CODE SEGMENT areas of memory. Code is program instructions that the microprocessor executes, such as addition, multiplication, and write/read data. The microprocessor writes and reads data in the STACK, HEAP and DATA SEGMENT areas.

The size of the STACK and HEAP areas varies as the process runs, while the other areas are fixed in size. The STACK area grows down in the address area, while the HEAP area grows upward if more memory space is needed, as shown in Fig. 3.6.

We will not explain all elements of the address block for a process shown in Fig. 3.6, but focus on the area of STACK, which we hereafter call "the stack", and where the drama of buffer overflow unfolds.

The stack contains a *stack frame* for each function call executed by the process. The frames lie next to each other "shoulder to shoulder" in memory or, put another way, on top of each other like a stack, hence the name "stack". A stack frame consists of data parameters used by the operating system to keep track of the function calls in the process. Among other things, the stack frame consists of local variables, where a certain amount of memory space, i.e. a *buffer*, is set aside to write data to the variable. If your application has a *bug* (an error), the application may write more data to a variable than is allocated space for in the buffer. This can happen, for example, if an application takes input from a web form where a specific length of data is expected, such as a street name that can have a maximum of 20 characters. Imagine now that the application allows the user to write text of arbitrary length, so the user can type 100 characters in the input field. The result is a so-called "buffer overflow", where excess data overwrites adjacent memory directly above in the frame, as shown in Fig. 3.7. If a buffer overflow occurs, in principle there is no limit to how much memory can be overwritten, so the other frames above on the stack can also be overwritten.

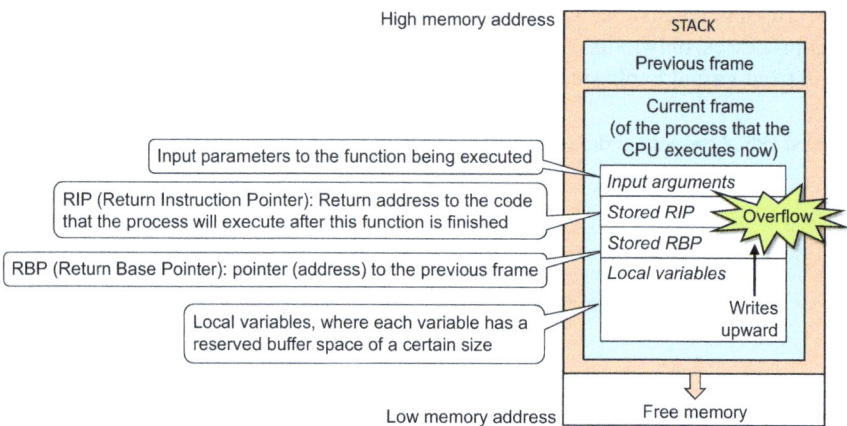

Fig. 3.7 Buffer overflow on the stack

The result of buffer overflow is most often that the program crashes. That is easy to understand, because when the pointer to the code (RIP) and the pointer to the frame (RBP) of the previous (calling) function are overwritten, the operating system will lose control on the execution of the process. When we use an application and experience that the program "freezes", this is typically what happens.

Smart hackers know how to exploit buffer overflow to take control of a system, and there are many advanced tricks that would take up too much space to describe here. The basic principle of creating an exploit that can exploit buffer overflow is that the attacker calculates exactly how to overwrite memory, for example by writing a new pointer in the RIP field that points to malicious code that the attacker wants to execute. The operating system is easily fooled and slavishly follows the pointer located in the RIP field, thus allowing the attacker to gain control of the system.

Manufacturers of operating systems and microprocessors have developed a set of countermeasures that make it difficult for attackers to create exploits that do what the attackers want. Some countermeasures are listed below.

- *NX (No eXecute)*. In early versions of microprocessors (up until about 2000), it was possible to let the stack contain code. Then attackers only needed to embed malicious code directly through the buffer overflow and run it from the stack. NX means that the operating system does not allow to run code residing on the stack. During the 2000s, all microprocessor and mamory manufacturers introduced support for NX by putting an NX bit on every address in the stack.
- *Stack Canaries* (also called "stack cookies"). By having each stack frame contain a checksum for part of the data in the frame, it can be verified that this data has not been overwritten as a result of buffer overflow. If the checksum does not match, the process is terminated. The principle of stack canaries was developed around the year 2000, and is supported by several compilers. The name "Stack Canary" comes from early submarines where live canaries were used to detect whether the batteries were producing toxic chlorine gas. A dead canary was a sign of chlorine gas and that the submarine had to rise to the surface.
- *ASLR (Address Space Layout Randomization)*. One method of creating an exploit is to reuse existing instructions in the SHARED LIBRARIES or CODE SEGMENT memory space illustrated in Fig. 3.6. In order to specify an address pointer to specific instructions, the attacker must know exactly where the code containing the instructions is located. In Fig. 3.6 these memory areas are neat and tidy, making it easy to locate code. ASLR jumbles memory addresses of software components to random locations in the address space each time the computer boots and processes start. Then it will no longer look neat and tidy as on Fig. 3.6, but the operating system keeps track nonetheless. That way, code locations become unpredictable for anyone other than the operating system, making it difficult, but not impossible, for an attacker to specify a pointer to a specific code address in memory. ASLR was introduced in the early 2000s and is in widespread use.

Despite these and other controls, attackers still manage to exploit buffer overflow to create exploits. It is one thing to prevent buffer overflow, where programming languages and compiler techniques play an important role. However, it proves very difficult to eradicate buffer overflow completely, so it still occurs frequently. Another thing is to prevent attackers from exploiting buffer overflow to create exploits, and here controls such as NX, Stack Canaries and ASLR play an important role.

The development of new techniques to create exploits and new controls to prevent them is an ongoing cat-and-mouse game between attackers and defenders of systems.

3.6 Virtualization

A "virtual system" generally means something that mimics the actual system, allowing the virtual system to function as the actual system. For example, VR (Virtual Reality) glasses show a world that mimics the real one, and we feel that we are in a kind of real world. In IT architectures, it is very convenient to implement virtual systems in software because physical hardware which is expensive and inflexible can be mimicked by software which is cheap and flexible. We sometimes say that services running in the cloud are virtual, e.g. we use a virtual editor when editing documents in the cloud, rather than in an editor on a local computer. Virtualization also is important for security.

3.6.1 Virtual Machines

Virtual machines offer a flexible way to install and run systems, where software mimics hardware. To interpret what a virtual machine is, we must first interpret what a machine is, which can be ambiguous. To distinguish the meanings from each other, we call them interpretation A and B. In interpretation A, we define "machine" to be physical hardware, which is essentially the microprocessor. In interpretation B, we define "machine" as consisting of the whole IT stack of microprocessor, operating system and installed apps/software as shown in Fig. 3.8.

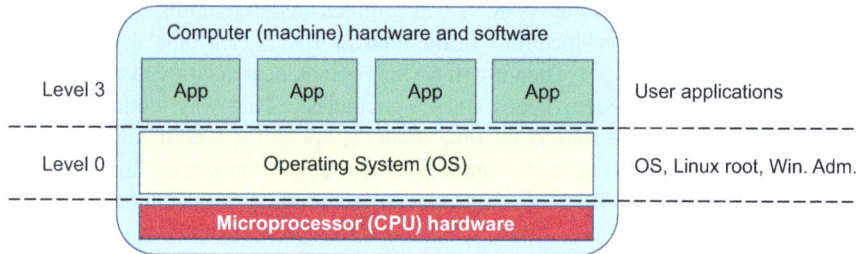

Fig. 3.8 IT stack of hardware and software in a computer

At this point, it is appropriate to introduce the main component of virtualization called hypervisor. This program module mimics the microprocessor such that the hypervisor receives microprocessor instructions from the operating system and software, so that for these it is as if the hypervisor is the microprocessor. However, the hypervisor does not execute the received microprocessor instructions, but forwards them to the real microprocessor which actually executes the instructions. The hypervisor also has other functions such as keeping track of multiple operating systems at the same time, as shown in Fig. 3.9, where three separate operating systems are installed on top of the hypervisor.

Interpretation A is such that if the microprocessor is the machine, then the hypervisor is a virtual machine. Interpretation B is such that if the whole IT stack of microprocessor, operating system and apps is the machine, then each separate stack consisting of hypervisor, operating system and apps is a separate virtual machine, as shown in Fig. 3.9.

Although interpretation A seems logical, it is not how people normally understand it. In the common IT-jargon, Interpretation B is usually assumed. When we

3.6 Virtualization

hear a phrase like *"installing a virtual machine"* it means installing a new operating system over the hypervisor, including apps whenever needed. Further in this book we follow the common IT jargon and interpretation B of the term "virtual machine", as illustrated in Fig. 3.9.

Operating systems that run over a hypervisor are typically called guest operating systems, abbreviated as guest OS, and are essentially ordinary operating systems. They are said to be *virtualized* because they run over a hypervisor which, after all, is a virtual microprocessor. Thus, with virtualization, it is possible to run many separate virtualized guest OSs with associated apps/software on the same physical hardware.

3.6.2 Virtualization Architectures

Two main types of virtualization are Type 1, aka. "native", and Type 2, aka. "hosted". These are briefly described below, with their respective implications for security.

Type 1 is *native virtualization*, or "on-the-metal virtualization". This is the most direct and efficient way to install and run virtual machines. With native virtualization, the hypervisor runs directly on the microprocessor, i.e. closest to the physical hardware, or "metal", as shown in Fig. 3.9. On top of the hypervisor, several different operating systems and associated apps/software can be installed, which constitute separate virtual machines.

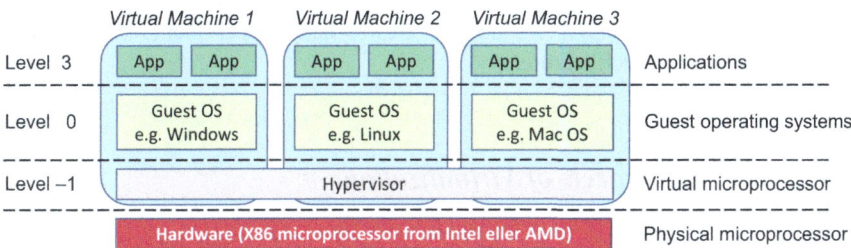

Fig. 3.9 VM type 1: Native virtualization architecture

Levels −1, 0 and 3 in Fig. 3.9 are, as explained in Sect. 3.4 about privilege levels, which determine what permissions each program module has in the system. Native virtualization provides a logical structure of privilege levels in that the hypervisor is more privileged than the guest operating systems that it controls. At the same time, the guest operating systems run at privilege level 0, for which they are designed, so that no customizations are needed for them to perform the necessary functions.

Type 2 is *hosted virtualization*, which means that the virtual machines run on top of a host operating system which in turn runs directly on the microprocessor. In hosted virtualization, the hypervisor is installed as a normal app, together with other apps, on the host operating system, as shown in Fig. 3.10. Over the hypervisor, several different operating systems and associated apps/software can be installed, which constitute separate virtual machines.

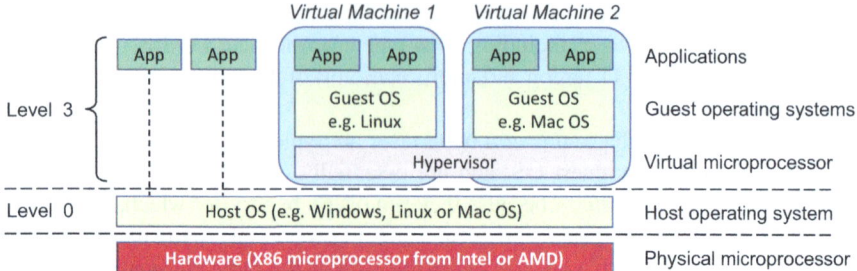

Fig. 3.10 VM type 2: Hosted virtualization) architecture

Levels 0 and 3 in Fig. 3.10. are as explained in Sect. 3.4 about privilege levels. Note that privilege level-1 is not used in this architecture, and that both the hypervisor and guest operating systems run with Level 3 as normal apps. Operating systems are basically designed to perform a set of privileged functions that require level 0 and that are not allowed for level 3 processes. To work around this constraint when they are guest operating systems, such functions must be called indirectly through the host operating system, which causes additional delay in processing. Therefore, hosted virtualization is a less efficient architecture than native.

If you have a system with a standard operating system and want to use virtualization without high demands on efficiency, you can use hosted virtualization because a hosted hypervisor is easy to install as a regular app. The hypervisor needs access to certain features that are typically not available for level 3 processes. Therefore, it may be necessary to enable a flag in BIOS/UEFI when a hypervisor is installed in a hosted virtualization architecture.

3.6.3 Security Aspects of Virtualization

In many ways, virtualization provides increased system security because an attack against a guest operating system does not affect the hypervisor or other guest operating systems. An infected guest operating system can be easily deleted, thus eradicating the infection. For example, dangerous malware can be tested in a virtual machine without compromising the security of the system itself. When analysing malware, it is common to run it in a virtual machine, which makes it easy to step through instructions one by one to see how the malware behaves. The hypervisor can store memory dumps and snapshots of the state of the malware and the guest operating system at certain points when running the malware.

Briefly mentioned below is a set of security requirements that must be met to have a virtualization architecture with adequate security.

- Guest operating systems must not be able to access each other's code or data, and must not be affected by each other.
- The guest operating systems must not be able to affect the hypervisor.

- Guest operating systems must not be able to detect that they are virtualized, that is, the should perceive their context as if it were microprocessor hardware.

These requirements can never be 100% satisfied, and virtualization architectures have been attacked in the wild. Like any other software, hypervisors also need security patching and updating.

3.7 Trusted Computing

Trusted computing means that aspects of the security of a system are anchored in hardware in some way. The idea of trusted computation is that hardware is significantly more robust to security threats than software. There are many different models for trusted computing, where some examples are:

- *Secure boot with* UEFI. This is described in the next section.
- *Trusted Platform Module (TPM)* is a coprocessor integrated into systems such as shown on Fig. 3.1. TPM are built into many types of systems, such as servers, laptops, smartphones, etc. The TPM supports three specific security features: (1) secure boot, (2) remote attestation, and (3) sealed storage. Secure boot with TPM differs from secure boot with UEFI.
- *TEE (Trusted Execution Environment)*, aka. *confidential computing*, means that data and calculations to a process are protected with hardware technology in the microprocessor, not just by security features in the operating system.
- *Tamper-resistant physical encapsulation* means that hardware is mounted with a protective enclosure that is difficult for an attacker to penetrate. An encapsulation is tamper-proof if in addition it can detect attempted physical tampering and, if necessary, automatically delete sensitive information, such as cryptographic keys.

The following explains only the Secure Boot principle with UEFI, as it is the most important and widely used method of trusted computing. If startup of a system is insecure, no security mechanism can restore security.

3.8 Secure Boot

The most common method of secure boot of operating systems is based on *UEFI (Unified Extensible Firmware Interface)*. UEFI replaces the traditional BIOS (Basic Input/Output System), which is no longer widely used. UEFI supports a secure boot sequence by ensuring that only correctly digitally signed and correctly verified drivers and boot loaders can be loaded into memory and executed by the microprocessor.

The process behind this might appear complicated but is in reality quite simple. It uses the principle of digital signature which is described in Sect 4.7. Without going into detail here, digital signature requires a pair of private and public cryptographic keys where the private key is used for signing, and the public key is used for verification.

This pair of public and private keys is generated by the computer manufacturer. During the configuration of a new computer at the factory, the public key, called *PK (Platform Key)*, is stored in the firmware ROM chip together with UEFI program code for initial system start-up. The firmware ROM chip is mounted on the motherboard so that the UEFI code is directly available for the microprocessor at start-up. The computer manufacturer also signs other program modules such as drivers and the boot loader with the corresponding private key, and stores these on the hard disk together with their digital signatures. During start-up the public PK is used to verify the digital signatures on program modules signed by the computer manufacturer. The PK and UEFI code form a so-called *trust anchor* because they are used to verify and enforce the security of the boot sequence.

When the system boots, the CPU (Central Processing Unit), i.e. the microprocessor, starts by retrieving and executing initial UEFI code from the firmware ROM chip. This code also retrieves the PK from the ROM chip and uses the key to verify that UEFI drivers and OS loaders stored on disk have correct digital signatures. If a digital signature is incorrect, the boot sequence is terminated. If all digital signatures are correct, UEFI drivers and OS loaders are loaded into memory. The microprocessor (CPU) starts running these program modules, which eventually load the OS kernel and the entire operating system. Then the system startup is completed, and the system is ready to serve the user. This startup sequence is illustrated in Fig. 3.11.

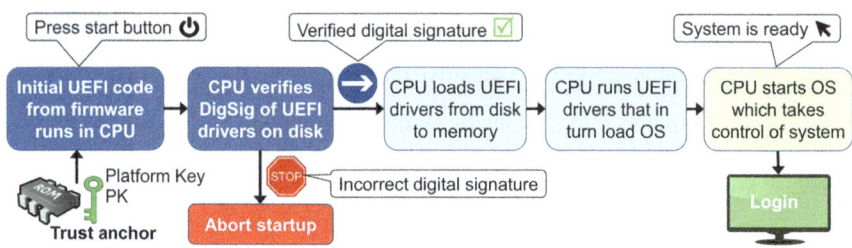

Fig. 3.11 Secure boot sequence

When we say that the CPU is running a program, it strictly means that the program is running using the CPU, because the instructions that execute are specified by the program. The analogy is that a car driver (program) runs using a car (CPU). In other words, it is the car driver (program) who is in charge and decides where they go, while the car (CPU) is the driver's tool to get where they want to go.

Secure boot thus depends on verifying digital signatures on program modules. The logic is that only the manufacturer of the computer is able to generate the correct digital signatures because only the manufacturer holds the private key, thus guaranteeing that these program modules are installed on the computer by the manufacturer and are not spoofed by attackers.

The abbreviation UEFI (Unified Extensible Firmware Interface) can be difficult to understand intuitively. "Unified" means that several companies that previously had their own proprietary standards joined forces to create a common standard. "Extensible" means that arbitrarily many and different UEFI drivers can be included. "Firmware Interface" means that this technology is the interface between a firmware chip with UEFI code attached to the system's motherboard and the system's operating system stored on disk. UEFI is a method of providing the system with a trust anchor in the form of a firmware chip.

Using Secure Boot can be controversial for several reasons. At the slightest error when verifying digital signatures, the system will not start. A less stringent variant of secure boot is therefore "authenticated boot", which means that the boot sequence is not terminated by an incorrect signature, but that the user is informed that the boot sequence was incorrect.

Another aspect of enforcing secure boot is that the user cannot easily install a completely new operating system because UEFI drivers and the OS bootloader dictate which operating system to load. For users who want to install their own operating system, the UEFI standard specifies the possibility of writing the user's own public PK (Platform Key) to the UEFI firmware chip. In other words, the user must generate a separate pair of public and private keys, and digitally sign UEFI drivers and the OS loader with their own private key. Alternatively, the UEFI can be programmed to boot without verifying the digital signatures, thus making it a non-secure boot. These aspects can be of interest to organizations that require strong security management in their own systems, and who need to be independent of the manufacturer. It is, of course, a complex procedure, difficult for individuals or small businesses to perform, which thus have little opportunity to install their own host operating system. The procedure of installing custom operating systems is easier for larger organizations that have the necessary resources for it. It is interesting to note that these technical aspects at the bottom of the IT stack which are the trust anchor of systems, are influenced by market forces and policies in organizations. When these systems are used as platforms for critical applications, it is understandable that organizations want to control the root of trust.

3.9 Intel Management Engine

Intel ME (Management Engine) is a small, low-power subsystem integrated with Intel microprocessors. Intel ME runs its own operating system called MINIX, which is always active as long as the microprocessor has power, i.e. regardless of whether the platform's main operating system (Windows, Linux, MacOS, hypervisor) is running normally, is dormant, or is turned off completely (but the power is on). Intel ME has full access to system hardware, including system memory, monitor, keyboard, camera, microphone, peripherals, and networking. MINIX is a mini OS with a limited set of features, which means that it has low power consumption and small footprint (physical size) in embedded systems.

According to Intel, ME is used to perform "various tasks," but Intel gives few details about what those tasks are. One of these tasks is to support AMT.

Intel AMT (Advanced Management Technology) is a scheme for remote management of workstations. It can be used on servers and client computers running Intel processors. Users of Intel AMT are usually large organizations, not home users. AMT must be enabled in UEFI/BIOS to be usable and is disabled by default when servers are delivered to the customer. AMT remote management can turn on, configure, control, or delete the OS on computers running Intel processors. Unlike typical platform management schemes that require OS support, AMT works even if no operating system (Windows, Linux, MacOS, hypervisor) is installed.

ME has full control over an Intel processor, and is therefore the most privileged feature of an Intel-based computing platform. AMT enables access to ME remotely controlled. A relevant threat scenario is, for example, that attackers manage to take control of ME via remote access based on AMT. This threat scenario is important for organizations to be aware of.

The company AMD is the second major manufacturer of microprocessors for servers and workstations, besides Intel. AMD offers *ARM TrustZone*, which has similar functionality to Intel ME, and at the same time is a technology for trusted computation in their ARM-based microprocessors.[2]

3.10 Side Channels and Covert Channels

Side channels and covert channels are two similar principles that can allow information to be transferred in violation of system security policies, such as leakage of secret information. The difference is subtle, but at the same time important to understand. The definitions below will clarify the difference. We'll start with the definition of side channel:

> *A side channel is an unintended channel that emits information resulting from the physical implementation of a system, and not from weaknesses in the theoretical model of the system.*

A typical side channel from a system can be evidence of time spent performing a program function or instruction in a microprocessor. It is often possible to guess the value of local variables in a function or instruction from the time it takes to execute the function. If the microprocessor performs encryption, an attacker can, for example, guess the value of bits in the encryption key simply by observing the time it takes to compute. Other typical side channels are power consumption,

---

[2] arm: Advanced RISK Machines, where RISK means Reduced Instruction Set Computer.

electromagnetic radiation and sound. Stopping electromagnetic radiation from a device can be done by placing it in a Faraday cage described in Sect. 7.1.1.

Advanced analytics of detected signals, e.g. with machine learning, can reveal sensitive information. Security controls against side-channel attacks include implementing equipment in a way that does not emit such signals, or placing IT equipment in a shielded environment where such signals cannot be intercepted by attackers. Side channels are related to emission security and TEMPEST described in Sect. 12.4.5.

We move on to covert channels, which is the second type of channel, with the definition below.

> *A covert channel is a mechanism that was not designed for communication, but that can nevertheless be misused to transmit information in a way that violates the system security policy.*

An example of a covert channel could be if the implementation of MAC (Mandatory Access Control) policy, described in Sect. 9.3.2, has a weakness. A simple explanation of MAC is that you should not be able to read a document with a hight security classification than your security clearance. This policy includes that you should not even know whether such a document exists, so of you try to list all documents, then those with a higher classification should not be displayed. However, if you attempt to create a document with a higher classification than your clearance, and the document already exists, the file system could notify you with "File name already exists". This would be a covert channel, because it lets you know that the document exists, which is a policy violation.

The communication capacity (aka. bandwidth) of covert channels is typically very low, but the fact that such a channel exists in a system at all is a vulnerability that should be removed.

3.11 Tasks

1. Trust anchor
   What is meant by the term "trust anchor" in system security? Cite an example.
2. Virtualization
   (a) What is the difference between a "real thing" and a "virtual thing"?
   (b) What is a virtual machine when talking about computers?
   (c) How can virtualization support security?

3. Privilege levels

Search and find the talk "Replace Your Exploit-Ridden Firmware with Linux" by Ronald Minnich, at the Embedded Linux Conference & Open Source Summit Europe 2017, e.g. on YouTube https://www.youtube.com/watch?v=iffTJ1vPCSo

   (a) Give the privilege levels of (i) a user, (ii) an administrator/root, and (iii) a virtual machine.
   (b) What is UEFI?
   (c) What is Intel ME (Management Engine)?
   (d) In what way can one say that UEFI and Intel ME have privilege levels, and what are they?

Chapter 4
Cryptography

Cryptography is typically bypassed, not penetrated.

Adi Shamir, Israeli cryptographer

Before the modern age, cryptography was only used for encryption which is to use a secret key to convert readable text (cleartext) to unintelligible nonsense text (ciphertext). Decryption is then to use the same secret key to recover the text. While encryption still is an important application today, cryptography has several other uses, especially to ensure the authenticity of entities (systems and users), and to ensure the authenticity and integrity of information. This is built into communication protocols on the Internet and other communication technologies. Modern cryptography uses advanced mathematical theory, where the security strength is based on computational hardness assumptions. While it is theoretically possible to break a well-designed cryptographic algorithm, it is infeasible in practice to do so because it would take millions of years with current computers.

In this chapter, you will learn what cryptography is, what types of cryptographic algorithms exist, how they are used, and what kind of security they support. Next, you will learn about the history of cryptography from ancient Egypt to the present day. Finally, you will learn about how future quantum computers could potentially make today's asymmetric encryption algorithms insecure, and how this has already triggered the development and standardization of post-quantum cryptography.

4.1 What Is Cryptography?

Cryptography is the science of secret writing, the purpose of which is to conceal the meaning of a text/message, most often in connection with communication. The etymology of the word comes from the Greek: kryptos ("hidden") and graphein ("writing").

An *encryption algorithm*, or crypto algorithm, is a method of encrypting information. Traditionally, it is called a *cipher* from the word "chiffre" in French. Encrypting is traditionally called "ciphering," while decrypting is called "deciphering." These words are becoming old-fashioned, so today, as a rule, the terms *encrypt* and *decrypt* are used today. The term "code" is sometimes also used to denote a cryptographic algorithm.

The combination of cryptography and cryptanalysis is called *cryptology*, and researchers, as a rule, work on both aspects. Trying to *cryptanalyse* (to crack) an encryption algorithm is the best way to test whether it is sufficiently strong.

Cryptography supports the following security goals/purposes:

- **Confidentiality:** Make data unreadable to entities that do not have the correct cryptographic key, even if they have the encrypted data.
- **Entity authentication:** An entity that communicates can gain certainty about the identity of other entities in the communication.
- **Data authenticity/integrity:** The recipient of data can verify the identity of the sender and that data is correct and has not been altered, either intentionally or unintentionally.
- **Digital signature:** Third parties (all entities) can verify the identity of the publisher/sender of data and that data is correct and has not been altered, whether intentional or accidental.
- **PKI:** An efficient and scalable infrastructure for distributing cryptographic public keys.

4.2 Cryptographic Features

Cryptographic functions can be broken down into categories, each of which fills specific roles in the implementation of cryptographic solutions, as shown in Fig. 4.1.

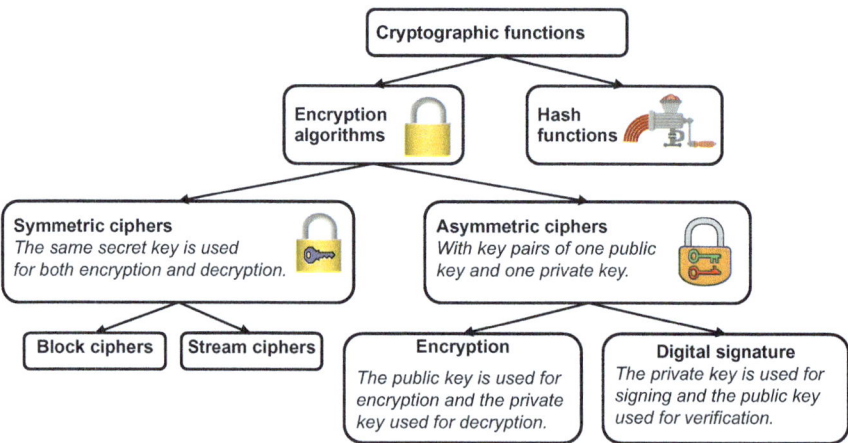

Fig. 4.1 Categories of cryptographic functions

The different categories of cryptographic features are described in the following sections.

4.3 Symmetric Algorithms

An algorithm is called symmetric when the same key is used for both encryption and decryption. A block cipher encrypts a block of plaintext into a block of ciphertext in each encryption operation. *AES (Advanced Encryption Standard)* is the most widely used block cipher, which supports most of the encryption over the Internet. AES has three alternative key sizes of 128, 192, or 256 bits and a fixed block size of 128 bits.

A stream cipher encrypts a stream of data by encrypting each bit separately. Strem ciphers are used, among other things, for encryption of traffic in 2G mobile communications. Figure 4.2 shows the principles of block ciphers and stream ciphers.

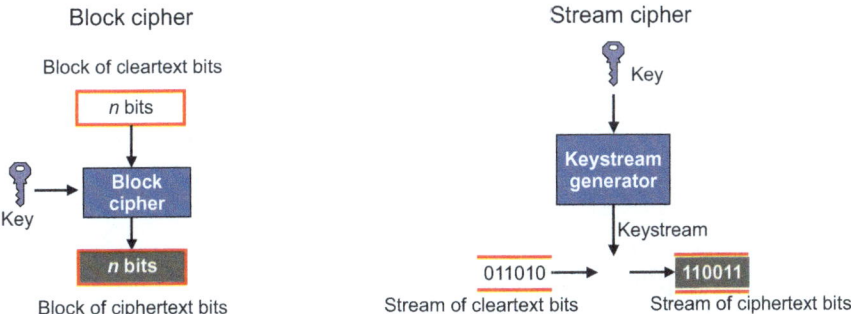

Fig. 4.2 Principle of block cipher and stream cipher, two types of symmetric algorithms

Let a clear text message be expressed as M and let its corresponding cipher text be expressed as C. The encryption function is expressed as E with input parameters message M and secret key K. Similarly, the decryption function can be expressed as D with input parameters, ciphertext C and secret key K. Formally, encryption and decryption can be expressed as below.

- Encryption: $C = E(M, K)$
- Decryption: $M = D(C, K)$

The process for encrypting and decrypting with a symmetric cipher is illustrated in Fig. 4.3, where Alice is the sender and Bob is the recipient of the message.

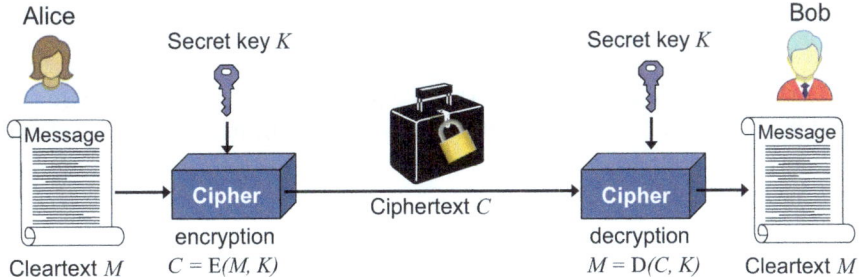

Fig. 4.3 Principle of symmetric encryption and decryption

Alice and Bob are, of course, processes/entities and not human beings. It is common jargon in the crypto community to use the terms Alice and Bob to represent entities that exchange encrypted information.

Two factors are important for the theoretical strength of a block cipher:

- **Key size:** One way for the attacker to find the secret key is to do an exhaustive search (complete search) of the entire key space. The time it takes for an exhaustive key search depends on the key size. Typical key size for symmetric encryption is 128 or 256 bits, as in AES. For a key size of 128 bits, an attacker would have to try an average of $(2^{128})/2$ keys to find the key, which would take millions of years.
- **Algorithm strength:** For example, cryptanalysis to find cryptographic keys exploits statistical patterns in the ciphertext, or knowledge both the plaintext and ciphertext of a message. To resist cryptanalysis, the encryption process must erase all statistical patterns in the plaintext so the ciphertext contains no statistical patterns, i.e. it appears random. An example of statistical patterns in normal text is shown in Fig. 4.15. The effect of removing statistical patterns is achieved, for example, with a block cipher based on Shannon's SP network shown in Fig. 4.17.

4.4 Block Cipher Modes of Operation

Block encryption itself is only a primitive cryptographic transformation (encryption or decryption) of a block with a certain number of bits. An *encryption mode* describes how the cipher is used to encrypt data files and messages larger than a block. Fig. 4.4 illustrates that secure encryption modes are important.

Cleartext   Ciphertext produced with ECB mode   Ciphertext produced with secure mode

Fig. 4.4 Example of insecure and secure encryption mode

The simplest encryption mode for a block cipher is ECB (Electronic Code Book). With the ECB, each block is encrypted separately with the message block and key as input parameters. However, it turns out that this method is insecure, as shown in Fig. 4.4.

4.4 Block Cipher Modes of Operation

The problem with ECB mode is that duplicate plaintext blocks are always encrypted as duplicate ciphertext blocks. Figure 4.4 Displays clear text in the form of an image of a penguin having large areas of similar color. Because plaintext pixels of equal color are encrypted into cipher text pixels of equal color, it is logical that the penguin is easily recognizable even if the image is encrypted.

A secure encryption mode ensures that duplicate plaintext blocks are encrypted to different ciphertext blocks. An example of a secure encryption mode is *counter mode*, abbreviated as CTR (Counter Mode). Other encryption modes that are also considered secure are e.g. CBC (Cipher Block Chaining), CFB (Cipher Feedback) and OFB (Output feedback). We only describe the ECB and CTR in the sections below.

4.4.1 ECB: Electronic Codebook

The simplest of the encryption modes is *ECB (Electronic Codebook)*. The name comes from traditional physical "code books" made of paper, where identical words or characters were always encrypted into the same cipher word or cipher.

To encrypt with the ECB, the message is divided into blocks of typically 128 bits each, and each block is encrypted separately, as shown in Fig. 4.5. The decryption is done by running the cipher algorithm in reverse and using the same key.

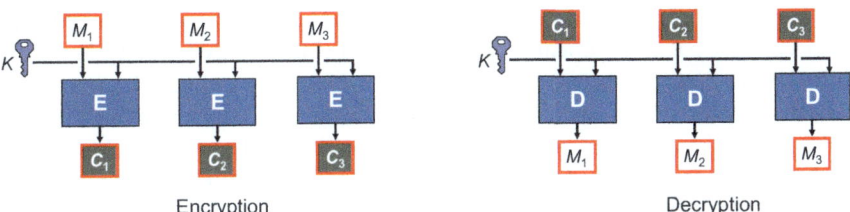

Fig. 4.5 ECB mode: Electronic Codebook Book

Cleartext blocks are denoted as M_1, M_2, M_3 in the figure, and the corresponding ciphertext blocks as C_1, C_2, C_3. Let M_i be a cleartext block and C_i the corresponding ciphertext block with index i. The same key K is used for both the encryption function E and the decryption function D. The terms below give a mathematical formulation of ECB encryption and ECB decryption of a single block.

- Encryption: $C_i = E(M_i, K)$
- Decryption: $M_i = D(C_i, K)$

ECB is very easy to implement, but as this encryption mode is insecure, it is normally not used in practice.

4.4.2 CTR: Counter Mode

With CTR (Counter Mode), block encryption is made into a form of stream cipher encryption, similar to the encryption method on the right side of Fig. 4.2. A CTR implementation includes a counter function, where the counter value is the same size as the block, typically 128 bits. Each counter value is encrypted to produce a counter cipher block of the same size. Encryption is performed by encrypting each counter value to calculate a counter cipher block that is XORed with a cleartext block, as shown in Fig. 4.6. The decryption is done in the same way, that is, with XOR addition of the same counter cipher block to the corresponding ciphertext block.

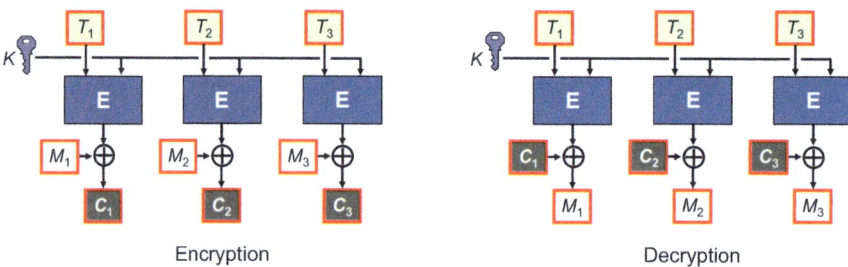

Fig. 4.6 CTR: Counter mode

Cleartext blocks are denoted as M_1, M_2, M_3, and the corresponding ciphertext blocks as C_1, C_2, C_3. Additionally, counter values are denoted as T_1, T_2, T_3. Let M_i be a cleartext block and C_i be the corresponding ciphertext block with index i. Also, let T_i be the counter value with index i. In CTR mode only the encryption function E is used, i.e. the inverse decryption function D is not used. The terms below give a mathematical formulation of CTR encryption and CTR decryption of a single block. XOR addition is denoted as $\oplus$.

$$\text{Encryption}: C_i = \text{E}(T_i,\ K) \oplus M_i$$

$$\text{Decryption}: M_i = \text{E}(T_i,\ K) \oplus C_i = \text{E}(T_i,\ K) \oplus \text{E}(T_i,\ K) \oplus M_i = 0 \oplus M_i$$

The explanation for why decryption is possible without using the decryption function D is that binary addition twice with the same counter cipher block $\text{E}(T_i, K)$ produces a string of 0-bits which gives the *null transformation*, so that the plaintext block is revealed.

The counter can be any function that produces a very long sequence that does not repeat during encryption of the cleartext. The most common is simple incrementation of the counter for each block, which gives the longest possible sequence for a given block size. The counter is initiated with a random value. CTR mode is relatively easy to implement and is widely used.

4.4.3 CBC: Cipher Block Chaining

With CBC (Cipher Block Chaining), the encryption of each block is a function of the previous ciphertext block, so that the encryption of multiple blocks forms a chain, as illustrated in Fig. 4.7. Since there is no block before the first block, an initialization vector *IV* is used for encrypting the first block. The decryption is done in a similar way, where each decrypted cleartext block is a function of the previous ciphertext block.

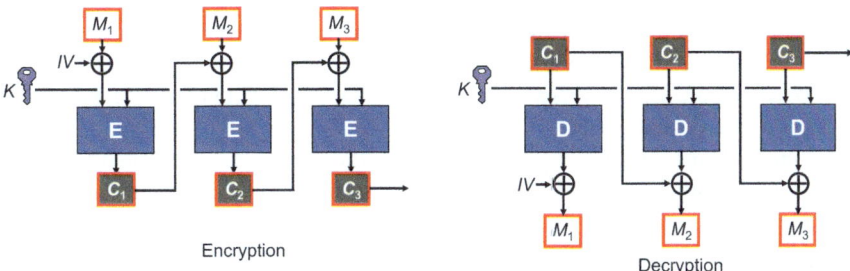

Encryption

Decryption

Fig. 4.7 CBC mode: Cipher Block Chaining

Cleartext blocks are denoted as M_1, M_2, M_3, and the corresponding ciphertext blocks as C_1, C_2, C_3. Additionally, the initialization vector *IV* is used for encryption and decryption of the first block. Let M_i be a cleartext block and C_i be the corresponding ciphertext block with index i. The terms below give a mathematical formulation of CBC encryption and CBC decryption of a single block. XOR addition is denoted as $\oplus$.

$$\text{Encryption}: C_i = \text{E}\big((M_i \oplus C_{i-1}), K\big)$$

$$\text{Decryption}: M_i = \text{D}(C_i, K) \oplus C_{i-1} = \text{D}\big(\text{E}((M_i \oplus C_{i-1}), K), K\big) \oplus C_{i-1}$$
$$= M_i \oplus C_{i-1} \oplus C_{i-1} = M_i \oplus 0$$

The initialization vector *IV* has different security requirements than a key, so it usually does not need to be secret. However, it should be random and never reused.

4.5 Hash Functions and MAC

A cryptographic hash function is a mathematical algorithm that converts arbitrary size data (often called the "message") into a fixed-size hash. Hashing is a one-way function, that is, a function that is practically impossible to invert, although theoretically it is possible. Ideally, the only way to find a message that provides a specific hash value is to try every possible message to see if they produce the hash, or use a lookup table with pre-generated hashes and messages. Cryptographic hash functions play an important role in modern cryptography, especially in the context of data authentication and digital signature.

A hash function should have the following characteristics:

- It is deterministic, which means that the same message always produces the same hash value.
- Calculating a hash value is very fast.
- It is practically impossible to generate a message that produces a predetermined hash value (i.e. reversing the hash function), although it is theoretically possible.
- It is practically impossible to find two different messages that give the same hash value, although it is theoretically possible. This is called "collision resistance". Note that there are two types of collision resistance (weak and strong), but which we will not describe here.
- A small change in a message should change the hash value so extensively that a new hash cannot be correlated with the old hash, which is called the "avalanche effect".

Cryptographic hash functions have many applications for cryptographic security, especially for message authentication code (MAC) and digital signature. They can also be used as common (non-cryptographic) hash functions, to index data in hash tables, for digital tagging, to detect duplicate data, to identify files uniquely, and as checksums to detect unintentional data corruption.

A MAC function is a cryptographic hash function that takes a secret key K as an input parameter in addition to the message itself M to be hashed. We therefore write the MAC function as Hash(M, K). We say that the MAC function produces a *MAC* (Message Authentication Code). If we assume that Alice and Bob share a secret key K, and Alice sends a message to Bob, a *MAC* enable Bob to verify that the message was sent by Alice and that it has not been altered along the way. Figure 4.8 shows how Bob uses *MAC* for message authentication. The *MAC* is computed with a hash function, formally expressed as MAC = Hash(M, K). A MAC function Hash(M, K) can be seen as a hash function controlled by the secret key parameter K. Bob receives both *MAC* and the message he first calls M' (because he is initially uncertain if it is the real message). Bob computes MAC' = Hash(M', K) for the message M' and verifies that it is identical to *MAC* that was received along with the message. Assuming that only Alice and Bob know the secret key K, this constitutes proof that no one else but Alice sent the message.

4.6 Diffie-Hellman Key Exchange

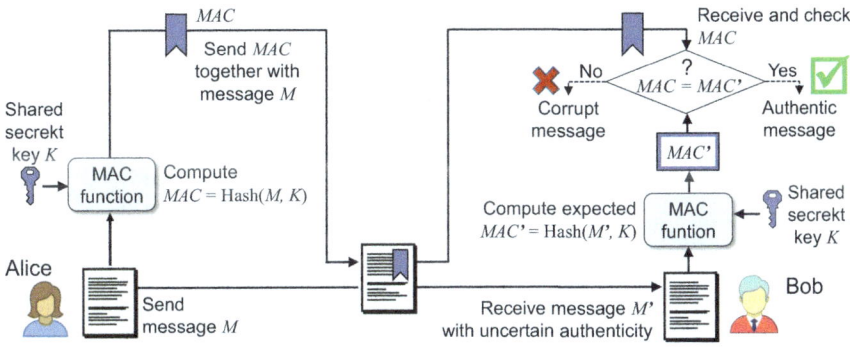

Fig. 4.8 Principle of message authentication with MAC

It is important to understand that *MAC* only gives certainty to Bob that Alice sent the message. Bob is unable to prove this to a third party because the key *K* is secret between Alice and Bob, as shown in Fig. 1.12 in Sect. 1.10.1.

4.6 Diffie-Hellman Key Exchange

Both symmetric encryption and message authentication described in the sections above require that secret encryption keys are exchanged in advance. Secure exchange or the establishment of secret keys is a fundamental problem in cryptography, where many different solutions are used. Diffie-Hellman represents an elegant method of exchanging symmetric session keys over an insecure channel.

4.6.1 Traditional Diffie-Hellman

Diffie-Hellman (DH) is a protocol for the establishment/exchange of secret keys between two parties over an openly insecure network. The name comes from Whitfield Diffie and Martin Hellman, who in 1976 published the article *New Directions in Cryptography,* describing the method [14]. However, the method had already been discovered in 1974 by Malcolm John Williamson, who worked at GCHQ (Government Communications Headquarters) in the UK, but there the method was classified as TOP SECRET [15]. The method is based on modular arithmetic, and is relatively simple.

Suppose Alice and Bob agree on a set of global parameters: a large prime number p and an integer g [mod p], which is a primitive root modulo p, i.e. g generates all integers in the interval $[1, (p-1)]$ by raising g (exponentiate g) to the power of k. To compute a number modulo p means to compute the residual value when the number is divided by p. If the number is less than p, the number is not divided, for the number is then already a residual value. As a toy example with small numbers, we choose prime numbers $p = 7$ and primitive root $g = 3$. It can be verified that only 3 and 5 are primitive roots modulo 7, i.e. 3 and 5 generate all numbers in the interval

[1, 6], while 2, 4 and 6 generate only a subset of integers in the interval [0, 6]. Below we show how g generates all integers in the interval [1, 6] by raising g to different powers k, and that the cycle repeats after 6 times, i.e. the period is 6.

3^1	= 3			[mod 7]
3^2	= 3·3	= 2		[mod 7]
3^3	= 2·3	= 6		[mod 7]
3^4	= 6·3	= 18	= 4	[mod 7]
3^5	= 4·3	= 12	= 5	[mod 7]
3^6	= 5·3	= 15	= 1	[mod 7]
3^7	= 1·3	= 3		[mod 7]

The sequential exponents of each number in the left-hand column are the discrete logarithms (with base 3) of the corresponding residual values. The second column shows for each row the residual value of the previous row multiplied by 3, which is the same as incrementing by 1 the exponent for each row. When p and g are sufficiently large, it is practically impossible to compute the discrete logarithm because there is a complex relationship between the order of the residual values and the sequential order of exponents in the left-hand column. This complexity is the basis for the security of the Diffie-Hellman algorithm. The parameters p and g define an *algebraic group*.

Alice and Bob each choose their own private keys: Alice chooses a and Bob chooses b. Then they each compute their respective public key values: Alice computes g^a [mod p] and Bob computes g^b [mod p] which they send to each other. Subsequently, both Alice and Bob are able to compute the same secret symmetric key $g^{ab} = (g^a)^b$ [mod p] = $(g^b)^a$ [mod p].

The security of this protocol is based on the difficulty of computing the discrete logarithm of large integers in an algebraic group of integers modulo p. For an attacker to discover the secret symmetric key, the attacker first would need to find one of the private keys, for example by computing Alice's private key $a = \log_g(g^a)$ [mod p]. If p (and g) are large integers, it is practically impossible to compute discrete logarithm, although theoretically it is possible. The Diffie-Hellman Protocol is illustrated in Fig. 4.9.

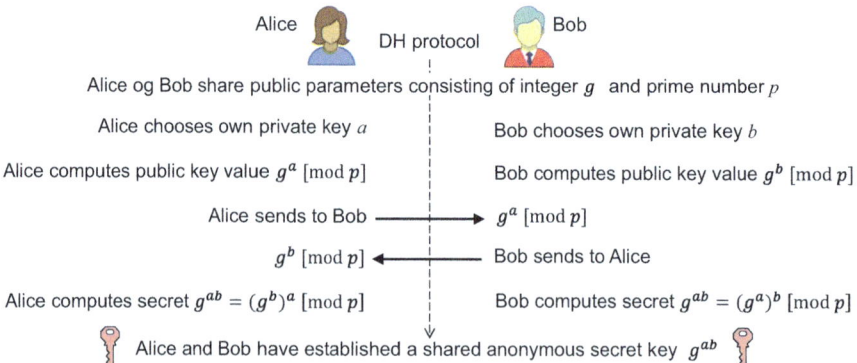

Fig. 4.9 Key exchange with Diffie-Hellman

4.6 Diffie-Hellman Key Exchange

At first sight, it may seem that the DH protocol can solve the problem of key exchange on the Internet by having all entities use Diffie-Hellman to establish common secret keys with each other. The problem is that Diffie-Hellman does not provide any kind of authenticity, which means the secret session key is *anonymous*. Metaphorically, it can be described as two people sitting in a dark cave having a confidential conversation with each other, but because it is pitch dark inside, they cannot see each other and therefore do not know who the other is. Authentication can be achieved by combining DH with digital signature, which assumes the existence of a PKI, which is described in the next chapter. Diffie-Hellman can also be combined with asymmetric encryption as described in Sect. 4.7.4.

4.6.2 ECDH: Elliptical Curve Diffie-Hellman

ECDH (Elliptic-Curve Diffie-Hellman) is a variant of traditional Diffie-Hellman described in the previous section. The mathematics is based on multiplication on discrete elliptic curves (EC). The description of ECDH in Fig. 4.10 is a simplified representation. The mathematical operator * represents discrete algebraic multiplication on an elliptical curve EC, and not ordinary multiplication.

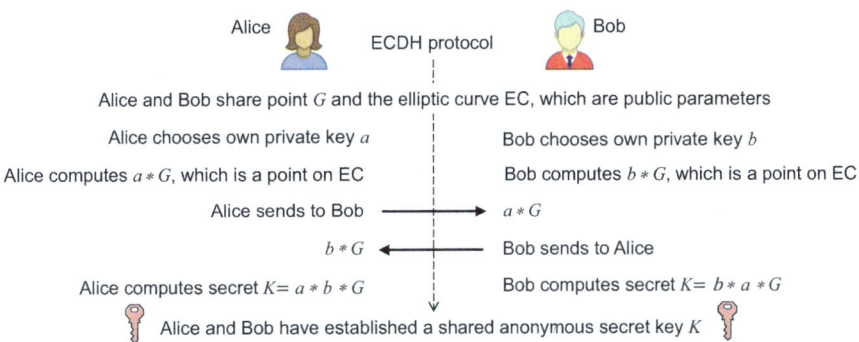

Fig. 4.10 ECDH—Elliptic-Curve Diffie-Hellman

As for DH, the session key computed with ECDH is anonymous, that is, the session key is not authenticated, which means that the protocol is vulnerable to man-in-the-middle attacks (MitM). The additional mechanism digital signature described in Sect. 4.7.4 is necessary for authenticity. A variant of ECDH that provides authenticity is to use a prespecified password, which both Alice and Bob know, when computing the session key, such as in the Wi-Fi WPA3 Dragonfly protocol described in Sect. 7.2.4.

As can be seen from Figs. 4.9 and 4.10, ECDH and DH are quite similar. The difference lies in the mathematics, where DH uses exponentiation which requires relatively intensive computation, while ECDH uses multiplication on discrete elliptic curves that is less computationally intensive. For large amounts of computations that take place on a high-traffic website, using ECDH instead of DH would mean significant savings in time and power consumption. You may not see a difference in a small website, but the more traffic, the greater the computation savings of ECDH.

4.7 Asymmetric Algorithms

Asymmetric encryption algorithms can typically be used for both encryption and digital signature, but some algorithms are only suitable for digital signature, such as DSA (Digital Signature Algorithm), which was standardized as early as 1991. The principles of encryption and digital signature with asymmetric algorithms are described below.

In simple terms, the principle of asymmetric encryption and decryption is that a key pair with a private and a public key instead of a single secret key as in the case of symmetric encryption. It is assumed that the private key is only known to the entity that will receive the message, which can be a person or a system. It is assumed that the public key is known to everyone. In addition, it is of course assumed that this public key is authentic, which is achieved through a *PKI (Public Key Infrastructure)*, which is described in the next chapter.

Imagine Alice sending an encrypted message to Bob. Let Bob's private key be written as K_{priv}^{B} and Bob's public key written as K_{pub}^{B}. The asymmetric algorithm has an encryption function E, and a decryption function D. The cleartext message is denoted as *M* and the ciphertext as *C*. Simple encryption and decryption of a message sent to Bob with an asymmetric algorithm can be formally expressed as below.

- Encryption: $C = E\left(M, K_{pub}^{B}\right)$
- Decryption: $M = D\left(C, K_{priv}^{B}\right)$

However, it is impractical to encrypt large amounts of data in this way because asymmetric algorithms are computationally inefficient. In practice, encryption takes place using a combination of both an asymmetric and a symmetric algorithm.

4.7.1 The RSA Algorithm

The RSA algorithm is named after Ron Rivest, Adi Shamir and Leonard Adleman, who publicly described the algorithm in 1977 [16]. A similar algorithm was developed by Clifford Cocks in 1973 at the British GCHQ, where the algorithm was classified as TOP SECRET [15]. RSA is the best known and the simplest asymmetric encryption algorithm, which also supports digital signature. The algorithm is based on modular integer arithmetic, as explained below.

4.7 Asymmetric Algorithms

Key generation:
- Choose two large prime numbers p and q, compute the modulus $n = p \cdot q$ and the totient function $\phi = (p - 1) \cdot (q - 1)$.
- Select the public key K_{pub} so that K_{pub} and ϕ are coprime numbers, i.e. they do not have any common factors.
- Compute the private key K_{priv} so that $K_{pub} \cdot K_{priv} = 1 \, [\mathrm{mod} \, \phi]$, i.e. so that the keys are the inverse of each other $[\mathrm{mod} \, \phi]$.

Encryption and decryption:
- Alice wants to send encrypted message M to Bob who has key pairs $\left(K_{pub}^B, K_{priv}^B\right)$.
- Alice encrypts plaintext M by computing the ciphertext $C = M^{K_{pub}^B} \, [\mathrm{mod} \, n]$.
- Bob receives and decrypts ciphertext C by computing the plaintext $M = C^{K_{priv}^B} \, [\mathrm{mod} \, n]$.

Signing and verification:
- Alice has key pairs $\left(K_{pub}^A, K_{priv}^A\right)$ and wants to send signed message M to Bob.
- Alice signs message M by computing the signature $S = M^{K_{priv}^A} \, [\mathrm{mod} \, n]$.
- Bob receives and verifies signature by computing $M = S^{K_{pub}^A} \, [\mathrm{mod} \, n]$.

Toy example for RSA encryption:
- Select two prime numbers $p = 3$ and $q = 11$, which results in the modulus $n = 33$ and the totient function $\phi = 20$.
- Select the public key $K_{pub} = 3$.
- Compute the private key $K_{priv} = 7$ so that $3 \cdot 7 = 21 = 1 \, [\mathrm{mod} \, 20]$.
- Specify the cleartext message $M = 9$.
- Encrypt M by computing the ciphertext $C = 9^3 = 729 = 3 \, [\mathrm{mod} \, 33]$.
- Decrypt C by computing the cleartext $M = 3^7 = 2187 = 9 \, [\mathrm{mod} \, 33]$.

If the prime number factors p and q, and thus module n, are sufficiently large, it is practically impossible to compute the private key K_{priv} just by knowing the public key K_{pub} and modulus n. The difficulty of computing the private key K_{priv} comes from the difficulty of factoring the modulus n when n is sufficiently large. As the computation power of modern computers is steadily increasing, it is probably possible to factor numbers with size 1024 bits in a few weeks with today's supercomputers. Therefore, it is currently necessary to use key sizes of 2048 bits or more for RSA, which is starting to get very large. See also Sect. 5.1.3 for key size requirements. Other asymmetric cryptographic algorithms, such as elliptic curves, require much smaller key sizes than RSA for the same cryptographic strength, and are therefore often preferred in modern implementations.

4.7.2 Hybrid Encryption

Suppose Alice and Bob want to use encryption for exchanging large amounts of data. Asymmetric encryption gives inefficient computation but has the advantage of relatively simple key distribution. Symmetric encryption makes key distribution difficult, but has the advantage of efficient computation. *Hybrid encryption* is a way to combine the best of two worlds, i.e. efficient computation and relatively simple key distribution, but has the disadvantage of not supporting *forward secrecy*, as explained later.

Hybrid encryption is done by Alice generating a symmetric secret key K that she and Bob will use to encrypt messages M to produce ciphertexts C. The challenge is to securely transfer the secret key K to Bob. This is done by Alice encrypting K with Bob's public key K_{pub}^B, resulting in the encapsulated symmetric secret key K^*. The key K^* is sent to Bob, who decrypts it with his private key K_{priv}^B, so that he has received the secret key K. Alice and Bob can now send encrypted messages between each other. The term *KEM (Key Encapsulation Mechanism)* is used for this method of encrypting the secret key K and transmitting the encrypted key K^*, which is thus an "encapsulated" secret key. The protocol is summarized below.

Alice does the following:

- Generates the secret key K.
- Encrypts (encapsulates) the secret key: $K^* = \mathrm{E}\left(K, K_{pub}^B\right)$
- Sends to Bob encapsulated (encrypted) secret key K^*

Bob does the following:

- Receives the encapsulated secret K^*
- Decrypts the encapsulated secret key: $K = \mathrm{D}\left(K^*, K_{priv}^B\right)$

The result is that Alice and Bob have exchanged a common secret key K. From now on, they can send encrypted messages to each other with symmetric encryption as shown in Fig. 4.3 in Sect. 4.3.

- Encrypt the message: $C = \mathrm{E}(M, K)$
- Decrypt the message: $M = \mathrm{D}(C, K)$

The scenario for hybrid encryption with asymmetric/symmetric algorithm is shown in Fig. 4.11.

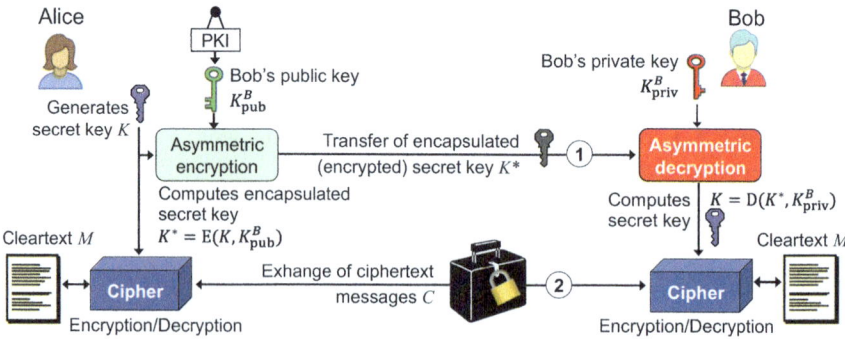

Fig. 4.11 Hybrid encryption with asymmetric and symmetric algorithms

4.7 Asymmetric Algorithms

In the scenario in Fig. 4.11 Alice is certain that she is communicating with Bob because she used Bob's public key to send the encapsulated secret key to Bob. However, Bob cannot know who he is communicating with. In order for Bob to authenticate Alice, the protocol would have to have additional elements for Bob to authenticate Alice. Such a protocol is described in Sect. 4.7.4.

The secret key K in Fig. 4.11 is a short-term key used only for one message or session. The key pair (K_{pub}^B, K_{priv}^B) are long-term keys that can be valid for several years.

When asymmetric encryption was invented in the 1970s, the fact that public keys could be forged was not considered a major problem. At that time, it was common for paper telephone directories of telephone numbers of every subscriber in a region to be sent physically to all subscribers in the same region, and it was suggested that cryptographic public keys could be listed along with the telephone numbers in the telephone directories. This was, of course, both impractical and insecure, and it took a few years before it became clear that authenticity is essential for the distribution of public keys. The solution for this was for public keys to be packaged in a certificate that is digitally signed by a *CA (Certificate Authority)*. If the user has an authentic copy of the CA's public key and trusts it, digitally signed certificates can be verified so that arbitrary public keys can be authentically received as certificates. This model is called *PKI (Public Key Infrastructure)*. Digital signature is described in Sect. 4.7.3 below, while PKI is described in the next chapter.

Hybrid encryption, as described above, revolutionized Internet security around 2000, when TLS 1.0 was adopted to encrypt communications between browsers and websites. The method was in use up to and including TLS 1.2 which was approximately until 2018. However, a potential problem with the secrecy of encrypted traffic in the long term into the future was realised. The problem is that if Bob's private key K_{priv}^B in Fig. 4.11 should it be leaked or cracked at some point in the future, the attacker would be able to decrypt historical traffic from all the way back to the time when that private key came into use. Ensuring forward secrecy of historical encrypted traffic is very important, therefore hybrid encryption as described in this section was eventually considered insecure. To solve this problem, a security protocol that provides *forward secrecy* was specified with TLS 1.3, which was introduced in 2018. This means that it should not be possible for an attacker to decrypt historical traffic even if the long-term key (private key) is leaked or cracked at some point in the future. TLS 1.3, which is described in Sect. 6.2.1, uses DH or ECDH described in Sect. 4.6 to establish session keys. A simple protocol for authenticated encryption with digital signature that provides forward secrecy is described in Sect. 4.7.4.

4.7.3 Digital Signature

Digital signature is based on asymmetric (encryption) algorithms, where the signer has a key pair consisting of a private and a public key. It is required that the private key is known only to the signing entity, which can be a person or a system. It can be assumed that the public key is known to everyone. It is of course necessary that everyone holds an *authentic* copy of this public key, which is achieved through a *PKI (Public Key Infrastructure)* described in the next chapter.

Let Alice be the sender og a signed message to Bob. Alice's private key is denoted K_{priv}^A as and her public key is denoted K_{pub}^A. The digital signature algorithm has a signing function denoted Sig and a verification function denoted Ver. The message is denoted *M,* and the signature is denoted *S*. Simple digital signature and verification of a message sent by Alice with a digital signature algorithm can be formally expressed as below.

- Signing: $S = \text{Sig}(M, K_{priv}^A)$
- Verification: $M = \text{Ver}(S, K_{pub}^A)$

However, it is impractical to sign large amounts of data in this way, because asymmetric algorithms are relatively inefficient, and should therefore not be used directly to sign large amounts of data. In practice, digital signing takes place using a hybrid method that involves a hash function along with a digital signature algorithm. Hash functions are quick even on large amounts of data, producing a hash value with a small and fixed size. A digital signature on the hash value of a message is the security equivalent of a digital signature on the entire message, and is efficient because the hash value is small.

Hybrid signing is done by the sender Alice creating a digital signature on the hash value of the message. The recipient Bob can then verify the digital signature on the hash. This principle allows for efficient digital signature generation and verification. This can be formally expressed as below.

Alice does the following:

- Computes the hash value of message *M*: $H = \text{Hash}(M)$
- Computes digital signature on hash value: $S = \text{Sig}(H, K_{priv}^A)$
- Sends to Bob cleartext message *M* and digital signature *S*

Bob does the following:

- Receives the signature *S* and message denoted *M'* because its authenticity is uncertain
- Computes the expected hash value of message *M'*: $H' = \text{Hash}(M')$
- Computes the hash value from the received signature: $H = \text{Val}(S, K_{pub}^A)$
- Compares hash values from the signature and the received message: $H = ? H'$

The logic is that if the hash values *H* and *H'* are equal, this is prof that Alice sent the message. This scenario is illustrated in Fig. 4.12.

4.7 Asymmetric Algorithms

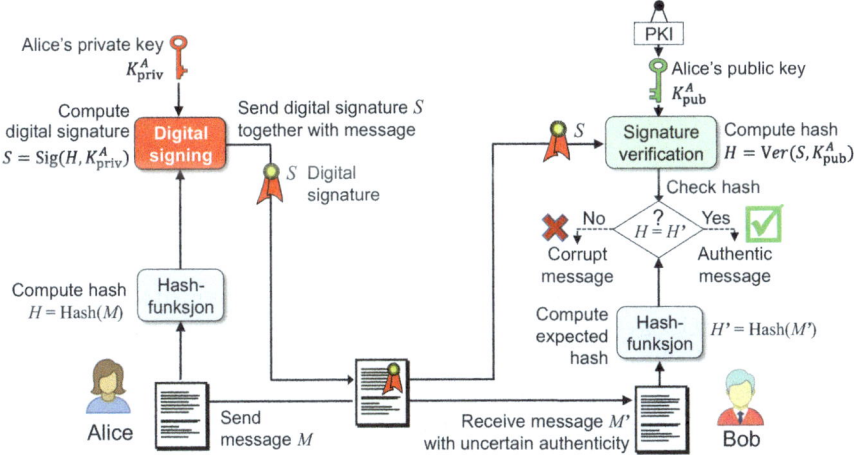

Fig. 4.12 The principle of digital signature generation and verification

Any entity that receives notification M and who has an authentic copy of Alice's public key K_{pub}^A, is able to verify the authenticity of the message, not just Bob, as shown in Fig. 4.12. When the authenticity of a message can be verified by any third party, the security service is called non-repudiation which is a strong form of data authentication. The difference between simple and non-repudiable data authentication is illustrated in Fig. 1.12 in Sect. 1.10.1.

For some asymmetric algorithms such as RSA, decryption and signing are essentially similar operations where both use a private key. Similarly, encryption and signature verification are often similar operations where both use a public key. Therefore, there are textbooks that describe signing as "decryption", because these operations are technically the same. These textbooks also describe digital signature verification as "encryption," because these operations are technically the same. Such descriptions are not incorrect, but can be confusing.

One purpose of digital signatures is that the meaning (interpretation) of a signed document cannot be changed after the signature has been applied. This objective is called WYSIWYS (What You See Is What You Sign). However, if the interpretation of the document involves complex processes with external dependencies, it may be possible to change the meaning of the document without changing the document itself and its signature. A simple example is that the visual representation of text using a computer screen can use different text fonts. How text visually looks on the screen depends on how the glyphs for each character are defined in a font installed on the system. A glyph is a character image, i.e. the visual representation of a sign. For example, if we swap glyphs for currency signs in a font on a system, an amount in US dollars might look like an amount in Japanese yen, which would cause a significant change of meaning. It undermines the digital signature objective if the same ASCII binary characters can look different on different systems. Another example of changing meaning is signing XML documents, where the interpretation of an XML document depends on external XML schemas that are not necessarily covered by the signature [17].

As long as data interpretation does not involve complex processes with external dependencies, we do not encounter challenges as described above. This applies, for example, when applying digital signatures to certificates in a PKI, as described in Sect. 5.2, or in cryptographic protocols, as described in Sect. 6.2.

4.7.4 Authenticated Encryption with Forward Secrecy

To achieve forward secrecy asymmetric algorithms must not be used for encrypting secret session keys, as is done with hybrid encryption described in the Sect. 4.7.2. Instead, Diffie-Hellman can used in combination with digital signature to exchange an authenticated secret session key K^{AB} to be used to encrypt messages between Alice and Bob. Figure 4.13 shows a simplified protocol based on these principles that provides authenticated encryption with forward secrecy. A similar protocol is used in TLS 1.3, described in Sect. 6.2. Forward secrecy means that the secrecy of historical encrypted messages is not threatened in the future even if the private keys of Alice and Bob are leaked.

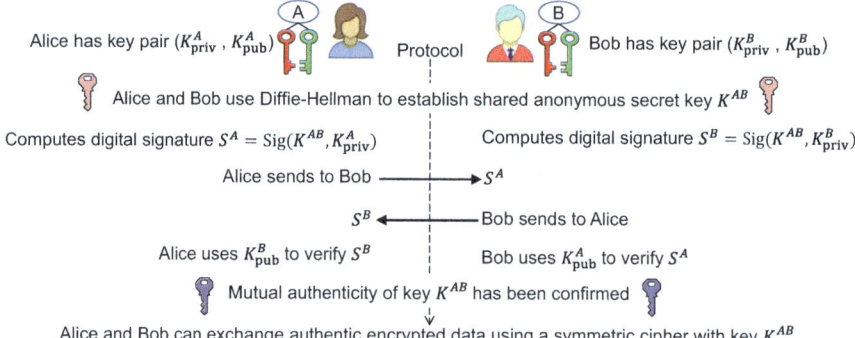

Fig. 4.13 Authenticated encryption with forward secrecy

When Alice and Bob send each other digital signatures of the secret *key K^{AB}*, it is proof of each other's identity. Of course, they do not send key K^{AB}, only the signatures S^A and S^B. Proper authentication through verification of the signatures depends on Alice and Bob having an authentic copy of each other's public keys. This requires a PKI described in the next chapter.

If a private signing key is leaked or cracked at some point in the future, the attacker could spoof the identity of the owner of the compromised key, which could threaten confidentiality in subsequent communication if the compromise is not detected and the key is not replaced. In contrast, compromising a private signing key will not undermine the confidentiality of historical messages, which is precisely the purpose of forward secrecy.

4.8 The History of Cryptography

The development of cryptography has occurred in parallel with, or as a result of, the development of cryptanalysis, which is the science of breaking cryptography. Cryptanalysis of a cipher can allow unauthorized persons to decrypt and read secret messages. Cryptanalysis has on several occasions changed the course of history, such as when the Allies managed to read the codes of Nazi Germany. Had the Allies not been able to read these codes, it would have been much more difficult and taken much longer to end the fighting and prevail in World War II.

4.8.1 Classical Ciphers

As early as 4000 years ago, in classical Egypt, scribes (the pharaohs' literate) used to replace hieroglyphics with others to hide their meaning, which is the earliest known form of cryptography. The best-known classical encryption method is the Caesar cipher, which was used in the classical Roman Empire. The Caesar cipher involves replacing each letter of the alphabet with the letter located a certain number of places down in the alphabet. For example, if the key is 3, the letter "A" is replaced by "D", "B" with "E", "C" with "F", etc. The word "ARM" is then encrypted to "DUP". Julius Caesar used this method to exchange secret messages with his generals. Figure 4.14 shows the Caesar cipher for the English alphabet set with key = 3.

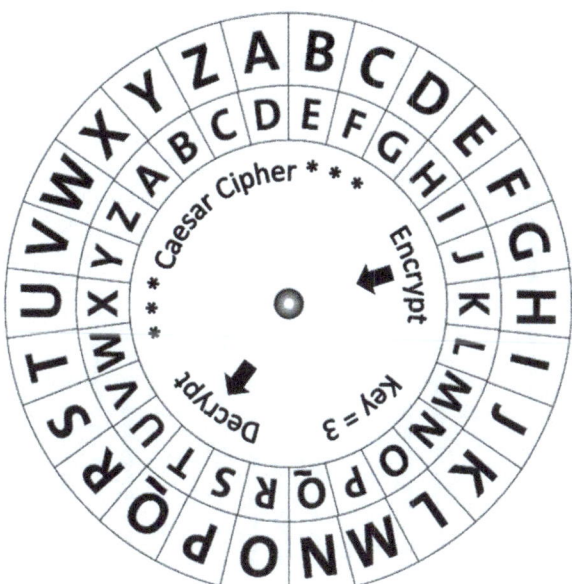

Fig. 4.14 The Caesar cipher

The Caesar cipher has as many combinations as there are letters in the alphabet, so cryptanalysis by experimenting does not take long to find the right key. Although the Caesar cipher is trivial, it is fun and educational to show the principle of a cipher.

4.8.2 Ciphers in the Middle Ages

Throughout the Middle Ages, polyalphabetic ciphers were typically developed. The basic principle of polyalphabetic ciphers is to use several different alphabets at the same time.

A variation of the Caesar cipher is to replace the innermost wheel in Fig. 4.14 with an alphabet where the letters are reshuffled, so that it becomes a new alphabet. With 26 letters, it can be defined $26! \approx 4 \cdot 10^{26}$ different alphabets. In addition, several such reshuffled alphabets can be used simultaneously, for example so that the first letter of the message is encrypted with the first alphabet, the second letter with the second, etc. in a repetitive sequence of a limited set of different alphabets. This type of cipher algorithm is often called Vigenère cipher after the French diplomat and cryptologist Blaise de Vigenère (1523–1596), who developed and used them. The huge number of combinations in a Vigenère cipher may give the impression that this provides strong encryption, but appearances are deceiving. The problem with the Caesar cipher and the Vigenère cipher is that they do not hide statistical patterns from the language or in the information being encrypted. Statistical patterns in the ciphertext make it easy to find the encryption key. Figure 4.15 shows relative frequencies of the letters in the English language, in which there is a distinct statistical unevenness in the frequency distribution.

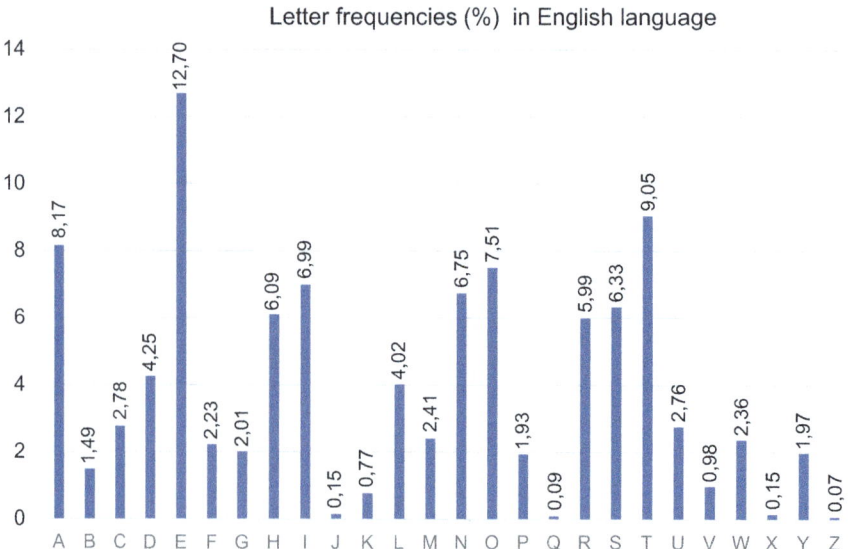

Fig. 4.15 Frequency of letters in the English language, where letter E constitutes 12.7% of average text

Cryptanalysis of a Vigenère cipher involves compiling statistics of letters in the ciphertext and matching these frequencies with the letter frequencies in the assumed language of the plaintext. With a little trial and error, the key can be quickly be recovered, and the message can be deciphered.

4.9 Ciphers Leading Up to World War I

Leading up to World War I, several important discoveries were made that still form the basic principles of modern cryptography. The first discovery is the formulation of *Kerckhoffs's principle*, and the second is the *one-time pad* as the perfect cipher algorithm. These are explained below.

4.9.1 Kerckhoffs's Principle

Auguste Kerckhoffs (1835–1903) was a Dutch cryptologist who published articles on cryptography. In one of his articles [18] he described six basic principles for the construction and use of ciphers. The most famous principle is simply called *Kerckhoffs's principle*, which reads like this:

> *"The design of a system should not require secrecy, and compromise of the system should not inconvenience the correspondents."*
> (The original principle in French: *"Il faut qu'il n'exige pas le secret, et qu'il puisse sans inconvénient tomber entre les mains de l'ennemi"*)
> (Kerckhoffs's principle)

Kerckhoffs's principle means that cryptographic security should not depend on "security by obscurity", i.e. that the algorithm must be kept secret. In other words, only secrecy of the encryption key should be required. This principle still applies today. The most well-known encryption algorithms are public standards. In a civil and commercial context, it would also not be practicable to keep crypto algorithms secret because the algorithms could easily be uncovered through reversal of software and hardware where the algorithms are implemented.

Until the 1970s, crypto algorithms were generally kept secret, because such equipment was not widely used. The breakthrough for public access to cryptographic algorithms occurred through the official encryption standard (DES) in 1977 and the publication of methods of asymmetric encryption in 1976.

Security by obscurity nevertheless has its benefit because it makes cryptanalysis much harder. However, secrecy requires strict discipline at all stages, both in the design phase, in implementation, during distribution and delivery of equipment, in use and in phase-out. As a rule, only military and government organizations can maintain sufficient discipline for the secrecy of crypto algorithms, and in such organizations, in fact, the secrecy of cryptography algorithms is common.

4.9.2 One-Time Pad

One-time pad (OTP) is a cryptographic algorithm in which the plaintext is encrypted character by character, or bit by bit, with a random key or "pad" equal in length to the plaintext and used only once. The one-time pad was invented in 1917 and patented in 1919. It is often called Vernam cipher, since Gilbert Vernam (1890–1960) was one of the inventors of the system.

If the key is truly random, secret and only used once, the one-time pad is perfect in the sense that cryptanalysis is theoretically impossible. This is called "perfect security." For classical messages with characters from an alphabet of size N, the one-time pad (key) will consist of a series of random integers in the interval $[0, (N-1)]$, which for the English alphabet is a series of random numbers between 0 and 25. Encryption of each character follows the principle of the Caesar cipher in that each character is encrypted with a key equal to the respective number in the series. This is essentially a kind of polyalphabetic Vigenère cipher. The difference is that the sequence of different alphabets is never repeated with a one-time pad, while it is repeated with a Vigenère cipher.

For binary messages, the one-time pad consists of a random series of bits with a value of 0 or 1. Let the message M be a strong of bits where M_i denotes message bit i, let K be the one-time pad (key) consisting of random bits where K_i denotes key bit with index i, and let C be the ciphertext as a strong of bits where C_i denotes cipher bit with index i. Encryption and decryption occur with the XOR operation denoted by the symbol $\oplus$ according to the following simple mathematical operations.

$$\text{Encryption}: C_i = M_i \oplus K_i$$

$$\text{Decryption}: M_i = C_i \oplus K_i = M_i \oplus K_i \oplus K_i = M_i \oplus 0 = M_i$$

One-time pads for encrypting text of normal characters were typically distributed as strings of numbers written on sheets of paper, often in the form of a pad of paper so that the sheets could be torn off and destroyed after use.

The Telex line between the Kremlin and the White House during the Cold War was based on rolls of punched tape in which the sequence of holes in the tape was randomly determined by a radioactive isotope and a Geiger counter. This equipment was invented and manufactured by STK (Standard Telefon og Kabelfabrik), located in Oslo, Norway. Hence, Norway has put its mark on crypto history. STK was acquired by Alcatel Telecom and later acquired by Thales Norway, which still produces crypto equipment—not only for the Norwegian Armed Forces, but also for other NATO countries.

Although the one-time pad provides perfect security, this algorithm is not widely used because it requires that keys have the size of data being exchanged, which poses major key distribution challenges.

4.10 Ciphers Around World War II

In the period before, during and after World War II, complex mechanical cipher machines were developed, which were also difficult to cryptanalyze using the methods of the time. A fundamental weakness in the construction of these devices was the lack of a scientific theory of cryptographic strength, that is, to assess their resilience to cryptanalysis. Enigma is a well-known cipher machine from that era, as is the story of how Enigma was cryptanalyzed by Polish and British mathematicians during World War II. Claude Shannon was an American mathematician who established information theory as a science discipline, and who defined basic mathematical principles to construct strong cipher algorithms, which are still valid. These interesting topics about cryptography from around the time of World War II are described in the sections below.

4.10.1 World War II Rotor Ciphers

During World War II, rotor machines were the most advanced in crypto. The belligerents in the war each had their own cipher machines. Most of these were cryptanalyzed by the enemy to a greater or lesser extent.

The Allies used the Swedish Hagelin M-209, Germany used the Enigma, and Japan used its own version of the Enigma, all of which are shown in Fig. 4.16.

Enigma (from Greek: "riddle") is the name of an electromechanical cipher machine used by German military forces for secret communications during World War II. The German side (erroneously) believed that Enigma would be secure for the foreseeable future. Enigma was easy to use, which also contributed to its prevalence. There are many copies of Enigma in museums around the world, and the picture in the middle in Fig. 4.16 shows that it was an elegant machine.

Hagelin M-209 (Swedish)   Enigma (German)   Japanese version of Enigma

Fig. 4.16 World War II rotor machines (The Hagelin photo is public domain. The Enigma photo is by Magnus Manske, and the photo of the Japanese rotor machine is by Mark Pellegrini, both under CC BY-SA.)

The Allies eventually gained full knowledge of the details of the construction of the Enigma. Polish and British mathematicians, including the ingenious Alan Turing (1912–1954), designed an electromechanical machine called "the Bombe", which was used to find the encryption keys (configurations of wheels and plugs on the Enigma) relatively quickly, allowing them to continuously read secret messages even if the keys were changed daily. A series of procedural errors in the use of Enigma also contributed to the effectiveness of the Bombe, allowing the Allies to decrypt Enigma messages during much of World War II.

After World War II, Enigma machines were used by some countries in the British Commonwealth (including India and Pakistan) without users knowing that the messages could be read by the British. This did not become known to the public until the early 1970s [19].

4.10.2 Shannon's Theory of Cryptography

Claude Shannon (1916–2001) founded information theory as a scientific discipline and defined the binary bit (0/1) as the smallest unit of information. He also defined information entropy as the measure of the amount of information.

Shannon is also considered the founder of mathematical cryptography. His article "A Mathematical Theory of Cryptography" written in 1945 was first classified by the U.S. government, but later published as a shorter version entitled "A communication theory of secrecy systems" in the Bell System Technical Journal [20]. Shannon singled out the security objectives of confidentiality and data integrity as the main purposes of cryptography. In the article, Shannon also proves that a theoretically perfect cipher must have the same properties as the one-time pad, which means that perfect ciphers are necessarily impractical.

To construct ciphers that are both practical and secure, Shannon defined basic design principles for symmetric block ciphers that are still valid today. A block cipher encrypts a block of bits in one operation, unlike a stream cipher that encrypts each bit separately.

Shannon found that secure block ciphers must contain functions for *substitution* and *permutation* which are repeated a certain number of times. Such a design is called an SP network (for Substitution and Permutation), as illustrated in Fig. 4.17. The starting point is that the plaintext consists of blocks with a specific number of bits, typically 128 bits, where each block is encrypted separately. Substitution with the S-function means that the block is divided into subsets of bits, where each subset is replaced by a different set of bits, defined by a substitution function with high entropy. Permutation with the P-function means that each subset of the block is moved to a new location within the block, defined by a high-entropy permutation

4.10 Ciphers Around World War II

function. For each round, different parts of the encryption key are added with the ⊕-function to each subset e.g. as XOR addition. Alternatively, parts of the key can be included as a parameter in the substitution and permutation functions. Computing one set of ⊕, S and P-functions is called a *round*. The rounds are repeated a certain number of times, typically 10 to 20 times, with new parts of the encryption key included in each round so that the entire encryption key is used after all rounds have been completed. The effect of repeating many rounds is that statistical patterns in the plaintext get gradually erased in the cipher text, and that the connection between plaintext and cipher text becomes practically impossible to recognize.

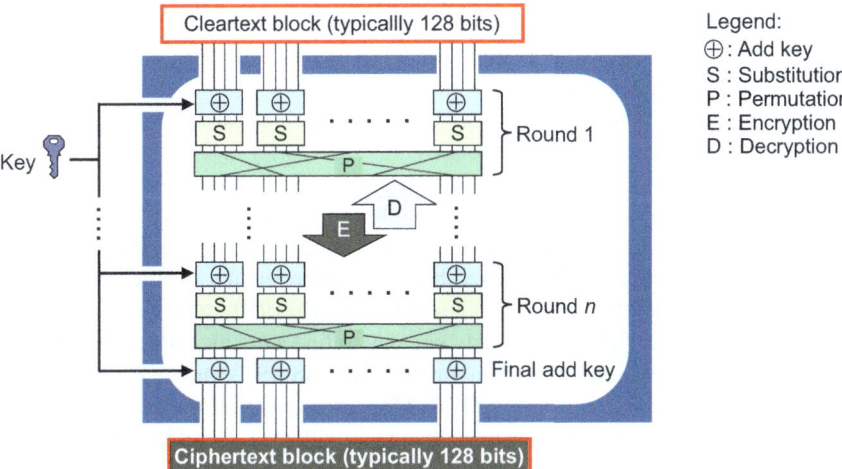

Fig. 4.17 SP network for block ciphers

All SP functions usually need to be invertible to be able to decrypt. It should be mentioned that the so-called Feistel architecture, which is also an SP network, allows non-invertible SP functions, but with the disadvantage that encryption and decryption take twice as long. The Feistel architecture was used in the earlier DES (Data Encryption Standard). Note that in some contexts it is not necessary for a cipher to be invertible to decrypt. For example, both encryption and decryption in CTR mode (counter mode) are done only with the encryption function E of a block cipher, as described in Sect. 4.4.2.

Shannon's contribution to cryptography is significant. The Advanced Encryption Standard (AES) algorithm, which is the world's most widely used cipher today, is based on Shannon's SP network. All SP functions are invertible in AES, which provides the highest possible computational efficiency. AES has block size 128 bits and three possible key sizes: 128, 192 and 256 bits. AES uses 10 rounds with 128 bit keys, 12 rounds with 192 bit keys and 14 rounds with 256 bit keys.

4.11 Ciphers Up to the Year 2000

Until the 1970s, all crypto algorithms were symmetric, meaning that encryption and decryption use the same key. Secret means that the key must be exchanged with secrecy between sender and recipient before messages can be encrypted and sent. Secret key distribution eventually became a major bottleneck, which triggered clever minds to look for better solutions.

James H. Ellis (1924–1997) of the British GCHQ (Government Communications Headquarters) proposed in 1973 the principle that messages could be encrypted with a non-secret key, and that only decryption required a secret key. That way, non-secret keys could be easily distributed, but Ellis never found a practical method to implement this principle. But so did Clifford Cocks (b. 1950), who was a colleague of Ellis's at GCHQ. Cocks needed only one day to find a practical method, which is the same as the well-known RSA-algorithm. Also in 1973, another colleague at GCHQ, Malcolm J. Williamson (1950–2015), spent only half an hour to design a convenient method for two parties to exchange a symmetric secret key over an open insecure channel. This method is the same as the well-known Diffie-Hellman algorithm for key exchange. Because Ellis, Cocks and Williamson worked for the top-secret GCHQ, their discoveries were naturally classified as secret. The tragedy for them is that their inventions were reinvented by academic researchers and published a few years later. It was only in 1997 that GCHQ announced that Ellis, Cocks and Williamson had actually made these inventions as early as 1973 [15].

In 1976, Whitfield Diffie (b. 1944) and Martin Hellman (b. 1945) published the article "New directions in cryptography", which describes both the idea of Ellis, which they call *public-key cryptography,* and the invention of Williamson, which became known as *the Diffie-Hellman algorithm* [14].

Inspired by the article by Diffie and Hellman, Ron Rivest (b. 1947), Adi Shamir (b. 1952) and Leonard Adleman (b. 1945) published a report in 1977 and an article in 1978 entitled "A Method for Obtaining Digital Signatures and Public-Key Cryptosystems", describing Cocks' invention and patenting what became known as the RSA algorithm [16].

The RSA and Diffie Hellman algorithms are called asymmetric algorithms because each uses different cryptographic material and the sender and recipient sides to establish encrypted communication.

These publications triggered an enormous research and development activity in cryptology, by which other algorithms, such as elliptic curves, were invented. Although the mathematics surrounding these algorithms is simple, it was challenging to create practical solutions, and it was not until around the year 2000 that these ideas were implemented and widespread in our modern ICT infrastructure. Web encryption with HTTPS, which is based on the Diffie-Hellman, RSA and elliptic curve algorithms, was introduced around the year 2000.

4.12 Ciphers After the Year 2000

With the transition to the 2000s, encryption with block ciphers received a long-awaited modernization from DES to AES. The same applied to hash functions, which were greatly modernized from MD5 to SHA-2 and SHA-3. At the same time, anticipation of future quantum computing made people in the crypto community warn of dark clouds on the horizon for traditional asymmetric crypto algorithms. This is briefly described in the following sections.

4.12.1 Block Cipher: Replace the Old, Tired Horse with a Young, Strong One

The DES algorithm, which had been the workhorse of encryption in the commercial sector since 1977, was limited by a key size of only 56 bits and a block size of only 64 bits, eventually making brute-force cryptanalysis practicable with ordinary commercial computers. For that reason, NIST (National Institute of Standards and Technology) announced in 1997 a competition for the design of a new block cipher that could replace DES.

Many proposals for a new block cipher were received, and after joint efforts with analysis done by NIST and by the contributors, the algorithm *Rijndael* was declared the best among them. AES (Advanced Encryption Standard), published in 2001, is the standard based on Rijndael, so AES and Rijndael are basically the same. Rijndael was developed by Vincent Rijmen (b. 1970) and Joan Daemen (b. 1965) from Belgium. For the AES standard, NIST chose three versions of the Rijndael algorithm, all with a block size of 128 bits but three different key sizes: 128, 192, and 256 bits.

4.12.2 Hash Functions: New Sponges Needed

The MD5 (Message Digest 5) hash function, published as RFC (Request For Comment) in 1992, had 128 bits hash size and was widespread until the early 2000s, when attacks were found that completely cracked it. SHA-1 (Secure Hash Algorithm 1) published by NIST in 1995 had 160 bits hash size, and was also widespread until the early 2000s, when attacks were found that made it weak. SHA-2 published by NIST in 2001 is specified with two versions, with 256 or 512 bits hash size respectively. As no serious attacks have been found, SHA-2 is still considered secure and is the most widely used hash function today. However, NIST feared that serious attacks against SHA-2 would emerge, and in 2006 announced a competition to design an even stronger hash function. Many proposals for a new hash function were received, and after joint efforts with analysis done by NIST and by the contributors, the hash function Keccak was proclaimed as the best among them. SHA-3 (Secure Hash Algorithm 3) published in 2015 is the standard based on Keccak, so SHA-3 and Keccak are in principle the same. SHA-3 is specified with four versions, with 224, 256, 384 or 512 bits hash size respectively. SHA-3 is based on a so-called sponge function, which metaphorically means that it sucks up data from the message to be hashed, and that the hash value is eventually squeezed out of it. As of 2024, the usage of SHA-3 is limited because SHA-2 is still considered safe. In the event of serious attacks against SHA-2, it can easily be replaced with SHA-3. In this way, SHA-3 can be considered a backup hash function.

4.12.3 Postquantum Crypto: A Lifeboat in Case the Ship Sinks

Quantum computing is the use of quantum physical phenomena such as superposition and entanglement to perform computation on data. Computers that perform quantum computation are called quantum computers. Small versions of quantum computers already exist, and huge sums are being invested in research and development because such computers can, in theory, do certain types of calculations much faster than traditional computers. In theory, quantum computing could solve certain mathematical problems such as integer factorization (which would crack the RSA algorithm) and discrete logarithm (which would crack the Diffie-Hellman algorithm) much faster than traditional computers. Today's quantum computers with capacities of around 50 qubits pose no threat to RSA and Diffie-Hellman. A capacity of more than one million qubits would be needed for a quantum computer to factor integers of size around 2048 bits, thus cracking RSA, or compute discrete logarithm on integers around 512 bits, thus cracking Diffie-Hellman.

The question is whether we humans, or future artificial intelligence, will be able to construct a practical quantum computer with a capacity of one million qubits and

4.12 Ciphers After the Year 2000

more. The quote *"Prediction is difficult- particularly when it involves the future"* (Mark Twain) might sound ironic, but is very much relevant here. Some experts believe that such quantum technology will become practically available by 2040–2050, and if we are to take such predictions seriously, it is necessary to think about the consequences. Others think that quantum computing is just hype that will never materialize as a practical computing paradigm.

Traditional asymmetric algorithms such as RSA, ECC and Diffie-Hellman are the foundation for today's digital signature and PKI technologies, which in turn are essential for security on the Internet and in sectors such as finance and defence. If RSA, ECC and Diffie-Hellman are broken, security could collapse completely, which in turn could destabilize the global civilization.

Although the likelihood is small that quantum computers may one day be able to factor large numbers and calculate discrete logarithm on large numbers so that RSA and Diffie-Hellman can be cracked, the consequences would simply be too serious for this threat to be ignored.

Fortunately, quantum computers do not pose a major threat to symmetric encryption with block ciphers or to hashed functions. Thus, AES, SHA-2 and SHA-3 will not be cracked, so that quantum computing will not rob us of everything that cryptography supports. Figure 4.18 illustrates the quantum threat to various crypto algorithms and the services they support. Confidentiality and authenticity/integrity based on symmetric algorithms are not threatened. Digital signature, confidentiality and PKI based on traditional asymmetric encryption are threatened.

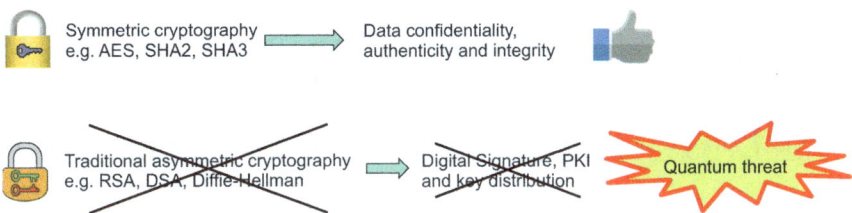

Fig. 4.18 Threat to crypto services from possible future quantum computers

The perception that this threat is realistic led NIST in 2016 to announce yet another competition, this time to propose new asymmetric algorithms that are not vulnerable to the computational power of quantum computers. The purpose is that they can replace existing asymmetric algorithms that are vulnerable to quantum computation-based factorization and discrete logarithms. That way we will be able to keep the valuable services digital signature and PKI. Such "quantum resistant" cryptography is called "post-quantum cryptography" (PQC), i.e. cryptography that can survive in a future world where it is presumed that powerful quantum computers will be available.

Quantum resistant algorithms are nothing new, they have been around for a long time. The reason why they are not used is that they are based on rather complex mathematics that is difficult to analyse and challenging to implement correctly in hardware and software. Mathematical candidates for post-quantum crypto are based on advanced abstract algebra and have exotic names such as *lattice-based,* multi-variate, *hash-based, and code-based algorithms.* The job that needs to be done, then, is to design good specifications based on the theoretical and mathematical basis, so that it can be implemented in a practical way.

NIST received many proposals for new asymmetric post-quantum algorithms in 2017, which were assessed and analysed by NIST and by its contributors until 2022. In 2024, NIST presented a set of draft PQC standards of which one algorithm is for encryption and key establishment, and two algorithms are for digital signature. After standardization during 2025, these are being implemented in solutions. However, NIST was not entirely satisfied with the selected digital signature algorithms and wanted additional proposals for algorithms. Therefore, NIST announced a new call for proposals which had a deadline in 2023. New submitted proposals for PQC digital signature algorithms go through a painstaking new process of analysis and evaluation over many years. This development is shown in Fig. 4.19.

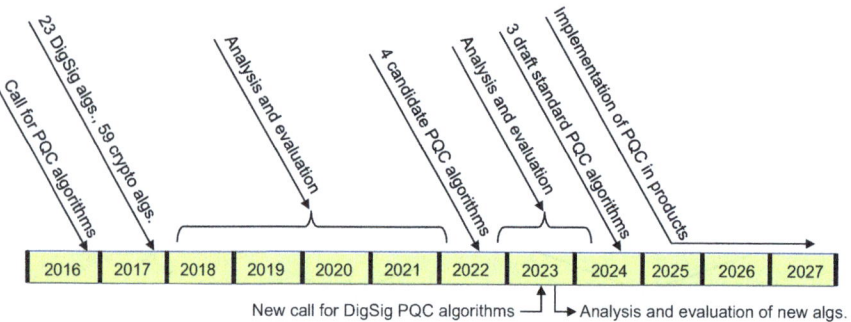

Fig. 4.19 Timeline for post-quantum cryptography development and standardization

The development of standards for PQC algorithms is a thorough and painstaking process that takes many years. It is necessary to be thorough since digital security throughout the global community may depend on these algorithms in the future.

For now, it seems far-fetched and unlikely that practical quantum computers will ever have a capacity of one million qubits, thus threatening traditional asymmetric cryptography. But if that were to happen, post-quantum cryptography standards and technology must stand ready to take over from the RSA, ElGamal and Diffie-Hellman algorithms. There is also the threat that adversaries in the future could get access to confidential information protected with current asymmetric crypto today, so that we need to start using PQC as soon as possible. Metaphorically speaking, post-quantum cryptography is a lifeboat in case the ship with traditional asymmetric cryptography sinks.

4.13 Cryptography and Energy Consumption

All digital computing uses electrical energy. At the same time, we want digital technology to be sustainable regarding energy consumption. Therefore, an interesting question is whether the use of cryptography leads to a noticeable increase in the consumption of electrical energy. This question can be considered from the perspective of energy consumption of a single device or worldwide.

4.13.1 TLS Everywhere

Until about 2000, almost all Internet traffic went unencrypted. When SSL (early version of TLS, described in Sect. 6.2.1) was introduced around the year 2000, encryption was only used to protect sensitive sessions, such as to protect passwords during login. In 2014, the campaign *"TLS everywhere"* started in the Internet industry, with the goal of making HTTPS the default for all web traffic. That campaign had essentially reached its goal by the year 2020, as the vast majority of websites had adopted HTTPS, and most web browsers had started to warn users or even block websites that do not support HTTPS. While this is good for security, it is important to consider the impact it has on energy consumption.

Studies show that the energy consumption for the encryption itself in TLS is proportionately much lower than the energy consumption for additional data transmission as part of the TLS Handshake and TLS Record protocols. At the same time, the energy consumption for both encryption and additional data transmission in TLS is proportionately much lower than the energy consumption for transmitting the data content, when the amount of transmitted data is greater than 500 KB [21]. Thus, although the use of TLS results in an increase in energy consumption, it is not dramatic when the amount of transmitted data is relatively large and when the device has a permanent power supply. However, the increase in energy consumption in TLS can be relatively large if the amount of data transferred in each session is small and the device is battery powered. This could be a problem, for example, for IoT devices.

The Internet of Things (IoT)), means that all kinds of devices connect to the Internet, typically through the 4G and 5G networks, but also through other networks such as Wi-Fi and Bluetooth. When these devices are not only smartphones and tablets, but also, for example, sensors in smart homes, in industrial equipment and on humans and animals, we can create advanced and useful services that were never before possible. At the same time, this significantly increases the threat surface, so security in IoT is essential. TLS or equivalent strong cryptographic protocols should be the first choice for encrypted data transmission and authentication of IoT devices. However, IoT devices often get power supply from battery that may be supposed to last for several years. At the same time, the amount of transmitted data in IoT applications is often relatively small, so that the energy consumption for TLS is relatively large in relation to the energy consumption of the data transfer itself. Thus, it is relevant to consider where power can be saved by using cryptography in IoT applications. TLS can be configured so that session keys last a long time, making it unnecessary to enforce the TLS handshake protocol for each data transfer. Another approach is *Lightweight cryptography*.

The purpose of lightweight cryptography is to minimize resource requirements with respect to physical size, storage capacity, data rates and energy consumption. Reduced resource requirements must always be weighed against security requirements. To satisfy both types of requirements as far as possible, research is being conducted on designing new cryptographic algorithms and protocols that use less energy than standard algorithms and protocols, but at the same time provide adequate strength in limited application areas.

4.13.2 Cryptocurrency

A blockchain is a set of data blocks that are logically linked to each other by digital signatures. A blockchain starts with a *genesis block* which is digitally signed by the first *signer*, who is one of the participants in the blockchain. This digital signature is added to the next block, which is typically signed by another signer whose identity is their anonymous public key. A block can contain arbitrary types of information. Each blockchain application typically stores a specific type of information in the blocks, which for cryptocurrency are transactions for transferring crypto coins between two parties. Other uses of blockchain are described in Sect. 5.4.

Cryptocurrency is based on transactions being described in data records that are signed and placed in data blocks, as shown in Fig. 4.20. In a cryptocurrency blockchain, each block contains a set of data records, with one of the data records describing the issuance of a new crypto coin to the signer of that block as a reward for signing. The other data records describe transactions for transferring crypto coins between participants in the blockchain. In addition, each block contains a digital signature on the preceding block. All subsequent blocks are logically bound to each other with digital signatures, as shown in the figure. All participants in the network have a copy of the entire chain, which is updated with each new block added.

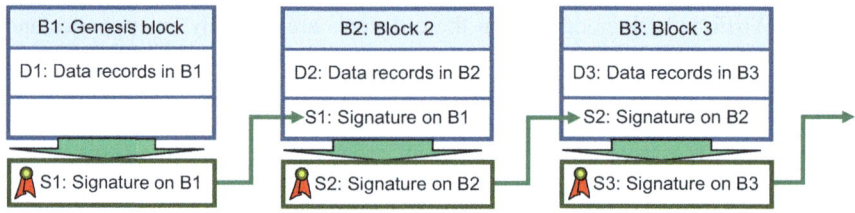

Fig. 4.20 Blockchain

4.13 Cryptography and Energy Consumption

A fundamental element of cryptocurrencies is the *consensus protocol*, that is, how participants in a blockchain network of a cryptocurrency should agree on who will be the signer of the next block. Consensus, i.e. agreement between all participants, is essential to maintain integrity and to prevent counterfeiting, such as using the same coin twice. In most cryptocurrencies, such as bitcoin, a new coin is issued to the signer of each new block as a reward for invested efforts. The two main categories of consensus protocols are *permissioned* and *permissionless*. The term "permissioned" means that certain entities have the authority to determine how consensus is formed. The term "permissionless" means that no entity initially decides, but that consensus is formed through proving some form of *effort*. Energy consumption of permissioned consensus protocols is negligible, while the energy consumption of certain permissionless consensus protocols can be astronomical.

In permissionless consensus protocols, consensus is usually established by PoW (Proof of Work). Bitcoin, for example, uses PoW, where the specific consensus algorithm is called *Nakamoto Consensus*, after the pseudonym of the author of the article describing bitcoin [22]. Another form of permissionless consensus is PoS (Proof of Stake), but as PoS has weaknesses, it is less widely used.

PoW effectively means "proof of electricity consumption", in the sense that the entity that uses the most electricity in the shortest period of time is most likely to become the signer of the next block, and thus the owner of a new bitcoin. The power consumption is spent looking for a hash value for the next potential block, where the hash value must have a number of consecutive zeros. Because the probability of generating a hash value with many zeros is very low, billions upon billions of hashes must be generated before such a hash value can be found. Here, the principle of "first come, first served" applies, so the stakes in hardware and power consumption among crypto miners can be absurdly large. Whoever first finds the hash value wins, and the price of one bitcoin determines how much resources it economically pays to invest. The total electricity consumption for bitcoin generation in 2021 was about 100 terawatt hours (TWh), which was around 0.4% of the entire world's electricity consumption and about the same as the Netherlands' electricity consumption. By comparison, Google's data centers collectively used "only" around 12 TWh in 2019, which is about 1/8 of the electricity consumption of bitcoin mining in 2021. The increase in the price of bitcoin and the resulting global race for mining is turning the cryptocurrency market into one of the world's most polluting sectors [23]. The use of cryptography for this type of cryptocurrency is obviously harmful to the environment, and is thus not sustainable.

While blockchain technology has some useful applications, it also has negative side effects. The two main applications of cryptocurrencies are speculation and anonymous/criminal money transfer, which can be considered harmful both from ethical and legal perspectives. If speculation and criminal use of bitcoin and similar cryptocurrencies are allowed to continue unregulated, it will eventually seize a significant portion of the world's electricity production. It is clearly a market failure when massive amounts of electric energy are wasted on harmful activities, and takes energy away from meaningful needs. Regions without regulation and where electric energy is cheap will be hardest hit. The free market is unable to fix this problem, and hence regulation is needed. An important question is how such regulation can best be implemented in each country, regionally and worldwide. Politicians around the world must wakeup themselves to this challenge. China has already clamped down on mining and trading in cryptocurrency. In 2021 the Chines government issued a *"Notice on Further Preventing and Disposing of the Risk of Hype in Virtual Currency Trading"* which bans all cryptocurrency-related activities.[1]

4.14 Tasks

1. <u>The Caesar cipher</u>
   Assume a substitution cipher algorithm similar to the Caesar algorithm in Fig. 4.14, but where the letters of the innermost circle are reshuffled. The letters can have an arbitrary order. The order of the characters of the reshuffled alphabet on the inner circle constitutes the key. There are a total of 30 characters, consisting of 29 letters and a hyphen '-'.

   (a) Give a simple mathematical expression of how many different keys this cipher algorithm has, and find the numerical answer with a calculator, e.g. online: https://www.calculatorsoup.com/calculators/discretemathematics/factorials.php
   (b) What is the key size for 30 characters expressed in bits? Consider whether the key size is sufficient to withstand exhaustive searches throughout the key space.
   (c) What would the key size be if the alphabet had 34 characters? Is the key size sufficient to withstand exhaustive searches throughout the key space? Explain the answer.
   (d) Can the algorithm withstand statistical cryptanalysis? Explain the answer.
   (e) Briefly explain how to cryptanalyze a ciphertext encrypted with this cipher algorithm.

2. <u>Digital signature</u>
   Alice wants to send message M with digital signature Sig(M) to Bob. They have each other's public keys, and have agreed on a cryptographic hash function Hash and a signature algorithm that operates in signing mode Sig (equivalent to decryption mode D) or in verification mode Ver (equivalent to encryption mode E).

---

[1] https://www.coindesk.com/policy/2021/09/24/china-tightens-crypto-mining-crackdown-bans-trading/

4.14 Tasks

(a) Describe the steps Alice must follow to send M with digital signature.
(b) Describe the steps that recipient Bob must follow to verify the authenticity of M.
(c) Explain why the digital signature proves to Bob that the received message is authentic and why Bob is able to prove to a third party that the message is authentic.
(d) How can a sender provide a plausible reason to deny (repudiate) having sent a signed message?
(e) Discuss the semantic interpretation of "digitally signed message," which could mean (i) that Alice agrees with the content of the message, or (ii) that Alice sent the message without necessarily agreeing with its content?

3. Stream ciphers

Imagine that a binary stream cipher (which uses a pseudorandom key string or one-time key, One-Time-Pad) has been used to encrypt an electronic funds transaction. Suppose no other cryptographic mechanisms are used. Explain how an attacker can change the transfer amount without knowing anything about the key being used. (You may assume that the attacker knows the format of the cleartext message used to transfer the money.)

Chapter 5
Key Management and PKI

> *When one lock can be opened by three different keys, it's a bad lock. But when one key can open ten different locks, it's a master key.*
>
> Confucius, ancient Chinese philosopher

All crypto systems need keys to operate. The strongest crypto system will give zero security protection if a secret or private key falls in the hands of attackers. Hence, the secure handling of cryptographic keys needs equally much attention as the design and implementation of crypto systems. While crypto systems operate automatically in hardware and software, key management also involves policy, user training, organizational processes and interactions, and coordination between all of these elements. This is to ensure that the generation, exchange, storage, use, destruction, and replacement of keys are done in a secure way. With the billions of Internet-connected systems and devices that all require cryptographic communication, the key management challenge is daunting. A public key infrastructure (PKI) is a solution which significantly reduces the complexity of this challenge by providing a secure basis for the exchange of cryptographic keys.

In this chapter, you learn what key management is and why it is important. You also learn about PKI which is an infrastructure for distributing public keys in the form of digitally signed certificates. You will understand how PKI is a prerequisite for the practical application of asymmetric algorithms for encryption and digital signature, which were topics in the previous chapter. PKI is also a prerequisite for network security which is described in the next chapter.

5.1 Key Management

The security of a cryptographic solution depends not only on key size and the algorithms' strength against cryptanalysis. Security also depends on practical aspects of how cryptographic solution are implemented and used, and on how keys are generated, distributed and managed.

Key management provides the basis for secure generation, storage, distribution and destruction of cryptographic keys, and is essential for cryptographic security. Weak key management can easily lead to breaches of security systems and of cryptographic data protection.

5.1.1 Key Types

A cryptographic key should only be used for a single purpose, such as either encryption, authentication, key wrapping, random number generation, digital signature generation. Keys must also be specified to be used either for single sessions, or for long-lasting use. The reason is that using the same key for two or more different types of applications exposes the key to multiple threats that pose a greater risk of compromising the key, and that a compromise of the key in one application will spread to the other applications.

It is important to consider threats that may lead to a violation of the confidentiality of the key or to a violation of the security of its application. Three important factors for such threats are:

- The more a key is used, the more it is exposed to attacks that can lead to a breach of confidentiality, i.e. that the key gets compromised.
- Research on cryptanalysis and increased processing speed in supercomputers cause algorithms and keys to "erode" over time, i.e. they become weaker against new and more advanced attacks. The erosion can occur steadily over time, or suddenly through a breakthrough in cryptanalytic research or computational technology such as quantum computers.
- The longer a key is used and the more applications it is used for, the greater the impacts can be in case the key is compromised, or the algorithm is broken.

The moral is that the use of a key must be limited both in time and in type of application.

The same principles apply to passwords, e.g. a password should never be reused for multiple different services. The more a password is used, the more it is exposed, hence, a password that is frequently used for accessing a sensitive or important service should be changed regularly, see Sect. 8.3.2. The classification of cryptographic keys is based on several aspects, such as:

- whether the key is symmetric (secret) or asymmetric (public/private),
- encrypting general data or encrypting secret cryptographic keys,
- authentication of general data or authentication of public keys,
- use over a long period of time (static keys) or for a single session (ephemeral keys).

NIST's standard "Recommendation for key management" describes 19 different key types, with recommendations on how long each key type can be used before it should be replaced [24]. A key's period of application is called *cryptoperiod*.

5.1.2 Crypto Periods

Crypto algorithms with keys have the two functions encryption and decryption which are the inverse of each other, and similarly for signature algorithms where signing and verification can be considered the inverse of each other. Logically, the first function provides protection (confidentiality or authenticity), while the second function represents active processing of encrypted/signed data. For the use of keys, it makes sense to distinguish between time periods in which the respective inverse functions are performed, because it has implications for the security of keys and algorithms.

The period during which a symmetric key is used to encrypt data is called the *protection period* (aka. Originator-Usage Period). The period during which a symmetric key is used to decrypt data is called the *processing period* (aka. Recipient-Usage Period). The periods in which a pair of public and private keys is used for encryption and decryption are also called protection period and processing period, respectively. Similarly, the periods in which a pair of private and public keys is used for digital signature generation and verification can also be called protection period and processing period, respectively. That is, signing is "to protect" the authenticity of data, while signature verification is to "process" the authenticity of the data.

As a rule, the periods of protection and processing are partially overlapping in general. This means, for example, that the use of a symmetric key for encryption should typically cease after a relatively short period of time, while the same key can be used for a long period afterwards to decrypt the data. The same principle applies to digital signatures, so the use of a private key for digital signing should cease after a relatively short time, while verification of the signatures with the corresponding public key can be done for a long period afterwards.

The cryptoperiod of a key includes both the protection period and the processing period, as shown in Fig. 5.1.

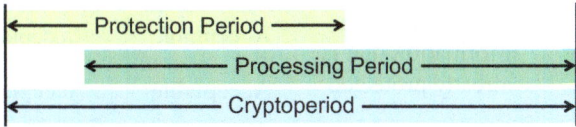

Fig. 5.1 The crypto period includes the protection and processing periods

The protection period necessarily begins earlier or simultaneously with the processing period, which in turn ends after or simultaneously with the protection period. Thus, the cryptoperiod lasts from the beginning of the protection period to the end of the processing period. A key should not be used outside the specified periods. The following two aspects must be considered when specifying crypto periods:

- The greater the sensitivity of the encrypted information, the shorter the crypto period should be to limit any harmful impacts of key compromise.

- Short crypto periods can be counterproductive, especially if availability is an important security goal. Key replacement entails additional work (overhead) and can lead to service interruption if complications occur. There may also be vulnerabilities in the key replacement process, especially if the replacement involves complex procedures.

The cryptoperiod is therefore a trade-off between key security requirements on the one hand and availability requirements on the other, for the application where the key is used.

5.1.3 Key Sizes

Increase in computational power and advances in theoretical cryptanalysis over time results in the decrease of cryptographic strength. As a consequence, key sizes must be increased over time. In its "SP800-57 Recommendation for Key Management" [24], NIST makes recommendations on key sizes for symmetric algorithms with a perspective of 20 years, as shown in Fig. 5.2.

Security Strength		Through 2030	2031 and Beyond
< 112	Applying protection	Disallowed	
	Processing	Legacy use	
112	Applying protection	Acceptable	Disallowed
	Processing		Legacy use
128	Applying protection and processing information that is already protected	Acceptable	Acceptable
192		Acceptable	Acceptable
256		Acceptable	Acceptable

Fig. 5.2 Maximum crypto periods based on key size of symmetric algorithms (Source: 24])

From Fig. 5.2 we see that NIST discourages protection (encryption) with symmetric keys less than 112 bits for current applications, but at the same time specifies (of course) that previously encrypted data can be processed (decrypted). Keys with size between 112 and 128 bits are recommended to be used for protection (encryption) until the year 2030, but not thereafter, when only processing is recommended. Keys that are 128 bits or larger are recommended to be used for both protection and processing beyond the year 2030. These recommendations can be useful for planning the use of cryptosystems and associated long-term key management.

5.1 Key Management

In order to make similar assessments for asymmetric keys, NIST provides recommendations on key sizes corresponding to the strength of symmetric keys with AES, for other asymmetric algorithms, such as Diffie-Hellman, RSA and ECC, as shown in Fig. 5.3. The abbreviations are relatively technical and are only briefly explained below:

FFC: Finite Field Cryptography (related to discrete logarithm)
DSA: Digital Signature Algorithm (signature algorithm based on Diffie-Hellman)
DH: The Diffie-Hellman Protocol for key exchange
MQV: Menezes-Qu-Vanstone (authenticated protocol for key exchange based on DH)
IFC: Integer Factorization Cryptography (related to factorization of large integers)
ECC: Elliptic Curve Cryptography (related to elliptic curves with large integers)

Security Strength	Symmetric Key Algorithms	FFC (DSA, DH, MQV)	IFC* (RSA)	ECC* (ECDSA, EdDSA, DH, MQV)
128	AES-128	$L = 3072$ $N = 256$	$k = 3072$	$f = 256\text{-}383$
192	AES-192	$L = 7680$ $N = 384$	$k = 7680$	$f = 384\text{-}511$
256	AES-256	$L = 15360$ $N = 512$	$k = 15360$	$f = 512+$

* The security-strength estimates will be significantly affected when quantum computing becomes a practical consideration.

Fig. 5.3 Recommendations on key sizes of similar strength for different algorithms. (Source: [24])

Figure 5.3 uses as reference the key sizes for AES, where the options are 112, 192 and 256 bits. Combined with the cryptoperiods in Fig. 5.2 we thus find recommendations on cryptographic keys for asymmetric algorithms with a perspective of 20 years.

It is interesting to note that 128 bits strength for AES equals 3072 bits strength for RSA. Around the year 2024, the usual key size for RSA is still 2048 bits, but according to NIST's recommendations, the size of RSA keys should be increased to 3000 bits and more. It looks better for ECC keys, where the recommendation is still relatively modest at 256-383 bits.

It is even more interesting to note the remark below Fig. 5.3, which states that there is a caveat that these recommendations should change significantly in case powerful quantum computers become practically available, see Sect. 4.12.3.

5.1.4 Key Lifecycle

Cryptographic keys are generated, exchanged, stored, activated, used, deactivated, and destroyed, making for a relatively complex cycle. Figure 5.4, which is based on NIST SP800-57 Recommendation for Key Management, shows states and phases of the key lifecycle. The numbered green circles show transitions between states, and the four major phases in the lifecycle of a cryptographic key are shown on the right hand side of the figure.

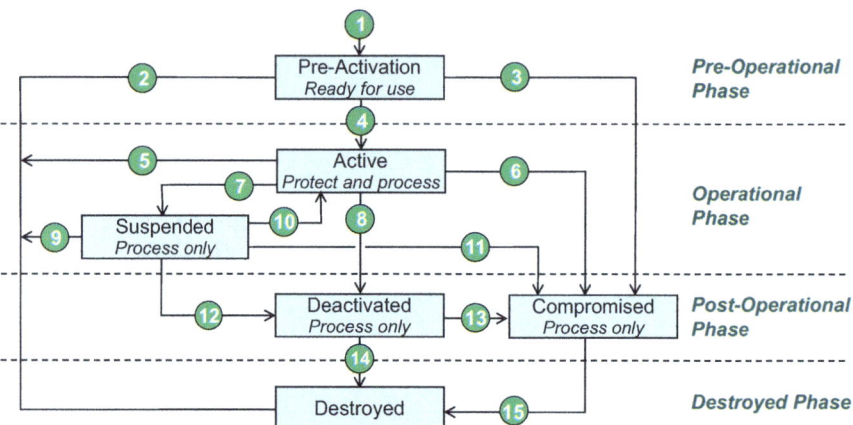

Fig. 5.4 Lifecycle of cryptographic keys

Key generation (transition 1) is the most sensitive of all steps because it is a dynamic situation where there are potentially many different vulnerabilities in the equipment and in the generation process. Special apps or hardware devices are often used to generate keys and initialization vectors for algorithms (see Sect. 4.4.3).

It is essential that keys and initialization vectors are random (unpredictable) and selected from the entire key space with uniform probability distribution. This almost sounds obvious, but we cannot take it for granted that keys are always generated randomly. It was revealed by Edward Snowden in 2013 that the NSA (National Security Agency) had bribed RSA Security Inc. (prominent security company) with US$10 million to implement and sell on the commercial market a compromised random number generator in their BSAFE products. The fact that the generator was compromised meant that it generated numbers that were predictable to the NSA. It was like a backdoor. When customers and users of BSAFE generated secret keys, the NSA was able to regenerate the same secret keys, thus decrypting traffic encrypted with those keys.[1] This was, of course, a scandal when this was revealed, which damaged the reputation of RSA Security Inc.

---

[1] https://www.wired.com/2013/12/what-we-really-lost-with-the-rsa-nsa-revelations/

Compromise of keys is when it is known or suspected that an unauthorized entity has obtained a secret/private key. Once a key is thought to be compromised, it should be immediately put into a compromised state in Fig. 5.4. Compromised is not the same as disabled, which is considered a normal state for example when a key's lifespan expires with the crypto period. Note that a compromised key can still be used for processing previously protected information. At the same time, users must be made aware of the risks of using a compromised key for processing.

The most serious form of key compromise is when it happens without it being detected by the user. Measures against unknown key compromise are the use of several security controls or several forms of encryption at the same time, so that a single incident does not necessarily result in a serious security breach. Another control, of course, is to limit the cryptoperiod for keys, which also helps limit impacts in case the key is compromised.

5.2 PKI

PKI (Public Key Infrastructure) is an infrastructure based on technology, policies, processes, organizations, and trust that together make it possible to distribute public keys in the form of digitally signed certificates. The purpose of the certificates, and thus the purpose of PKI, is to facilitate strong authentication of entities on the Internet and in other networks, as well as encrypted transfer of information between entities for any type of business processes on the Internet. PKI is a pillar of Internet security.

5.2.1 The Challenge of Key Distribution

Secure key distribution has been a challenge throughout the history of cryptography. It may seem paradoxical that a secure communication channel for key exchange is needed for cryptography to provide a secure communication channel for information exchange. Why could not the secure channel of key exchange also be used for information exchange? The problem is that secure key exchange is a relatively onerous and expensive process, so it would be impractical to exchange large amounts of information through the same channel. Therefore, it is economical to let secure exchange of small cryptographic keys happen first, which in turn enable secure communication channels for arbitrary amounts of information.

Secure key exchange is nevertheless a challenge because it can be both cumbersome a process and can potentially involve large numbers of keys.

The challenge can be illustrated by envisioning an open computer network (such as the Internet). In order to have flexible security in a computer network, each pair of nodes must be able to establish encrypted communications that support confidentiality and authenticity. This means that each pair of nodes needs a separate cryptographic key. Then the question is how many different keys are needed and how these keys should be distributed. Given a computer network with n nodes, then the formula for the number of node pairs and thus the number of keys that need to be exchanged, is expressed as $\frac{n(n-1)}{2}$. Note that the product n^2 appears in this formula, which means that the number of key exchanges grows quadratically with the number n of nodes. Figure 5.5 shows the number of key exchanges needed in a five-node network and a network with ten billion nodes (such as the Internet).

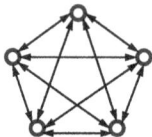

Fig. 5.5 The number of exchanges of symmetric keys as a function of the number of nodes in a network

5 nodes
10 key exchanges

10 billion nodes
50 billion billion key exchanges

Since the number of key exchanges increases quadratically, key exchange with symmetric keys does obviously not scale to large networks such as the Internet. An important motivation for the invention of asymmetric cryptography and public keys in the 1970s was precisely to solve the challenge of key exchange in large networks. The idea was that the public keys could be published through a public register, for example in the telephone directories of the time, which were printed on paper and distributed to all telephone subscribers. Eventually, it became clear that such records would be vulnerable to forgery of public keys, which meant that the idea was flawed.

Imagine that Alice wants to send a secret message to Bob encrypted with Bob's public key, but an attacker has replaced Bob's key with his own in the key register where Alice looks up the key. As a result, Bob cannot decrypt the message, but the attacker can. Furthermore, imagine that Alice signs a message with her private key and sends the signed message to Bob, but an attacker has replaced Alice's key with her own in the public key register. The result is that Bob retrieves the wrong public key for Alice, that the verification of the signature fails, and that Bob therefore rejects the message from Alice.

It is essential to understand that security requirements for symmetric keys and public keys are different. The security objective for the distribution of symmetric keys is confidentiality, while the security objective for the distribution of public keys is authenticity. In fact, exchanging an authentic public key is just as challenging as exchanging a confidential symmetric key. Luckily, there is a way for authentic distribution of public keys that scales relatively well to computer networks the size of the Internet, namely PKI (Public Key Infrastructure).

Cryptography can solve security challenges on the Internet, but at the same time creates challenges for key distribution. PKI can simplify key distribution, but at the same time creates trust challenges. Trust management is thus a fundamental component of PKI.

5.2.2 X.509 Certificates and PKI Components

The basic principle of a PKI is that public keys are distributed as *certificates* which are digitally signed data records containing the public key. Along with the public key, other attributes related to the key are also included, such as owner, issuer, validity period, application, key size, names of crypto algorithms (hash function and digital signature algorithm) and other things relevant to the key's application. When all these attributes are packaged together and signed, it is called a certificate. The digital signature is generated with the private key of a *certificate authority (CA)*. The certificate can then be validated by verifying its digital signature, which can be done by anyone holding an authentic copy of the CA's public key.

The standard for such certificates is called X.509 [25] and is published by the International Telecommunications Union (ITU). In addition, standards are needed for how certificates should be generated, issued, distributed, validated, stored, revoked, etc. These types of standards are published by the IETF (Internet Engineering Task Force) which produces RFCs (Request For Comments) and other standards documentation for the Internet infrastructure in general, and also on how to use and deploy X.509 in practice. In particular, RFC 5280 [26] specifies necessary extensions for X.509 that are required in modern Internet applications. Figure 5.6 provides a simplified representation of the structure of an X.509 certificate.

Fig. 5.6 X.509 certificate

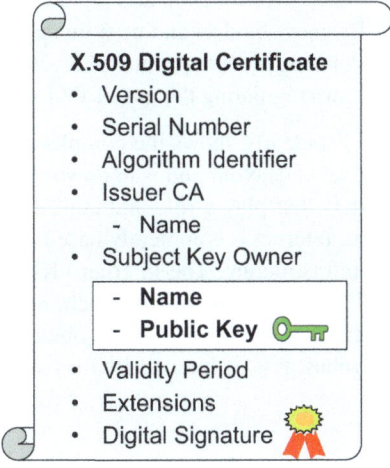

Most browsers (Firefox, Chrome, Safari, Edge) can display certificates for websites visited on the web by clicking the padlock in the address bar at the top of the browser window. Tools like OpenSSL can be used to easily generate X.509 certificates. Such tools support certificate generation by CAs (Certificate Authorities) as well as management and use of certificates in servers and clients. A simple high-level recipe for a CA to generate X.509 certificates goes as follows:

1. Receive the public key and name of the subject (owner) who requires a certificate.
2. Collect and express all attributes as a data record in the format of X.509.
3. Compute the hash value of the X.509 data record.
4. Sign the hash value with the private key of the issuer (CA).
5. Append the digital signature as the last attribute in the certificate.

The digital signature represents a logical binding between the public key and the subject's (owner's) name. This binding defines the authenticity of the public key, that is, the certificate is proof of who owns the public key. For Internet applications, the domain name of the subject (owner of the certificate) is always included as part of the subject name because a domain name is always unique, while a common organization name is not necessarily unambiguous worldwide.

A PKI needs much more than just a standardized structure of certificates. A PKI consists of the following components (non-exhaustive list):

- Standard for the X.509 data format.
- Standards for how certificates should be generated, issued, distributed, validated, stored, revoked, etc. For the Internet, this work was done by the IETF working group PKIX (PKI with X.509 certificates).
- Products that implement the standards.
- CA (Certificate Authority): organizations that issue/sell certificates.
- Operating system and browser manufacturers that distribute CA root certificates.
- Other governmental and private organizations that provide PKI-related services.
- Business models and trust models for organizations offering PKI services.
- Policies, procedures and policies for administering PKI services.
- Laws regulating the use of PKI services.

The list clearly shows the complexity of a PKI. The Internet PKI is largely a commercial endeavour and is not owned by anyone, but is managed jointly by all stakeholders that play a role and contribute in one way or another. The security of the entire Internet is completely based on its PKI, which can thus be considered a critical infrastructure. The Internet PKI is continuously under threat and regularly suffers from cyber incidents. Such attacks enable identity spoofing and the stealthy distribution of malware which obviously can have serious impact. Hence, sustaining the robustness of Internet PKI is essential.

5.2.3 Generating X.509 Certificates

A PKI forms a graph of certificates logically bound together with digital signatures with a root certificate as the starting point and subject certificates as endpoints. Figure 5.7 shows the sequence for generating certificates in a PKI, starting with the root certificate, then an intermediate certificate, and finally a subject certificate.

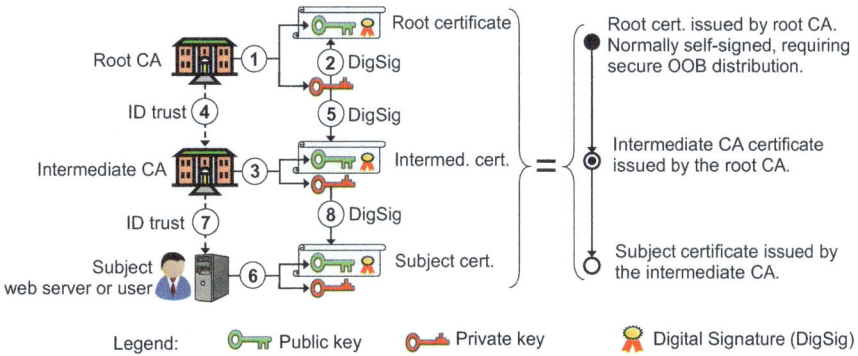

Fig. 5.7 Process for generating certificates in a PKI

The sequence of steps in Fig. 5.7 is briefly explained below.

1. The root CA generates a root key pair.
2. The root CA assembles the dataset for an X.509 certificate that includes its own public key and signs the certificate with the corresponding private key. This is called self-signing, which syntactically looks like a signed certificate, but logically is not because self-signing does not provide authenticity. For this reason, the distribution of root certificates requires a secure OOB (Out-Of-Band) channel, i.e. a different channel than the one the certificate is intended to protect.
3. The intermediate CA generates a key pair with a public and a private key, and requests the root CA to generate a certificate for the public key.
4. The root CA has confidence in the identity of the intermediate CA and in the authenticity of the public key received for certification.
5. The root CA generates and signs the intermediate CA's public-key certificate.
6. The subject entity generates a key pair and asks the intermediate CA to generate a certificate for the public key.
7. The intermediate CA has confidence in the identity of the subject and in the authenticity of the public key received for certification.
8. The intermediate CA generates and signs the subject's public-key certificate.

The meaning of self-signed root certificates can be difficult to understand, and the term "self-signed" is actually misleading because it can give a false impression of security. In reality, self-signing provides no security in the sense of authenticity. In fact, anyone can generate a self-signed certificate and specify an arbitrary name for the issuer and the subject, which potentially can be used in spoofing attacks. The only way to know if a self-signed root certificate is authentic is if you trust the OOB (Out-Of-Band) channel through which it was received.

However, two useful purposes of self-signing can be mentioned: First, it provides a checksum for detecting accidental breach of integrity. Second, populating the signature attribute field in X.509 certificates can be useful because an empty signature field can cause errors in applications that require content in the signature field.

Because self-signing does not provide any authenticity of the root certificate, it is necessary to distribute root certificates through a secure OOB communication channel. This is a relatively obscure topic in Internet security, which means that it is difficult to know exactly how clients and servers are receiving root certificates. The most common procedure is that downloading and installing browsers and operating systems comes with a set of root certificates included. In addition, new root certificates are downloaded through continuous software updates. Trust in root certificates thus depends on confidence in the online software installation and update process.

5.2.4 Validation of X.509 Certificates

Communicating over the Internet, especially through the web, often involves establishing a secure encrypted connection based on the server certificate that contains the server's public key. For example, the security protocol TLS 1.3 uses public-key certificates to establish authentic session keys, see Sect. 6.2.1. Before the public key can be used to establish a secure connection, the client must be assured that the key is authentic, which is done by validating the certificate. Figure 5.8 illustrates the principle of certificate validation.

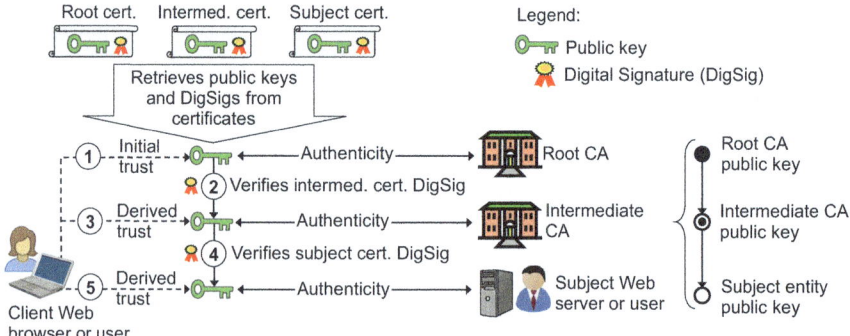

Fig. 5.8 Validation of certificates

5.2 PKI

The sequence in Fig. 5.8 is briefly explained below.

1. The client (e.g. browser) extracts public keys from all relevant certificates, i.e. from the root certificate, from intermediate certificate(s), and from the subject certificate.
2. The client has initial trust in the authenticity of the root certificate and its public key.
3. The client uses the root CA public key to verify the authenticity of the intermediate certificate.
4. The client uses the intermediate CA public key to verify the authenticity of the subject certificate.
5. The client has derived and established trust in the authenticity of the subject's public key, which can then be used to establish a secure connection to the subject entity (web server) with, for example, the security protocol TLS 1.3.

5.2.5 Trust Models for PKI

The graph structure of a PKI reflects an underlying trust model, and if trust is lacking, a PKI becomes dysfunctional. For example, if subjects and clients lack trust in a root CA, they should neither purchase certificates from that CA nor from underlying CAs, nor use that root certificate for validating subject certificates.

Many different trust structures can be used as a basis for a PKI. Figure 5.9 provides an overview of different PKI trust models that have been proposed and/or are in use.

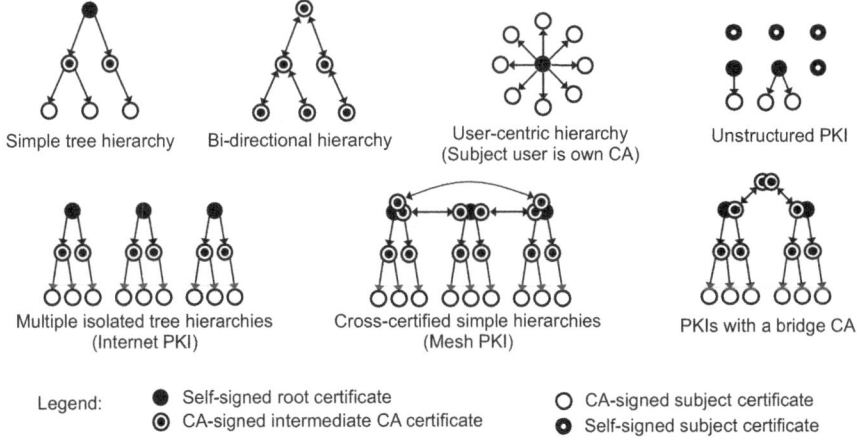

Fig. 5.9 PKI trust models

The PKI for the Internet consists of many isolated, strict hierarchies, as shown in the bottom left corner of the Fig. 5.9. Each browser comes with several hundred root certificates already installed, which means that organizations (subject entities) have a large selection of CAs they can buy or obtain certificates from.

The PKI for DNSSEC is another PKI structure as a single, strict hierarchy, as shown at the top of the Fig. 5.9.

The PKI for PGP (Pretty Good Privacy), also called GPG (Gnu Privacy Guard), is based on a user-centric model where each user is their own CA, as shown as number three on the top row of the Fig. 5.9.

The trust models also represent business models, because many CAs are commercial players. For these, a particular trust model can be advantageous for positioning themselves in the market. However, many CAs issue certificates at low cost or even for free, so selling certificates is not very lucrative.

CAB (CA/Browser Forum)[2] is a consortium for the CA industry, whose members are certification authorities and vendors of browsers, operating systems and other PKI-related products and services. CAB defines industry guidelines for handling X.509 certificates and how certificates are used in applications.

Through the eIDAS regulation, the EU has defined requirements and procedures for PKI and for the establishment of trust in CAs for the issuance of Qualified Website Authentication Certificates (QWAC).

5.3 Challenges and Solutions for PKI

PKI is a complex infrastructure where new challenges constantly arise that must be solved. This section briefly describes some well-known challenges and how they have been solved.

5.3.1 PKI Trust Scope Limited to Authenticity

It is important to understand that PKI and public-key certificates only support the establishment of trust to authenticity, that is, the binding between a public key and the name of the subject. For example, certificates say nothing about whether the subject entity is a law-abiding and serious organization—those types of assessments are not part of certificate validation. Figure 5.10 shows two simple scenarios where a client validates certificates from UiO.no (University of Oslo, a law-abiding organization) and from criminal-company.con (a fictitious criminal organization). Web browsers always display a padlock symbol in the address bar when a webpage's certificate has been validated and the connection is encrypted with HTTPS.

---

[2] https://cabforum.org/

5.3 Challenges and Solutions for PKI

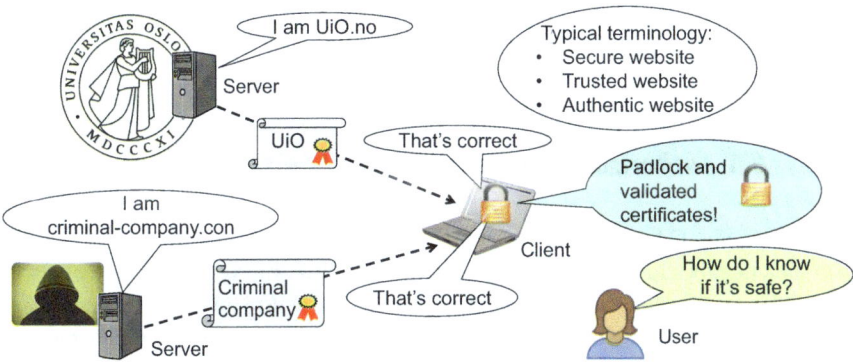

Fig. 5.10 The potentially misleading concept of certificate trust

For the client, there is no difference between the semantics of the two certificates—certificate validation only determines whether the domains UiO.no and criminal-company.con are the rightful owners of the respective public keys in the certificates, as they are in the example in Fig. 5.10.

Let's be clear: The padlock symbol in the address bar of the browser says nothing about whether there is a serious organization behind the website—that question must be determined by the user himself (a human) through other means. Alternatively, the validation can be combined with a whitelist of pre-trusted legitimate organizations that can be checked automatically during the certificate validation.

5.3.2 Certificate Revocation

Certificates can be revoked and invalidated before their validity period expires. This can happen for different reasons, such as in the case of suspected compromise of the private key of the CA or of the subject entity. It is a major challenge to give timely notification to the entire Internet that a certificate has been revoked.

Best practice for PKI dictates that the status of certificates should always be checked shortly before they are used. Otherwise, the client risks that a revoked and thus invalid certificate gets incorrectly validated. Previously, a CRL (Certificate Revocation List) was used, but it proved to require very large network capacity. From 2018, a method known as OCSP (Online Certificate Status Protocol) is used instead.[3] OCSP requires relatively small network capacity, which allows the status of certificates to be checked in near real time. An OCSP certificate is a signed and relatively recent status report for a subject certificate issued by an OCSP responder. The OCSP certificate should always be sent together with the subject certificate, which is called "must-staple" because it should not be the recipient client's job to

---

[3] https://datatracker.ietf.org/doc/html/rfc6960

obtain OCSP certificates for subject certificates received from, for example, web servers.

5.3.3 CA Authorization and Certificate Transparency

Until 2018, any CA could issue certificates for arbitrary domain names. This was risky practice, because it only took a single CA to be compromised for attackers to issue fake certificates for arbitrary domains on the entire Internet. Several such attacks occurred, in which fake certificates were issued for prominent organizations. The solution to this problem was to introduce *CA Authorization* (CAA),[4] which means that the owner of a domain name can specify in DNS which CA is authorized to issue certificates for that domain.

However, there was still the problem that any CA technically could issue certificates for a domain without being authorized. The solution to this problem was to introduce *certificate transparency* (CT).[5] This means that all certificates issued must be registered on a CT server, of which there are several on the Internet. Clients receiving certificates should always receive a digitally signed confirmation from a CT server that the certificate has been published. In that way you get busted if you issue certificates without authorization, i.e. you get busted both by publishing it and by not publishing it.

5.4 Use of Blockchains

The structure of blockchains is briefly described in Sect. 4.13.2. The most important feature of blockchain is to guarantee data integrity through a distributed chain of digitally signed data blocks (records). The traditional method of guaranteeing data integrity is a simple digital signature, which depends on needing trust in the signer. In a blockchain you do not need to trust any individual participant, but you trust that the community of participants together make sure that the content of the data blocks is correct, and/or that the content of data blocks does not change. This is done by linking blocks of data with digital signatures, where all participants, or a set of participants, can have the role of signers. The selection of the signer for each block is done with a consensus protocol that gives participants an inherent incentive to sign only the correct blocks. An attack against the consensus protocol could aim to become a signer of a corrupt block. If that were possible, for example, the attacker could generate counterfeit digital coins and falsify documents. An attacker who controls 51% of the global computational capacity for mining new bitcoins, would have the power to mount such an attack.

---

[4] https://datatracker.ietf.org/doc/html/rfc8659
[5] https://datatracker.ietf.org/doc/html/rfc6962

Potential applications for blockchain technology are precisely when it can be useful to guarantee data integrity in a distributed manner by many participants together. The placement of blocks of data in a blockchain also provides a timestamp that cannot be forged. Relevant areas of application for blockchains include:

1. Integrity of documents exchanged in open infrastructures.
2. Integrity of contracts for purchase and sale, loans, and insurance.
3. Money transfer and financial transactions.
4. Integrity of data collection from sensors.
5. Digital Rights Management (DRM).
6. Logistics and tracking of goods.
7. E-voting.

EBSI (European Blockchain Services Infrastructure) is an initiative of the European Commission, started in 2018, which is managed by EBP (European Blockchain Partnership), in which many European countries participate. EBP's vision is to leverage blockchain technology to create cross-border solutions for public administrations, businesses, citizens and their ecosystems, to verify and create trust in digital information and services. As an example, it is conceivable that educational institutions put digital signatures on educational information in EBSI, so that employees e.g. can easily verify the authenticity of university degrees of job applicants.

5.5 Tasks

1. Certificates
   (a) Give (four) reasons for having limited cryptoperiod for keys.
   (b) What is meant by "protection period" and "processing period" for cryptographic keys?

   Go to https://www.google.com/ using your favorite browser, click the padlock in the address bar to see the server certificate (X.509 public-key certificate), view the certification chain, and click on the root certificate. Check the details of the root certificate to see the validity period, which corresponds to the cryptoperiod of the public key.

   (c) What is the validity period (from/to) of the root certificate?
   (d) The private signature key was used to sign the intermediate CA certificate. Check the date of issue of the intermediate CA certificate. Would you say that the validity period of the private signature key in the root certificate follows the recommendations for the private signature key protection period according to NIST SP800-57 shown in Fig. 5.2?

(e) Assuming that powerful quantum computers will be practically available by 2030—does the validity period of the root certificate make sense?
(f) Use the service at https://www.ssllabs.com/ssltest/ to test certificate and TLS server quality on e.g. the google.com and fb.com sites. Compare the results.

2. Trust model
    (a) Describe the trust model for the PKI used by web browsers.
    (b) Mention the pros and cons of this model.
    (c) It is said that key certificates and PKI create trust. What is meant by this?

3. Security of cryptographic keys
    (a) Name factors that determine the strength of cryptographic security solutions.
    (b) Why is key management important for cryptographic security solutions?
    (c) Three important key categories are: (i) symmetric secret keys, (ii) asymmetric public keys, (iii) asymmetric private keys. Explain the type of security services/protection (i.e., confidentiality, integrity, and authenticity) required for each key category.
    (d) Describe security controls that can be used to protect cryptographic keys.

Chapter 6
Network Security

> *The Internet gave us access to everything; but it also gave everything access to us.*
>
> James Veitch, British comedian

Network security is about protecting and controlling the communication between systems and devices. Network security is essential for conducting transactions and communications among businesses, government agencies and individuals. The concept of security protocol means a specification for exactly how data shall be transmitted in a secure way between systems, and typically involves encryption and digital signatures. The most prominent communication platform is of course the Internet, for which there are many different security protocols. Other network technologies are e.g. Bluetooth, mobile, industrial networks, in-car networks, and satellite links, for which there are also various security protocols. Two different aspects of network security are the security of the data transmission itself, which is called *communication security*, and the security to protect computer networks of organizations from attacks, which is called *computer network security*.

The learning objective of this chapter is first to understand how the Internet and computer networks are structured, as a basis for the main learning objective which is to understand principles of network security. In particular you will learn principles and technologies for communication security between nodes in networks, and principles and technologies to protect an organization's computer network against attacks.

6.1 Computer Networks and the Internet

Computer networks are infrastructures that allow systems to send and receive data, i.e. for data communication. Systems that communicate through a computer network are called nodes or hosts.

Nodes in a computer network are systems consisting of hardware and software that can send and receive data through some form of network connection. The Internet consists of millions of separate computer networks that are connected to a greater or lesser extent, so that it is possible to send data between them. The connection point for a computer network is called, for example, an access node, an interface or a gateway. In the name Internet, "inter" means precisely "interconnection" of different computer networks.

Nodes that exchange data need a set of rules for how the data is formatted in *packets* and how the data packets are sent in sequence. A data packet is simply a chunk of bits grouped together in a structure. The set of rules for transferring data packets is called a *network protocol*. The Internet is based on many different network protocols that each fulfil its own role. It is important to understand how different network protocols build on top of each other with hierarchical levels of abstraction. Defining levels of abstraction allows the various functions to be separated in a consistent way so that it is easier to design, implement and configure all the components of a computer network into a well-functioning whole. The levels of abstraction are also pedagogical, without them it would have been impossible to learn and understand the complexities of the Internet.

Ultimately, the data packets must be transmitted as physical signals which can be radio signals through air, electronic signals through metal cables, or optical signals through glass fiber cables. Hence, airspace, metal cables and fiber cables are physical media for data communication.

6.1.1 The Domain Name System

To locate something in a physical or virtual space, a method is needed for naming and reaching locations in that space. On a topographic map, the combination of latitude and longitude uniquely designates each geographical location. On a street map, the locations of buildings are designated by their street addresses. In the public telephone system, each subscriber can be contacted by their telephone number. On the Internet, virtual locations have numeric IP addresses (Internet Protocol addresses), as well as names—called domain names. The purpose of having both an IP address and a domain name for a location on the Internet is that IP addresses can be easily changed and are used by nodes for routing traffic across the Internet, while domain names are stable and generally more easily remembered (by humans) than IP addresses. The DNS (Domain Name System) consists of special servers on the Internet—called name servers—that translate domain names into the corresponding IP addresses.

There are two types of IP addresses: IPv4 (IP version 4) defined in 1983 and IPv6 (IP version 6) defined in 1998. An IPv4 address is a 32-bit integer, while an IPv6 address is a 128-bit integer. With 32 bits, only about $2^{32} = 4.3 \cdot 10^9$ (i.e. billion) IPv4 addresses can be defined. It was probably considered a huge address space in 1983, but as the Internet spread to every corner of the globe, it turned out to be too small. With 128 bits it is theoretically possible to define $2^{128} = 340 \cdot 10^{36}$ (i.e. sixtillion) IPv6 addresses. That address space seems huge today, but one day it might also be too

6.1 Computer Networks and the Internet

small. IPv4 addresses and IPv6 addresses can be used interchangeably on the Internet.

IP rotation is a technique for rapidly and continuously changing IP addresses. IP rotation can have multiple purposes, both legitimate and nefarious. IP rotation can for example be used for load balancing, so that a website can spread the traffic it receives to multiple servers. IP rotation can also be used by threat actors to avoid IP address blacklisting by a victim company they try to attack, e.g. when performing a password spraying attack against that company.

The DNS was developed in the early 1980s when the Internet was only used for non-commercial services and the World Wide Web had not yet been invented. The DNS was intended it to make it easy for users and applications to identify computers on the Internet. The naming system was designed to have the structure of an inverted tree as illustrated in Fig. 6.1.

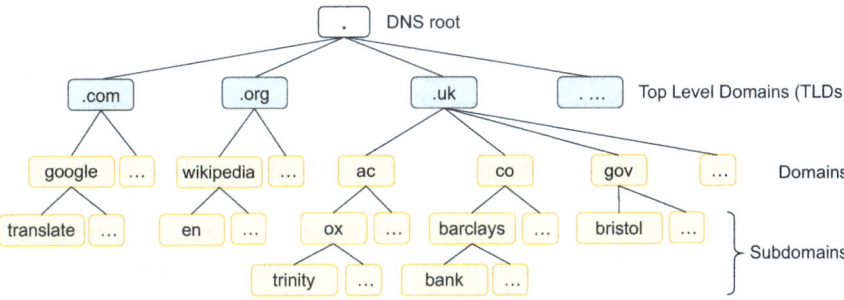

Fig. 6.1 Structure of domain names

The original intention was to group domain names under various TLDs (Top Level Domains) that represent categories of organizations such as .com for commercial, .org for non-commercial, and .uk for organizations in Great Britain (United Kingdom). The expectation was that the tree would be relatively balanced and that the domain names of most organizations would contain subdomains and be relatively deep. However, with the explosive commercial use of the Internet, .com quickly became the most popular TLD with 46% of all domain names under it, followed by .org (4.5%) and .ru (3.4%), as of December 2023.[1] Typically, there are relatively strict policies for country code TLDs (ccTLDs) such as .uk for Great Britain and .ru for Russia, but otherwise there is no strict separation of categories of organizations according to TLD. The borderless nature of the Internet also contributes to the convergence of domain names under the .com TLD. In contrast to the public telephone system where all telephone numbers are organized under national authorities, domain names (except ccTLDs) and IP addresses are largely borderless and out of reach for national jurisdictions. The responsibility of overseeing all the TLDs is held by ICANN (The Internet Corporation for Assigned Names and Numbers) which is a global multistakeholder group and nonprofit organization headquartered in the USA. So-called gTLDs (generic TLDs) are TLDs that are not

---

[1] https://www.statista.com/statistics/265677/number-of-internet-top-level-domains-worldwide/

country specific and that must have three or more characters. Any public or private organization anywhere in the world can apply to ICANN to establish and operate a new gTLD. The applicant must demonstrate the technical and financial capability to operate the gTLD according to specific requirements defined by ICANN. As of 2023 there are around 1500 TLDs, so there is a wide range of TLDs to choose from if you do not absolutely need a domain under .com or .org.

DNS servers are responsible for translating host domain names into numeric IP addresses that are used to route data packets to the right destination on the Internet. Each domain has at least one authoritative DNS server that provides information about that domain and other name servers of domains subordinate to it. A computer device receives the IP address of the closest DNS server through the DHCP (Dynamic Host Configuration Protocol) when connecting to a local network. For example, when connecting to a Wi-Fi network, the smartphone or computer receives the local DNS server's IP address from the Wi-Fi network.

Figure 6.2 illustrates DNS's role in support of routing traffic across the Internet, where the DNS server resolves (translates) the domain name en.wikipedia.org into the IP address 185.15.59.224 which is turn is used to route the request to Wikipedia's web server. It is common that the same IP address serves multiple hostnames. This is possible because when a client web browser requests a resource from a web server it includes the requested hostname as part of the request. The web server uses this information to determine which web page to return to the client. For example, Wikipedia has hostnames for every supported language, where the hostname en.wikipedia.org is the English version of Wikipedia. All host names under wikipedia.org have the same IP address, such as ja.wikipedia.org for the Japanese version and ar.wikipedia.org for the Arabic version.

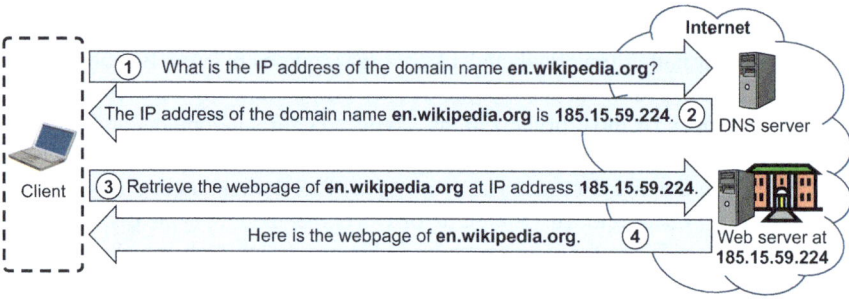

Fig. 6.2 The role of DNS for finding IP addresses to be used for Internet routing

To find the IP address of the domain name wikipedia.org, you can type the following command in the command window on Windows: `nslookup wikipedia.org`, which returns both an IPv4 address and an IPv6 address for Wikipedia's web server. To find your own IP address, you can type the following command in the command window on Windows: `ipconfig`. The command prompt window can be opened by typing "command" in the Windows search bar.

For the correct functioning of the Internet, it is essential that DNS information is correct, otherwise data may be routed to the wrong IP address. DNS hijacking and DNS poisoning are attacks aimed at spoofing IP addresses. The information in DNS is entered by a domain registrar, which is an entity with permission by an authority for a TLD to register domains for its customers. Every country has a national authoritative registrar for the respective ccTLD (country code TLD), and every other TLD also has an authoritative registrar. In turn the TLD registrars authorize sub-registrars to register and edit domain name information.

DNSSEC (Domain Name System Security Extensions) is a protocol for securing data exchanged in DNS. The protocol provides cryptographic authentication and integrity of data, but not confidentiality.

6.1.2 The Internet Stack

A hierarchical set of protocols is called a *protocol* stack, meaning a stack of different protocols that represent different abstraction layers of communication. A communication protocol defines formats of data packets to be sent and received by processes in a computer host. Different computer networks can be based on different protocol stacks. During the 1980s and early 1990s there were two competing protocol stacks for open global networks, where the Internet stack won, and the OSI stack was an also-ran. At the bottom of a protocol stack there is the physical layer, which is hardware for sending and receiving physical signals and which is not really a protocol. The next layer is *the link* layer, which formats the data in a way suitable for transmission through the physical layer and that interfaces with the hardware on the physical layer. There are different link protocols depending on the physical medium used, such as Wi-Fi, mobile broadband or a LAN cable. In case of the Internet stack, above the link layer is *the internet layer*, where *IP (Internet Protocol)* is the protocol used. IP is a peer-to-peer protocol for exchanging data between nodes on their way through the Internet. Above IP lies the *transport layer*, which can have different end-to-end protocols between host nodes that exchange data. *TCP (Transmission Control Protocol)* is the most common transport protocol, and during the development of the Internet, TCP and IP were developed together. Thus, it became common to say that TCP/IP is the protocol stack for the Internet. Other important transport protocols are *User Datagram Protocol (UDP)*, which is used for faster transmission at the cost of reduced reliability, and *ICMP (Internet Control Message Protocol)*, which is used to send operational messages between nodes on the Internet. Above the transport layer is the *application layer*, where there are also various application protocols, e.g. *HTTP (Hypertext Transfer Protocol)* for web traffic, IMAP (Internet Message Access *Protocol)* for sending e-mails, or *FTP (File Transfer Protocol)* for transferring files. The Internet stack thus consists of protocols for the link, internet, transport, and application layers. Apps (client and server applications) that talk to each other through the Internet are not part of the Internet stack, but they use the Internet stack to communicate.

The protocol stack is implemented as a set of program modules, one for each layer. Remember from Sect. 3.1 that a "process" denotes program module that is active and running in the microprocessor. Figure 6.3 shows the layered architecture of the Internet stack, with examples of apps and protocols. The functionality of each layer is handled by its corresponding protocol process.

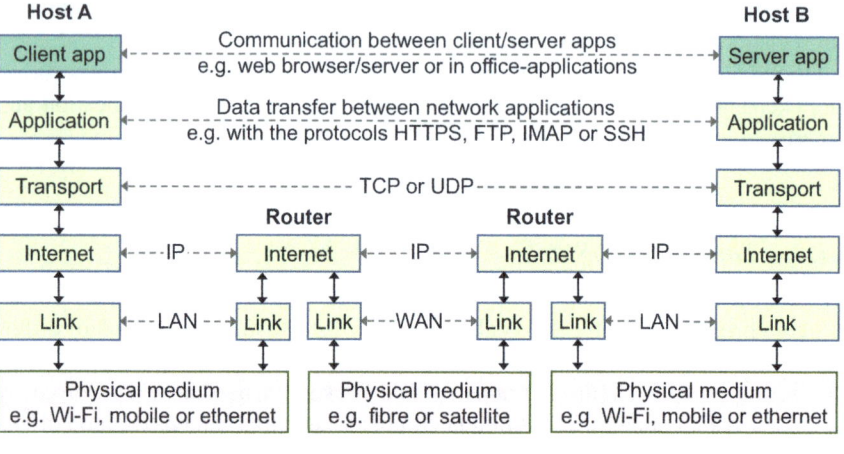

Fig. 6.3 Layered architecture for data transfer through the Internet

The functionality of the Internet stack in Fig. 6.3 can be described as follows.

Suppose a browser (client app) in host node A is going to transmit data to a web server (server-app) in host node B. First, host node A's browser process calls the application layer process and hands over the payload data. The application layer process calls the TCP (transport layer) process and relays the payload with a header as an application layer packet. In turn, the TCP process calls the IP (Internet Protocol) process and relays the payload with an additional header as a TCP packet. The IP process then calls the link layer process and relays the payload with an additional header as an IP packet. Finally, the link layer process adds a link-layer header to the data, activates host note A's physical transponder which sends the whole payload with packet headers as physical signals through a physical medium. Each packet header describes the content and where to forward the packet. The first router on the way will receive and unwrap the data packet up to the IP layer, before deciding where to forward the packet. In this way the data packet continues its journey through the Internet until it arrives at host node B on the receiving end. There, the headers are unwrapped from the packet layer by layer by the corresponding processes until the payload data arrives at the web server process (server app) at host node B. To send a reply, the web server at host node B returns payload data back to the browser on host node A using the same principle.

Apps that communicate with each other are called clients and servers, where the client typically sends a request to a responding server. This can be, for example, a browser (client) that communicates with a web server, but also other apps. For example, office applications can communicate directly with an Internet cloud server to translate text from one language to another. Operating systems and apps have built-in capability for communication over the Internet for various reasons, such as updating software or communicating with other apps. Operating systems and apps then use an application protocol such as HTTPS for communication over the Internet. Note that HTTPS can also mean HTTP, which is the unencrypted variant of HTTPS. See Sect. 6.2.1 for description of HTTPS. *Host nodes* are nodes that contain client and server applications, such as host nodes A and B in Fig. 6.3. *Routers* are network nodes that ensure that data packets find their way from host node A to host node B through the Internet.

The Internet stack was specified in the 1970s as the result of a research project by DARPA (Defense Advanced Research Projects Agency) in the US. Another protocol stack is OSI (Open Systems Interconnection), which was specified by the ITU (International Telecommunications Union) in the 1980s. The Internet is a simple protocol stack compared to OSI which is a relatively complex 7-layer protocol stack. While the popularity of the Internet grew slowly but steady, in the 1990s many national telecom operators, especially in Europe, made a major commercial effort to establish global data networks based on the OSI stack. It is fascinating technology history how the Internet ultimately took over the world despite OSI being supported by major national telecom operators. Ironically, the X.509 standard for digital certificates is an ITU standard that was intended to be used in the context of OSI but is now used on the Internet. It is interesting to note that OSI was specified from the very beginning with built-in security, that is, security functions were integrated into the protocols. In contrast, security was completely overlooked during the development of the Internet stack, which is the main reason why the Internet has been, and still is, relatively vulnerable to cyberattacks. Internet security had to be added afterwards, which has been difficult and the reason why the Internet still has fundamental security weaknesses. From a security perspective, it would probably have been best if OSI had won over the Internet in the quest for world domination. For example, the naming system in OSI was strictly based on country codes, similarly to the public telephone system where each country has jurisdiction over phone numbers with specific country prefixes. In this way the OSI network would have allowed stronger national jurisdiction than what is possible with the DNS and the borderless Internet.

This book uses only the protocol designations from the Internet stack. The OSI stack never gained widespread practical use, but is often cited as a reference model for data communication. It is e.g. common to read or hear people talk about "layer 7 protocols" or "layer 4 protocols". To understand what they mean, we need to look at Fig. 6.4, which shows the mapping between protocol layers of the Internet and OSI stacks. The figure also shows examples of protocols on each layer. For example, the link layer uses a MAC (Medium Access Control) protocol to activate and control hardware to send and receive electrical signals through cables, light signals through fiber, or radio signals through air and space.

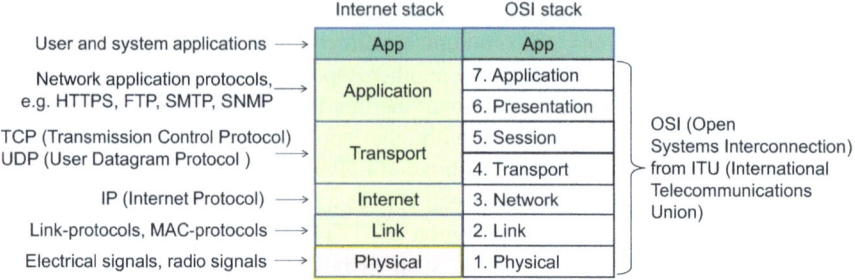

Fig. 6.4 Correspondence between the Internet stack and the OSI stack

Although the terms *app* and *application* sound similar, they are actually two completely different things in both the Internet and OSI stacks. The reason the application layer (layer 7 of the OSI stack) was denoted "application" is that back in the 1980's when the naming was invented, each type of communication was seen as a separate "communication application", which was not intended to be integrated into other applications such as text editors or spreadsheets. For example, telephony, file transfer, and message transmission were considered separate standalone applications. In those days, nobody could imagine that e.g. a spreadsheet app would send and receive data through data networks. Today, all forms of communication have largely merged on the Internet, where everything is just data transmission which is integrated with every app. The application layer protocol used by most apps is HTTP(s), but this protocol which was first developed in 1989 was initially intended for web surfing which became popular from around 1993. Web surfing was the killer app which caused the Internet to explode in popularity in the mid-1990's. Nevertheless, there are several different application protocols, each with its own specific use on the Internet, e.g. for controlling routers and controlling traffic.

The data packets transmitted between host nodes consist of payload data and headers. Each layer in the protocol stack adds its own packet header that contains information about what to do with the data, and where to send it. Figure 6.5 shows how payload data is combined with packet headers in the different layers of the Internet stack. The example shows the HTTP and TCP protocols in the top layers, but other protocols are also possible.

6.1 Computer Networks and the Internet

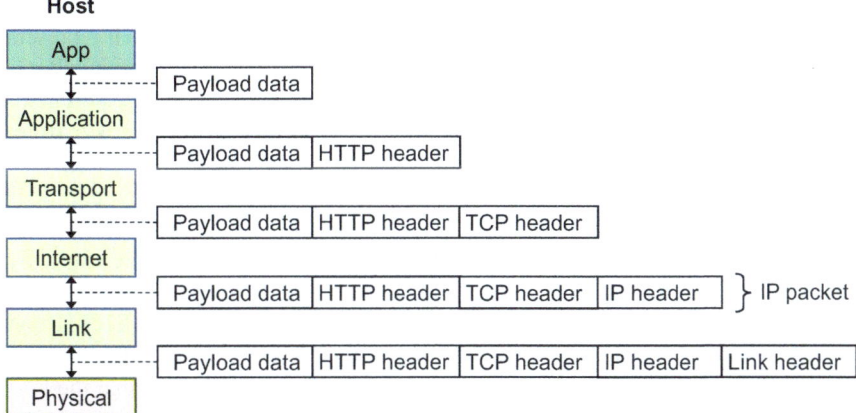

Fig. 6.5 Data packets in the protocol stack

The link header in data packets on the bottom layer contains the MAC address (Media Access Control) of the sending and receiving devices within a local physical network. It applies to physical equipment such as computers, printers, network cards, routers, switches, mobile phones, whether they communicate through LAN cables, Wi-Fi or Bluetooth. Each device listens for electrical signals on the cable, or for Wi-Fi radio signals in the air to detect if a link packet has a recipient address equal to its own MAC address. Only packets with the correct MAC address are processed, all others are ignored. In a Wi-Fi network, everyone listens to everyone, but only receives link packets with the correct MAC address.

The MAC address consists of 48 bits, where the first three octets (24 bits) identify the manufacturer of the equipment, and the next three (24 bits) are the manufacturer's own unique identification of the device. At the link layer, all addressing takes place with MAC addresses, therefore network equipment such as switches and routers must translate IP addresses into MAC addresses, which are placed in the link header before the packet is sent through a cable or as radio signals. In most types of equipment, the MAC address is fixed, but in some types of equipment it can be changed by the user. The MAC address can also be used for access control, so that only equipment with given MAC addresses can access a network or service.

The concept of *IP packet is* central in the Internet context because this is the unit of data that is routed through the Internet. The IP header contains the IP addresses of the sending and receiving host nodes, which allows each router node in the path to determine which link the packet should be forwarded through on its way to the receiving host node. The return address can be used to return an acknowledgement back to the sender to confirm that the packet has arrived. Dedicated protocols are used by the router nodes to keep up-to-date on the most optimal routes through the Internet.

The IP address represents the host node's near-physical location on the Internet. In addition, for each IP address there is a range of *port numbers* which represent specific network applications or process in a specific host node. The port number is required for a packet to reach the correct application on a host node. The port number is transmitted in the TCP header (or UDP header). A port number is a 16-bit integer, which means that it is possible to define 65,535 different ports in each host node. Many network applications have fixed port numbers. For example, port 80 is reserved for HTTP and port 443 is reserved for HTTPS. Firewalls are typically configured to block most ports by default, and only open ports that are explicitly needed. A few specific ports are usually always needed and thereby open, such as ports 80 and port 443 because online services largely send traffic through HTTP and HTTPS. The fact that a specific port is "open" means that packets with that port number in the TCP header is allowed to pass through the firewall or that the host node accepts those packets.

A *LAN (Local Area Network)* is a local computer network that connects nodes within a limited area, such as a residence, an open-plan office, an entire office building, or a limited outdoor space. Communication within a LAN does not require IP routing because the MAC address is sufficient to locate hosts within the same LAN. Ethernet and Wi-Fi are the two most common technologies for LAN. An ethernet LAN is based on metal or fiber cables that connect many nodes which can send and receive electrical or optical signals simultaneously. A Wi-Fi network covers an airspace where many nodes within the coverage area can send and receive radio signals simultaneously. Metaphorically, the term "wireless" is used when radio signals are sent through air, as opposed to through metal and fiber cables that represent "wires". A WAN (Wide Area Network) is a computer network over long distances. WAN is typically based on fiber cables that run point-to-point, i.e. between a pair of nodes. Satellite communications naturally also cover large distances and can broadcast radio signals one-to-many, i.e. signals from one satellite can be received by many receivers. Regardless of the type of medium, the physical signals can be modulated by the nodes so that they are interpreted as a stream of binary bits, i.e. as 0 or 1.

6.2 Communication Security

Data sent between nodes is exposed to threats against confidentiality (interception of data), against integrity (falsification of data) and against availability (prevention of data transmission). A simple analogy to communication security is physical transport security, such as for transport of cash money as shown in Fig. 6.6.

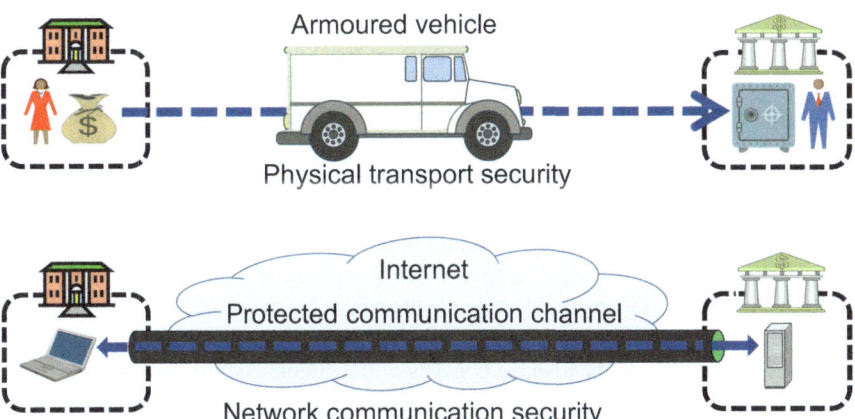

Fig. 6.6 Analogy between physical transport security and communication security

Communication security is typically based on cryptography, although physical shielding of cables also provides a form of communication security. This chapter describes only cryptographic methods. For nodes to understand each other, data communication must follow rules for how data packets should be formatted and sent in sequence. A communication protocol is one such set of rules. Communication protocols are standardized so that manufacturers can develop products for data communication that can communicate with each other. For secure data communication, there are many different security protocols, i.e. communication protocols where data packets are encrypted, or where data packets contain keys and certificates.

Security protocols can have different purposes, such as authentication, integrity, confidentiality, key exchange, e-voting, anonymity, etc. It proves surprisingly difficult to specify security protocols that are free of vulnerabilities, and unfortunately, serious vulnerabilities in prominent security protocols are constantly being discovered. Removing such vulnerabilities often requires a revision of standards, which can take years to fix.

An important security protocol is TLS (Transport Layer Security), which is used in conjunction with TCP to support end-to-end encryption of application data through the Internet. Another important security protocol is IPSec (Internet Protocol Security), which encrypts each IP packet separately and is typically used for encrypting data traffic between two computer networks. TLS and IPSec are described in the following sections.

6.2.1 TLS: Transport Layer Security

TLS (Transport Layer Security) is a prominent security protocol that, among other things, supports encrypted web communication with HTTPS. The first version of this protocol was called SSL (Secure Sockets Layer) and was published by the browser manufacturer Netscape in 1996. Due to serious security vulnerabilities in

SSL, a new, improved protocol called TLS was specified as an RFC by the IETF in 1999. The first versions of TLS also had some weaknesses, which have been corrected for each new version of the standard. The latest version is TLS 1.3, which was published in 2018. In addition to fixing vulnerabilities from the previous version, TLS 1.3 is designed for speed in establishing the session key for a session. Time spent for setting up a session is particularly dependent on the number of rounds back and forth of messages sent between client and server, because each round can produce significant time delay. In TLS 1.3, only one round of messages is needed to establish the session key, which is called the *handshake*, as shown in Fig. 6.7. A "round" means that one or several messages first are sent in one direction, followed by one or several messages being sent in the return direction. With reference to Fig. 6.7, the handshake consists of the client sending message 1 and the server responding with messages 2–6, which allows both client and server to compute the session key in step 7. This is a very efficient protocol design.

Forward secrecy is a desired feature of security protocols that prevents previous session keys from being compromised even if a long-term cryptographic key used for session key establishment is compromised sometime in the future. For TLS, the long-term crypto key is usually the private signing key of the server. Forward secrecy is achieved with the use of Diffie-Hellman, which is mandatory in TLS 1.3, but was only an option in TLS 1.2. The session key's confidentiality is based on Diffie-Hellman, while session key authenticity is based on the server's private key.

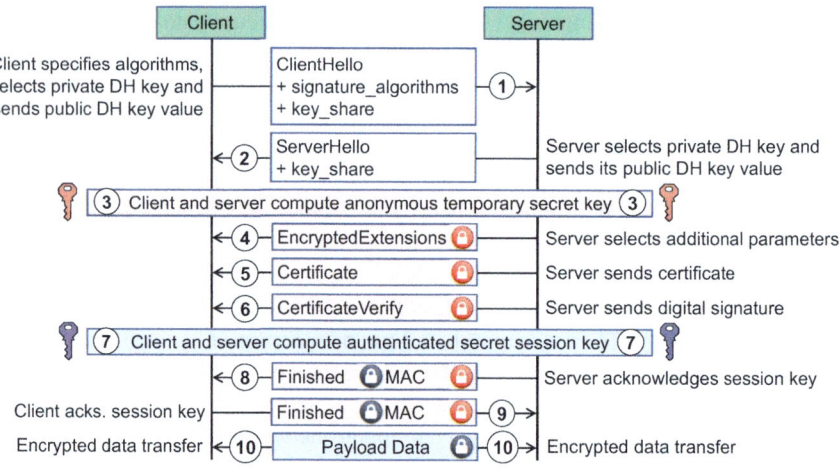

Fig. 6.7 Simplified TLS Handshake protocol message diagram

Below is a simplified description of each message and step in the message chart.

1. ClientHello: Client (e.g., a web browser) sends message to server (e.g., web server) based on assumptions that the server supports the DHE (Diffie-Hellman Ephemeral) algorithm with a specific algebraic group as described in Sect. 4.6. The client sends its DH public key computed under the algebraic group.

6.2 Communication Security

2. ServerHello: The server sends its DH public key which is computed under the same algebraic group.
3. The client and server compute a temporary (ephemeral) key as described in the Sect. 4.6. This key is anonymous, i.e. without any authenticity.
4. EncryptedExtensions: The server sends additional parameters as may be required. The message is encrypted with the ephemeral anonymous key.
5. Certificate: The server sends its certificate that contains the server's public key subject name, as described in Sect. 5.2.2. The message is encrypted with the ephemeral anonymous key.
6. CertificateVerify: The server sends a digital signature over message (5) as proof of the server's subject name. The message is encrypted with the ephemeral anonymous key.
7. The client and server calculate the session key based on the ephemeral anonymous key and other parameters sent by the server in messages 4–6.
8. Finished: The server sends a confirmation of the session key in the form of a MAC calculated with the session key, as described in Sect. 4.5. The message is encrypted with ephemeral anonymous key.
9. Finished: Client sends confirmation of the session key in the form of a MAC calculated with the session key. The message is encrypted with the ephemeral anonymous key.
10. Application Data: The authentic session key has been established and a secure session is created, in which exchanged data is encrypted with the authenticated session key.

TLS actually consists of several protocols that fulfil different functions. The sequence of messages 1–9 is part of the TLS Handshake protocol with the function of establishing an authenticated secret session key between the host nodes. Message 10 is part of the TLS Record protocol with the function of encrypting payload data.

The message diagram in Fig. 6.7 displays the simplest profile for establishing the session key with the handshake protocol. This figure only shows unilateral authentication, i.e. the client authenticates the server, and not vice versa. Unilateral TLS authentication is the most common situation in a web context where the browser authenticates the website (web server) with TLS, while the website authenticates the user with user authentication (e.g. password). TLS also allows mutual authentication called mTLS (mutual TLS), meaning that the client also sends a public key certificate to the server so that the session key can be computed based on authentication with both certificates. Other profiles are also specified for TLS Handshake, such as client and server having a predefined session key, or having a session key from a previous session.

TLS encrypts payload data in the application layer data packets, as shown by the shaded payload data fields in Fig. 6.8. The encryption is done by the TLS Record protocol, which can be thought of as a separate protocol layer that sits between the application and transport layers. TLS together with HTTP at the application layer turns into HTTPS.

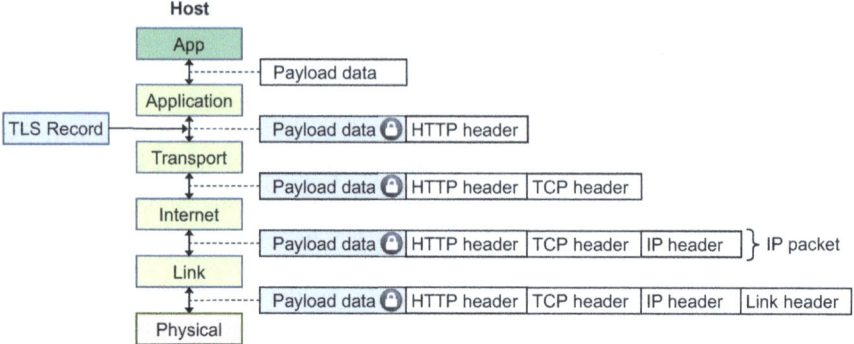

Fig. 6.8 Encryption of user data with TLS

TLS (and formerly SSL) is fundamental for communication security on the Internet. The "S" in the name HTTPS (Hypertext Transfer Protocol Secure) means that web traffic is encrypted with TLS. TLS/SSL was implemented in browsers and web servers from the year 2000. Previously, passwords were usually sent as plaintext during online login, which meant that passwords could easily be "sniffed" (stolen) on a large scale. In the 2000s, SSL was mostly used for encryption of sensitive web traffic. Since 2014, *TLS Everywhere* has been established as standard practice, meaning that web servers always use HTTPS regardless of whether the content is sensitive or not. This is undoubtedly good for security, but it can also provide greater CPU load and energy consumption, as described in Sect. 4.13.1. However, the burden of encryption is hardly noticeable in modern computer platforms, as CPUs are increasingly powerful and the transmission capacity through the Internet increases.

HSTS (HTTP Strict Transport Security) is a technology to force browsers to use HTTPS to websites that have introduced HTTPS across all their web domains. HSTS is a security control against TLS stripping, which is a type of MitM (Man-in-the-Middle) attack where, for example, an attacker sets up their own Wi-Fi network to trick users. The attacker's Wi-Fi network converts HTTP traffic from the user's browser into HTTPS that is forwarded to the real website. If the user does not notice that the connection is using HTTP instead of HTTPS, the attacker can e.g. read sensitive login credentials for identity theft. With HSTS, a browser will not allow sending HTTP traffic to an Internet domain that requires HSTS. After 2015, HSTS is used by an increasing number of websites and is considered to be standard practice today.

6.2.2 IPSec: Internet Protocol Security

IPSec (Internet Protocol Security) applies cryptography to IP packets to support confidentiality and authenticity. IPSec works at the internet layer and is a very flexible protocol with several different profiles. A useful profile is to encrypt IP packets

6.2 Communication Security

including the IP header, but then a new IP header is needed to route the packets through the Internet. This profile is called *tunnel mode* because it metaphorically creates an encrypted tunnel for routing IP packets through the Internet. Figure 6.9 displays the encryption of IP packets with IPSec in tunnel mode, where the shaded data fields in blue indicate encrypted data. The example shows HTTP and TCP protocols in the upper layers, but other protocols are also possible.

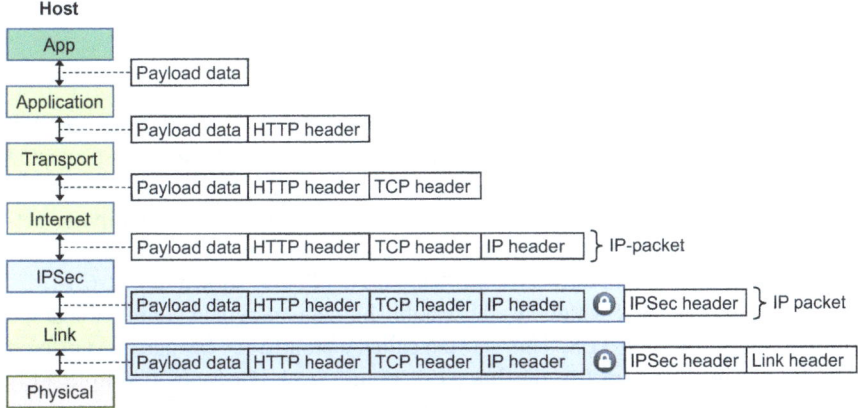

Fig. 6.9 Encryption of IP packets with IPSec in tunnel mode

IPSec in tunnel mode can be thought of as a new layer between the internet and link layers because a new header called *IPSec header* is added, as shown in Fig. 6.9. The sender and recipient address pair differ in the IP header and the IPsec header. Typically, the address pair in the IP header points to host nodes within private computer networks, while the address pair in the IPsec header points to gateway router nodes that form interfaces between private computer networks and the rest of the Internet. This keeps host nodes' IP addresses hidden, which is an important security feature for private computer networks. Figure 6.10 shows how IPSec can be used to create an encrypted tunnel between separate computer networks. The black tube in the figure symbolizes encrypted connection.

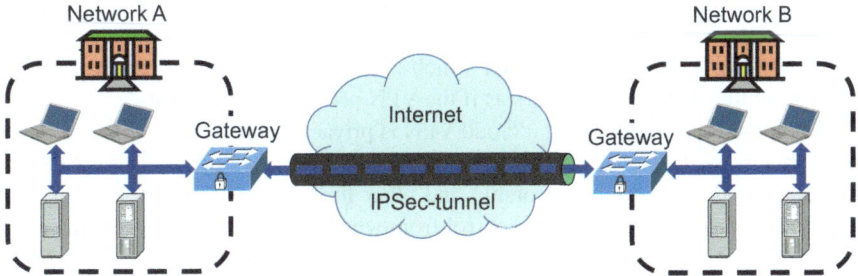

Fig. 6.10 IPsec tunnel between two computer networks

For routers on the Internet, it looks like the traffic is going between the two gateway nodes. In reality, the end-to-end traffic goes between host nodes within the respective computer networks, but IPSec keeps these hidden. A configuration as in Fig. 6.10 is also called a *VPN (Virtual Private Network)* because the IPsec tunnel is logically a private network connection between the two computer networks. There are different variants of VPNs, which are described in the next section.

6.2.3 VPN: Virtual Private Networks

VPN (Virtual Private Network) denotes encrypted (private) connections through open networks (Internet). Logically, it is as if a host node has its own protected computer network, but in reality, the open Internet is used. VPN is a very general term that covers many different architectures and configurations for computer networks. An encrypted connection at the IP layer between the gateway nodes of two separate computer networks forms an IPSec-based VPN, as shown in Fig. 6.10.

Another type of VPN widely used is establishing an encrypted connection at the transport layer based on TLS. This is often called a cloud VPN because it is offered as a service in the cloud (on the Internet). Figure 6.11 illustrates a simple VPN architecture where a user first sets up an encrypted connection to a VPN server, and through the VPN server sets up an encrypted connection to another website on the Internet which is the user's real target site. The black tubes symbolize encrypted connections.

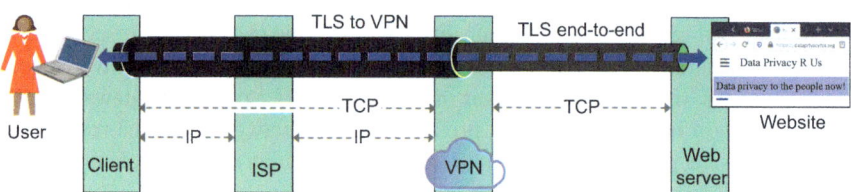

Fig. 6.11 Cloud VPN

To the user's ISP (Internet Service Provider), it appears as if the user has a connection to the VPN server, which is correct, but the ISP cannot see the connection from the VPN server to the website, which is the user's real target. From the perspective of the Website, it appears as if the VPN provider is the client.

A common reason for using a cloud VPN is privacy, such as when the user does not want e.g. the ISP or other parties to know which website they are visiting. Other reasons include bypassing censorship and other forms of content blocking on the Internet. Often, websites want to control from where in the world users can receive media content, which is done with *geo-blocking* based on the user's IP address received through the TCP connection. By using cloud VPN, users can hide their true

IP address and bypass geo-blocking because the website receives the IP address of the VPN server through the TCP connection and thus believes that the user has the VPN server's geolocation.

A potential weakness of VPN from the user's point of view is that the VPN provider may leak information or be forced to disclose information about the user's web activity. To hide the user's web activity even better, several VPN services can be used on top of each other.

TOR (The Onion Router) is precisely a VPN that wraps the traffic in three separate VPN connections on top of each other, metaphorically like the layers of an onion, and where the VPN nodes are chosen randomly. The development of TOR started at the US Naval Research Laboratory in the early 2000s and is now driven by the TOR project,[2] which is an independent organization.

If a business uses the IP address for location-based subject access control, then VPN represents an attack vector. With VPN, an attacker outside the geographical area of access can make it appear as if they are within the correct geographical area for access. Hence, location-based access control must be seen as a relatively weak mechanism in isolation.

6.2.4 *Steganography*

Steganography means to hide a (secret) message embedded in another message or physical object. The word steganography comes from the Greek, where *Steganós* means "covered", and *Graphia* means "writing". Steganography is not the same as cryptography. The difference is that an encrypted message is "discoverable" but impossible to interpret without the decryption key, while steganography makes the message "invisible" or difficult to detect, even if it is otherwise easy to interpret. It is naturally possible to combine both cryptography and steganography.

With digital steganography, a message can be hidden, among other things, in a media file such as a picture, music or video. Media files are ideal for steganographic transfer due to their large size. For example, a sender can use an image file and adjust the least significant bit on every ten pixels, so that an 8-bit letter can be transmitted by 80 pixels. The change is so subtle that it is not visible with simple visual inspection.

Steganalysis means analysing a message to detect the use of steganography. The easiest method to detect a modified file is to compare it with the original. Alternatively, statistical analysis may reveal the use of steganography.

However, there are steganography techniques that are resistant to statistical steganalysis — i.e., it is practically impossible to tell whether a file contains a hidden message or not.

---

[2] https://www.torproject.org/

Although steganography can be used to protect confidential information, the capacity is relatively low. The advantage of steganography over cryptography, as mentioned, is that the very existence of the message can be hidden with steganography.

Steganography can be abused by attackers with covert malware transmission, which is called *stegware*. Stegdetection is challenging, as mentioned above. An alternative method of eliminating the threat of steganography in general, and stegware in particular, is to transform media files in a way that destroys any hidden messages, which is called *Content Threat Removal*.

6.3 Computer Network Security

A computer network is always exposed to threats, both from within and from outside. An important security control is to prevent unauthorized external entities from gaining access to resources within the network, for example to commit data theft (espionage), misuse of computer network services or destruction/deletion of data and resources (sabotage). A simple analogy to computer network security is a medieval castle with defenses against intruders, as shown in Fig. 6.12. It is interesting to note that the two situations have many pairwise analog features, e.g.

Gatekeeper/drawbridge/moat	=	Gateway/external firewall
Inner gate	=	Internal firewall
Lookout tower	=	IDS (Intrusion Detection System)
Outer court	=	Outer segment
Inner court	=	Inner segment

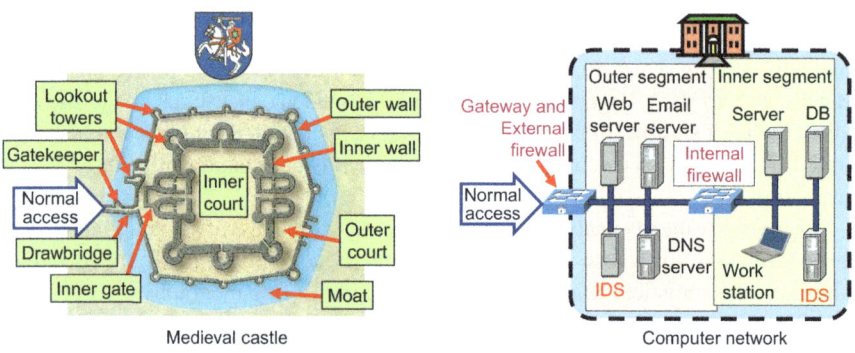

Fig. 6.12 Analogy between medieval castle defences and computer network security

There is also a kind of analogy between medieval society and cyberspace. In the Middle Ages, people were exposed to threats on country roads, but a castle or walled city provided protection. Similarly, cyberspace is full of threats, but a well-secured computer network provides protection.

Computer cybersecurity has traditionally followed the approach of *perimeter security*, where a perimeter represents a separation between different parts or segments of a computer network. The idea behind perimeter security is to protect against threat actors located outside the perimeter. Such a perimeter may be based on a firewall and/or authentication and access control for entities within a computer network. Eventually, ZTA (Zero-Trust Architecture) [27] has become a complementary approach to perimeter security. In the ZTA approach it is assumed that attackers have already infiltrated internal segments. "Zero trust" is about preventing attackers who have already infiltrated part of a computer network from infiltrating other parts of the computer network. Zero trust metaphorically means that segments or host nodes should "distrust" each other and therefore always check connection establishment and access between network segments with authentication and access control.

Firewalls and intrusion detection systems are essential components to ensure computer network security. These topics are described in the following sections.

6.4 Firewalls

The task of the firewall is to control traffic between computer networks. There is always a firewall between the Internet and a private/corporate computer network. Typically, one or more firewalls are also set up internally in the computer network. By controlling the traffic, the firewall can filter out unwanted traffic and traffic that forms of attacks against the computer network and its resources. A firewall must have rules for which traffic should pass through and which traffic should be rejected.

The term "firewall" is a rather bad metaphor because a physical firewall is a barrier where nothing gets through, while a digital firewall is a busy hub where lots of traffic passes through all the time. A better name would therefore be "checkpoint", because it intuitively gives the association that a security guard controls what can pass.

Properly configuring firewalls requires expertise and knowledge of network protocols. A small misconfiguration can allow the firewall to let malicious traffic pass through. The standard basic rule for firewalls is default-deny, i.e. any type of traffic is rejected unless otherwise specified. Rules must then specify what is allowed to pass through.

Firewalls can be implemented as a software module on a standard computer or server, or as a separate physical device with specially designed hardware. The most powerful physical hardware firewalls with the capacity to process many Tbps of traffic have prices up to one million dollars. In virtual computer networks in the cloud, firewalls are usually implemented as software that runs on standard servers.

There are different types of firewalls, where three main types are (1) stateless packet filter, (2) stateful packet filter, and (3) application firewall, which are described in the following sections.

6.4.1 Stateless Packet Filter

A stateless packet filter is the simplest type of firewall developed in the late 1980s. This type of packet filter inspects packet headers at the transport and internet layers, and from this determines whether an IP packet should be accepted or rejected based on attributes such as

- IP addresses of sender and recipient
- Port number for the application processes of sender and recipient
- Type of transport protocol (e.g. TCP, UDP or ICMP)
- Type of message for a specific protocol at the transport layer
- Which interface on the link layer a packet arrives at

The term "stateless" means that the packet filter is unaware of the state of a session between client and server. This severely limits the flexibility of what kind of filtering rules can be defined. Despite this limitation, such packet filters still fulfil important functions for pre-sorting IP traffic. In addition, stateless packet filters have very high performance.

6.4.2 Stateful Packet Filter

A stateful packet filter keeps track of the state of each connection or session between client and server passing through the filter. This means that each IP packet in an ongoing session is recognized as part of a specific session. This is done by keeping a record and saving traffic history for each ongoing session.

The packet filter can create temporary rules for each specific session. Typically, return traffic from an external server on the Internet will be allowed to pass through the firewall if the session was initiated by an internal client. When the session ends, the temporary rule along with the record of traffic history is deleted.

When the state of a session is known, it gives great flexibility to the definition of filtering rules, and makes it possible to define more precise rules that thereby increase security.

State-based packet filters also have high performance. One disadvantage is that memory is required to record traffic history of each ongoing session. A possible threat is that an attacker opens a large number of sessions with the intention of depleting the firewall's memory for traffic history. Thus, if the memory is full, the firewall will no longer be able to handle additional sessions, creating a denial-of-service attack.

6.4.3 Application Firewall

An application firewall is capable of inspecting payload data in addition to application layer packet headers, providing the greatest possible flexibility for defining filtering rules.

Rules can be defined based on specific application protocols such as HTTP, FTP, SMTP, and IMAP. Filtering each protocol is supported by a specific module implemented in software or hardware. Application firewalls can even be configured for filtering specific user apps, e.g. Facebook, YouTube and LinkedIn. It provides the ability to define filtering rules for detailed elements of each specific user application.

This flexibility comes at the expense of a very high processing load in the firewall. To handle a large volume of traffic, the firewall needs to be implemented with high-performance computing hardware, which also results in high cost.

Application firewalls can filter traffic in an end-to-end connection between client and server, or it can split the connection into two parts where the firewall plays the role of a proxy server between the client and the server. A proxy firewall is called a gateway, and is used, among other things, in VPN connections between separate computer networks, as explained in Sect. 6.2.2 about IPSec.

High-performance application firewalls are often referred to as *Next-generation firewalls*, although this type of firewall has been around since 2010. Another term is *Web Application Firewall (WAF)* for stopping attacks against websites, where DDoS is a common type of attack. DDoS means that a website is bombarded with fake traffic that causes the web server to become overloaded and can no longer serve legitimate traffic. DDoS traffic is typically generated with botnets that can consist of several million infected computers (see Sect. 2.2.4). Stopping DDoS traffic can be difficult and resource intensive. An advanced WAF firewall can use real-time machine learning to define filtering rules, so it can drop DDoS attack traffic before it reaches the web server in the shortest possible time.

6.5 Intrusion Detection

Intrusion into computer networks compromises security, e.g. by creating breaches of confidentiality, integrity and availability of resources.

Intrusion detection, as the name says, is the detection of intrusion attempts by analysing network traffic and system activity. An IDS (Intrusion Detection System) is precisely a system or application intended to warn of detected (attempted) intrusions. An IDS can also provide support to handle automatic responses around the network to resist the intrusion.

An IDS can be implemented on a host and is thus called *host-based IDS (HIDS)*, or implemented on a network node and is called *network IDS (NIDS)*.

6.5.1 Signature-Based Detection

Signature-based detection means that the IDS can identify a known pattern of network traffic and system activity previously identified to be part of an attack. The specific attack pattern is called a "signature". This means that the IDS must be configured to recognize signatures of known types of attacks, where an alarm is triggered if a signature is detected. The disadvantage of signature-based IDS is that it is not capable of detecting new types of attacks for which there is not yet a defined signature.

A signature-based IDS has good accuracy for known attacks, and thus produces only a small proportion of false-positive alarms. On the other hand, a potential weakness of signature-based IDS is that it can suffer from a high proportion of false-negatives because new types of attacks are not detected.

6.5.2 Anomaly-Based Detection

Anomaly-based detection was primarily introduced to detect unknown attacks. The basic approach is to use machine learning (Chap. 15) to create a model of normal (non-harmful) activity, and then compare current behavior against that model. In this way, abnormal behavior can be detected, which can thus be a new type of attack. Since these models can be trained according to applications and configurations, anomaly-based detection generally has greater potential for detecting attacks compared to signature-based detection.

The disadvantage of anomaly-based detection is that it can give a large proportion of false positives, because legitimate but previously unknown activity patterns get classified as harmful activity. Anomaly-based detection also provides a high processing load. Progress in designing more efficient machine learning and classification is needed to make the detection process more efficient and accurate.

6.6 Network Architecture for Security

Computer networks can be set up and configured in many different ways regarding network security. Some general basic principles are worth being mentioned. A computer network always has an external firewall that can also fill the role of gateway and router. The external firewall filters traffic between the Internet and the computer network. A computer network is usually divided into segments that are separated by internal firewalls. Figure 6.13 illustrates a simple network architecture separated in two segments by an internal firewall.

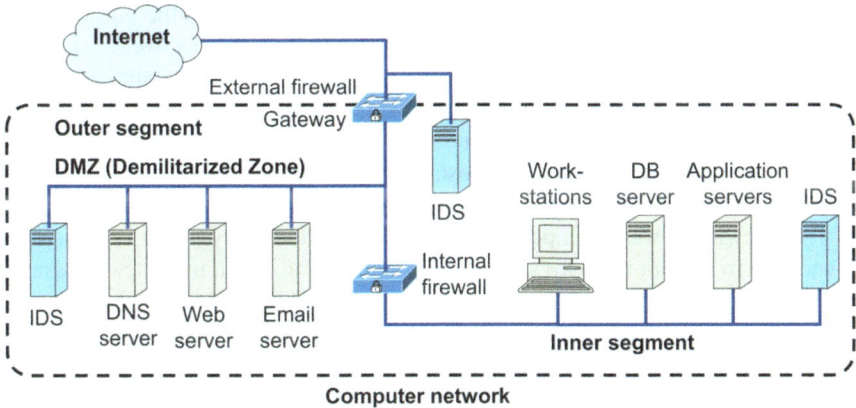

Fig. 6.13 Simple network architecture

The first segment is between the external and internal firewalls and is called the *DMZ (Demilitarized Zone)*, which is a metaphor for the outer segment that has strong exposure to the "enemy" (threats from the Internet). Systems in the DMZ are typically web, email and DNS servers. Beyond the internal firewall are internal segments where the production systems and workstations are connected. To detect attacks, IDS (Intrusion Detection System) can be used in both DMZ and internal segments, as shown in Fig. 6.13. It provides a better overview of possible attacks and security incidents by placing IDS in both segments, which can also be used for correlating observations in both segments. A honeypot can also be placed in different segments. A *honeypot* is an application intended to deflect and detect attempts at unauthorized use of systems. A honeypot is similar to a legitimate part of a system so the attacker should be lured into believing it is real. *Tar pit* is a related technology, which, in addition to imitating a legitimate system, also has a slow response time, with the intention of trapping attackers in slow-motion fake systems.

It is commonly said that different levels of "trust" between segments make it necessary to separate them by firewalls. The concept of *zero trust* means that no parts of the network "trust" each other, so that they are always segmented with firewalls or with mutual authentication. This is called zero-trust architecture or ZTA (Zero-Trust Architecture). The advantage of ZTA is that an intruder who has gained unauthorized access to one segment or system cannot easily access other segments or other systems in the network. The downside is greater complexity in configuring networks and applications to support legitimate traffic between segments.

6.7 TLS Inspection

HTTPS, which is based on TLS, provides strong end-to-end encryption between client and server for web services. From the user perspective, end-to-end encryption is a natural requirement for communication security on the Internet. However, many organizations want to be able to read all traffic passing through the interface between the Internet and the organization's computer network, even when the traffic is supposed to be encrypted end-to-end. From the organization's perspective, the motivation for inspecting the traffic is primarily to detect attacks and malicious traffic hiding as encrypted data. For example, implementing a ZTA (zero-trust architecture, see the previous section and Sect. 6.3), requires the ability to inspect connections and traffic to and from internal hosts [28]. For this, *TLS inspection* enables the organization to split end-to-end encryption so that encrypted traffic can be inspected.

The gateway firewall can operate as a *TLS termination proxy* which allows a TLS connection to be split into two separate TLS connections, with the gateway assuming the role of proxy between client and server. Then the client "believes" that it is connected to the server, and the server "believes" that it is connected to the client, but in reality, both client and server are connected to the proxy gateway node. Hence, the TLS-encrypted payload data gets decrypted in the gateway, and can be inspected by the network owner.

A proxy CA is installed on the gateway node for generating virtual server certificates on the fly. A virtual server certificate is a "fake" server certificate in the sense that it syntactically looks like the server certificate from the external website, but in reality comes from, and is issued by, the proxy CA without the website's knowledge. The proxy CA has a private key, and all internal clients in the computer network must be configured with the proxy CA's corresponding public key preinstalled, for example as a root certificate.

For an internal client to "believe" that it has an HTTPS connection with an external server, the proxy CA must generate and sign a virtual server certificate that appears to come from the remote server node, but this certificate actually represents the proxy server. This situation is shown in Fig. 6.14, where the example shows a computer network in which an internal client sets up a connection to an external server on the Internet.

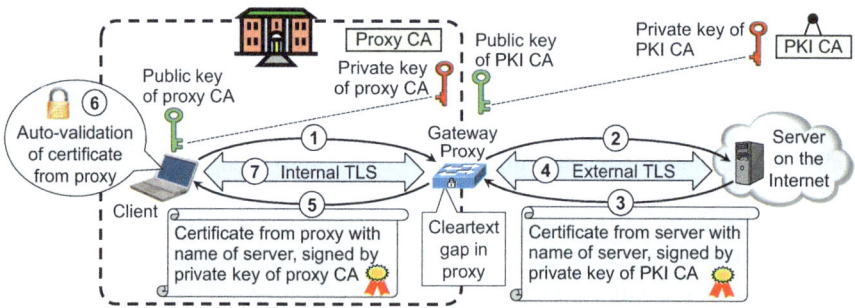

Fig. 6.14 TLS inspection

The sequence of the scenario in Fig. 6.14 is described below, with reference to the TLS Handshake protocol in Fig. 6.7 in Sect. 6.2.1.

1. The internal client sends a ClientHello message to set up an HTTPS connection with TLS to the remote server, but this message is stopped in the gateway proxy.
2. The gateway proxy sends a new ClientHello message to set up an HTTPS connection with TLS to the remote server.

6.7 TLS Inspection

3. The server returns a public-key server certificate signed by a PKI-CA on the Internet.
4. The HTTPS connection with TLS is established between the proxy client and the remote server.
5. The internal CA generates a proxy server certificate with the subject name of the external server, signed with the private key of the proxy CA. The proxy server sends the certificate to the client.
6. The proxy server's certificate is validated by the client using the proxy CA public key preinstalled on the client.
7. The HTTPS connection with TLS is established between the internal client and the proxy server.

Note that the traffic and payload data are unencrypted in the gateway proxy, which means that the traffic can be read in clear text. To the internal client user, it looks like an end-to-end HTTPS connection to the external server. It is very difficult for the user to detect that the encrypted TLS connection has been split with TLS inspection. Two possible ways to detect TLS inspection are the following:

- The server certificate validation path can be followed to the root certificate that contains the proxy CA public key. It provides an indication of TLS inspection if the issuer name and proxy CA subject name reflect what the root certificate really is.
- From a location outside of the organization's computer network, the user can get the external server's real certificate, copy the external server's public key (or its hash), and then compare that key with the public key in the certificate from the gateway proxy. It would give a strong indication of TLS inspection if the public keys are different.

TLS inspection is a highly invasive technology regarding privacy (GDPR) and financial legislation (PSD2) in the European Union. The NCSC in the UK has published a brief description of the issue [29]. In the EU, using TLS inspection in an organization requires the consent of employees of the organization to be compliant with the principles of privacy in the GDPR. For TLS inspection to work seamlessly with automatic validation of the proxy server certificate, all client terminals must be configured with the proxy CA certificate pre-installed as a root certificate. Equipment owned by the organization can be configured in this way, but it should not be done on private and personal equipment.

Many employees would find it unacceptable that their employer should be able to inspect traffic when, for example, they log into a personal online banking account. To make TLS inspection less invasive, an organization may choose to whitelist a set of external domain names that should be exempt from TLS inspection, which are typically online banks and government agencies for social security, tax, and health services.

TLS inspection is in principle incompatible with CAA (CA authorization) and CT (certificate transparency) described in Sect. 5.3.3, because the internal CA is not authorized by the external server to issue a server certificate for the server, and because the internal CA does not send the proxy server certificate to the CT server on the Internet.

The technology behind TLS inspection is very potent and can be a threat if misused. Even without preinstalling a proxy root certificate from an internal CA, TLS inspection will generally work. If a client receives a server certificate from a proxy that is not automatically validated, the user will receive a pop-up warning asking the user to accept or reject the certificate. We often encounter these warnings, which in most cases means that the validation of a legitimate server certificate fails for various reasons, such as the server certificate or a CA certificate having expired. Many of us have thus been conditioned to ignoring such warnings. Therefore, it can be assumed that many users will readily accept a rogue server certificate from an attacker and be tricked into setting up an HTTPS connection to a rogue TLS inspection proxy controlled by an attacker. These types of attacks have been reported and documented—for example, such attacks were carried out in cooperation between the PC manufacturer Lenovo and the former advertising company Superfish.[3]

TLS inspection represents both a tool for strengthening security, as well as a potential threat to security if abused and used in the wrong way. Organizations must therefore take care to follow good practice when implementing TLS inspection [30], and users must be aware of potential attacks based on abusing TLS inspection.

6.8 Tasks

1. Security protocols
   a. What is a network protocol in general and what is a security protocol in particular?
   b. Provide examples of security functions supported by security protocols.
   c. Name at least four well-known security protocols.
   d. At what layers of the Internet stack do TLS and IPSec operate? Why is a port number reserved/used for HTTPS (HTTP with TLS) but not for IPSec?

2. Firewalls
   Figure 6.13 shows a simple network architecture with firewalls.
   a. What is the purpose of DMZ? What services are typically located there?
   b. At which layers of the Internet stack does a packet filter firewall operate?
   c. What type of firewall can filter packets based on payload data in data packets?

3. TLS inspection
   a. Describe a legitimate reason for using TLS inspection.
   b. Describe a threat scenario that abuses TLS inspection.
   c. How can a user determine if TLS inspection is being used in an HTTPS web connection?

---

[3] https://en.wikipedia.org/wiki/Superfish

Chapter 7
Wireless Security

Our mobile phones have become the greatest spy on the planet.

John McAfee

In general, the term "wireless" denotes any type of data communication which does not require a cable (wire). More specifically, wireless communication means data communication via radio signals. An important aspect of radio signals is that they can be picked up by anyone within the range of the sender, and hence radio signals can easily be intercepted by attackers. Secure design of wireless technologies must take this aspect into account, typically by using encryption. There are numerous technologies for wireless data communication, where some are implemented in consumer devices that we carry with us. Early generations of these radio communication technologies typically had weak security, but that has improved over the years. Wireless technologies are also used in industrial applications such as satellite links, GPS (geo-positioning systems), and fixed point-to-point radio links, where security is also integrated to a greater or lesser extent.

In this chapter, you first learn basic concepts and principles of radio communication, which are often called *wireless communications* because no cables are needed. Next, you learn about the most prominent wireless communication technologies, their security challenges, and how security solutions are built into each type of technology. The technologies covered are Wi-Fi, Bluetooth, and mobile networks.

7.1 Radio Communications

Wireless communication means that no cables and wires are used to transmit data, but that data is instead sent as radio signals through space, air, or other lightweight materials. Radio signals transmitted from a radio transmitter can be picked up by receivers that are within the range the radio signals can reach. This has great potential because radio signals can easily reach many recipients.

The next section briefly explains how radio signals consist of electromagnetic radiation. Then follows a brief description of the radio spectrum, which means that different wavelengths of radio signals are reserved for different purposes such as Wi-Fi, Bluetooth, NFC and mobile communications.

7.1.1 Radio Signals

Radio signals are electromagnetic waves consisting of intertwined synchronized waves of electric and magnetic fields perpendicular to each other, as shown in Fig. 7.1.

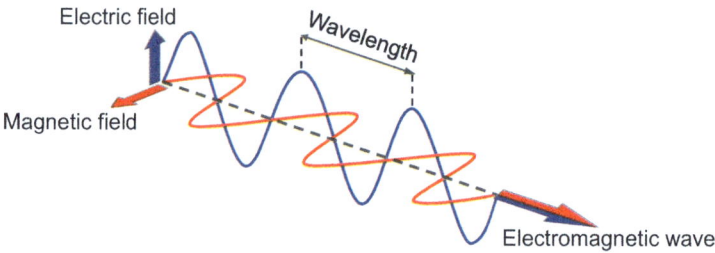

Fig. 7.1 Electromagnetic wave

A single electromagnetic wave can be thought of as a particle without mass, also called a *photon*. In physics, it is considered to be both a wave and a particle. When many waves/photons are sent as a continuous stream, it is called electromagnetic radiation.

Radio signals, light and X-rays are examples of electromagnetic radiation, but with different wavelengths. The shorter the wavelength, the more energy the wave/photon has. Photons with extremely short wavelengths, as in ultraviolet light, X-rays and gamma rays, can be dangerous to humans because it can damage cells and chromosomes. Radio waves, on the other hand, are considered harmless as long as the rays are not too powerful. By standing close to powerful radio beams, body tissues can be heated, in the same way that infrared radiation heats the skin. For example, microwave ovens heat food with powerful radio beams with a wavelength of 12.2 cm and frequency of 2.45 GHz, which has the property of efficiently heating the water molecule H_2O. This is about the same wavelength and frequency used for the 4G mobile network, but here the rays are very weak, so that one does not get hot from using mobile phones.

Sheet metal or fine-mesh metal mesh provides shielding from radio radiation. A *Faraday cage* is a room or container that is completely encircled with metal plates or wire mesh, and thus cannot receive radio signals from outside, nor let radio signals out. A microwave oven is thus a Faraday cage that shields the outside environment from the strong radiation inside the oven.

Regardless of the wavelength, electromagnetic waves will always propagate at the speed of light through void/vacuum and air, which is about 300,000 km/s. To put it into perspective, the distance from the Earth to the Moon is about 384,000 km, so it takes just over 1 second to send a radio signal or flash of light from the Earth to the Moon. Electromagnetic waves are formed when electrons undergo rapid acceleration. One way to accelerate electrons rapidly back and forth to generate radio waves is by having electrical voltage alternate with a given frequency in an antenna. Electrical signals in wires and antennas consist precisely of rapidly alternating electrical voltage. On the receiving side, radio waves hitting an antenna generate electrical signals in the receiving antenna. An antenna is the interface between electrical signals in metal and radio signals in air. In this way, electrical signals on the sending side can be converted into radio waves that are transmitted at the speed of light and converted into electrical signals again on the receiving side, which can be located a few meters or hundreds of kilometres away.

7.1.2 The Radio Spectrum

The electromagnetic spectrum is the total range of frequencies/wavelengths of electromagnetic radiation. This range starts from radio waves with long wavelengths of several kilometres, via visible light and X-rays with short wavelengths, to gamma rays, which have the very shortest wavelengths of the size order of atoms. Electromagnetic waves have frequency in the same way that sound waves also have frequency. The relationship between light speed, wavelength and frequency of electromagnetic waves can be expressed by the formula

$$c = \lambda \cdot \nu$$

where

$c = 300{,}000$ km/s $= 3 \cdot 10^8$ m/s is the speed of light,
λ (lambda) is the wavelength, and
ν (nu) is the frequency of the wave.
If we assume frequency $\nu = 3$ GHz $= 3 \cdot 10^9$ Hz (which is used in the 4G/5G network, for example), the wavelength will be
$\lambda = c / \nu = 3 \cdot 10^8 / 3 \cdot 10^9 = 0.1$ m $= 10$ cm.

Figure 7.2 below shows the central range of the spectrum for electromagnetic waves with a simplified distribution of well-known radio frequencies and other electromagnetic rays.

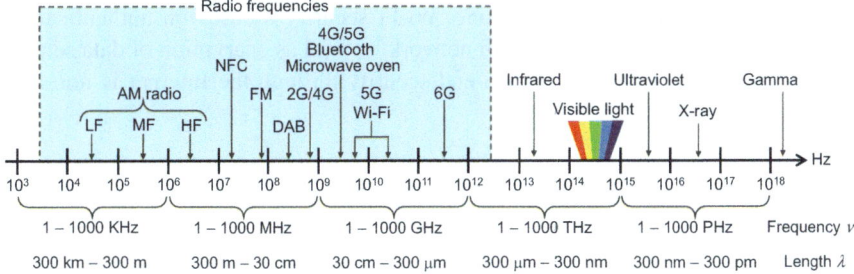

Fig. 7.2 The spectrum of electromagnetic waves

The transmission capacity of radio signals increases proportionally with the increase of frequency. Based on this consideration, it is advantageous to send radio signals at high frequencies because it allows more data to be transmitted. Unfortunately, the increase in frequency and shorter wavelengths also reduces the penetration depth of the wave, which in practice means that radio signals are hindered by walls and ceilings in buildings because they absorb the energy in the waves. Very high-frequency radio signals therefore require a clear line of sight between transmitter and receiver, as for example with the most high-frequency radio signals in 5G networks.

Finally, it must be mentioned that the frequency spectrum of radio waves is a limited natural resource because only one radio transmitter should transmit with a given frequency within the same geographical area to avoid interference. If many transmit at the same frequency at the same time, they will interfere with each other's radio signals so that it would be difficult for receivers to distinguish between the signals and thus difficult to receive the signals correctly. To create order, the use of the radio spectrum is regulated both internationally and nationally, depending on frequencies. ITU Radiocommunication Sector (ITU-R) is part of the UN and has the role to manage the international radio-frequency spectrum and satellite orbit resources. In addition, every country has its own government body for regulating the radio spectrum within the country borders. These bodies auction out chunks of the spectrum to private companies for billions of dollars, e.g. for operating mobile communication networks. This shows how valuable the radio spectrum is.

7.2 Security in Wi-Fi

Wi-Fi is a technology for radio-based local computer network based on the IEEE 802.11 family of standards. The English spelling "Wi-Fi" has no meaning other than that it refers to "Wireless", and was introduced to avoid calling it "IEEE 802.11". Through Wi-Fi, devices can transmit data wirelessly over a range of up to approx. 100 meters. Computers, mobile phones and other wireless devices can connect to a Wi-Fi network to access the Internet, or can send data wirelessly directly between themselves. Wi-Fi uses radio frequencies in the license-free 2.4 GHz and 5 GHz bands.

The challenge for security in Wi-Fi is mainly that radio signals from Wi-Fi typically cover open areas, so that radio signals within coverage areas of Wi-Fi antennas can be sent and received by everyone. Wi-Fi security focuses on authentication between Wi-Fi devices and the Wi-Fi network, as well as encryption of data sent as radio signals between these. End-to-end security through the Internet is not supported by Wi-Fi.

7.2.1 Basic Wi-Fi Concepts

Devices that can communicate with a Wi-Fi network are called mobile stations (STA). These typically connect to an access point (AP) for further connection with local area networks and the Internet. There are two operating modes for Wi-Fi networks: (1) infrastructure mode, and (2) ad hoc mode. In infrastructure mode, a device (STA) communicates with the access point (AP) to access the network, as shown on the left in Fig. 7.3. In ad hoc mode, devices (STA) communicate directly with each other, as shown on the right in Fig. 7.3.

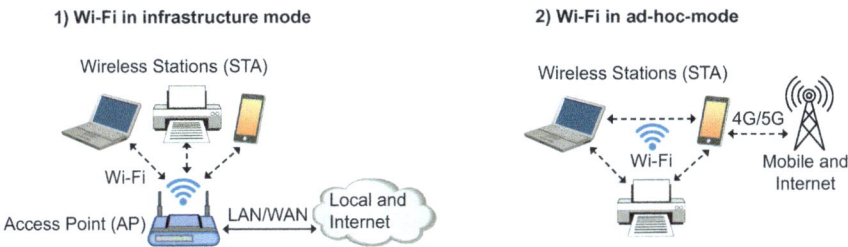

Fig. 7.3 Wi-Fi network modes

Infrastructure mode is most common for Internet access in office settings or homes where there is an infrastructure with a LAN/WAN connection to an ISP. Ad hoc mode is convenient outdoors or when there is no permanently installed Wi-Fi network you have access to.

7.2.2 Development of Security in Wi-Fi

Wi-Fi technology has gone through innovation and development since the first generation of Wi-Fi was presented in 1999. The six different generations from 1999 until today are simply called Wi-Fi 1, Wi-Fi 2, ... Wi-Fi 6, while the standards are formally called IEEE 802.11b/a/g/n/ac/ax. Each new generation offered greater performance in the form of increased transmission capacity, new functionality and, not least, stronger security. Figure 7.4 shows the timeline of the development of technologies for Wi-Fi, as well as the development of the security protocols WEP, WPA, WPA2 and WPA3.

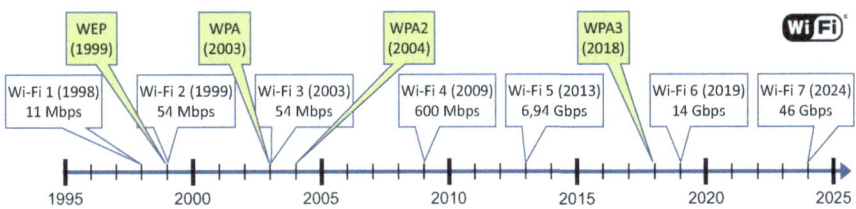

Fig. 7.4 Development of Wi-Fi and security protocols for Wi-Fi

A security protocol is a set of formally specified message formats that two or more parties exchange among themselves to achieve or support a security function. This can be, for example, to exchange session keys, to authenticate or to encrypt data. It is surprisingly difficult, even for experts, to specify security protocols without weaknesses that allow attackers to spoof messages and compromise the CIA (confidentiality, integrity and availability) of messages and transmitted data. The first version security protocol for Wi-Fi, called WEP (Wireless Equivalent Privacy), turned out to have so many vulnerabilities that it was trivial for attackers to break into Wi-Fi networks. The replacement was called WPA (Wi-Fi Protected Access), which also had serious weaknesses, and was therefore quickly replaced with WPA2 in 2004. As WPA2 still had some weaknesses, WPA3 was introduced in 2018 with new security features, including encryption of open Wi-Fi networks. All new equipment sold today should support WPA3.

7.2.3 Secure Network Access with WPA

WPA (Wi-Fi Protected Access) has through many years of development become a robust and complete standard for access authentication and secure radio communication for Wi-Fi. Between 2006 and 2020, WPA2 certification was mandatory for all new Wi-Fi devices. As of 2020, WPA3 certification is mandatory for all new Wi-Fi devices.

WPA in infrastructure mode mainly has two possible architectures for access authentication, called *WPA-Personal* and *WPA-Enterprise*, both of which are illustrated in Fig. 7.5. WPA3-Personal requires at least 128-bit cryptographic strength, while WPA3-Enterprise has 192-bit cryptographic strength.

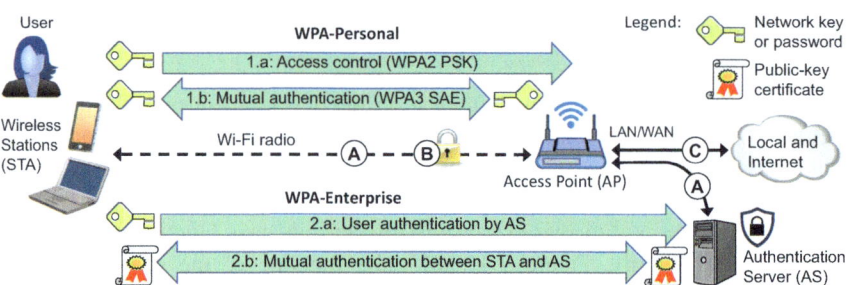

Fig. 7.5 WPA authentication and secure access with WPA-Personal and WPA-Enterprise

- **WPA-Personal**
  In WPA-Personal, the router/access point has a password or authentication key that all users must know and enter in order to access the network. WPA-Personal is designed for home and small office networks and requires no authentication server (AS). In WPA2, access authentication and session key creation were based

on the PSK (Pre-Shared Key) protocol, where the green arrow 1.a in Fig. 7.5 is a simplified representation. WPA3 uses the SAE protocol (Simultaneous Authentication of Equals), where the green arrow 1.b is a simplified representation. The session key for encrypting the radio connection is established by the SAE protocol based on the AP key/password and other network parameters combined with the ECDH described in Sect. 4.6.2. In contrast to PSK in WPA2, SAE in WPA3 provides mutual authentication between STA and AP, preventing e.g. man-in-the-middle attacks. SAE is described in Sect. 7.2.4 below. Web access with WPA-Personal is not really based on user authentication because all users/devices enter the same key.

- **WPA-Enterprise**
 WPA-Enterprise is designed for corporate networks with an authentication server (AS). This requires a more complicated setup but provides additional security. There are different variants of the Extensible Authentication Protocol (EAP) used in WPA-Enterprise. The green arrow 2.a in Fig. 7.5 is a simplified representation of Access authentication with password/key. It is also possible with mutual authentication between the wireless device STA and the authentication server AS based on certificates, as illustrated by the green arrow 2.b. Access with WPA is usually based on user authentication because users provide their own password/authenticator. It is also possible to configure access based on a shared password, similar to WPA-Personal.

Letter **A** in Fig. 7.5 denotes messages for session establishment. Once the session key is established using either of the methods described in WPA-Personal or WPA-Enterprise, encrypted radio communication between the STA and AP can begin, as denoted by letter **B** in Fig. 7.5. At the same time, the router/AP opens for communication with local area networks and/or the Internet, as shown by letter **C**.

WPA3 also introduced OWE (Opportunistic Wireless Encryption), which allows wireless devices to establish encrypted connections to open/public Wi-Fi hotspots or guest access from cafes and hotels, even without access authentication. In places that still use WPA2 without OWE, Wi-Fi connections are established without encryption with the risk that data could be read or manipulated by unauthorized persons, unless the data is additionally encrypted via HTTPS or VPN.

A weakness of OWE is that no authentication is done between user/device and AP/router. This means that an attacker can spoof a Wi-Fi network by giving it the name of a genuine Wi-Fi network nearby, which can trick users to believe that they are getting Internet access through a legitimate network. To prevent the theft of sensitive information through that type of attack, it is again essential that Internet connections are encrypted with HTTPS or VPN at the application layer.

7.2.4 SAE: Simultaneous Authentication of Equals

The SAE protocol is based on The ECDH protocol described in Sect. 4.6.2, but additionally includes a common long-term secret parameter x which in practice is the AP's key/password. The user is expected to provide a key/password to the STA. As the STA and AP have a shared secret parameter, this variant of ECDH provides an efficient way for STA and AP to establish *authenticated* session keys over an open insecure channel. Figure 7.6 provides a simplified presentation of the SAE protocol. STA and AP calculate a common secret point G (called password element PE) on EC based on AP key/password and other network parameters. The mathematical operator * represents discrete algebraic multiplication on an elliptical curve EC, and not ordinary multiplication.

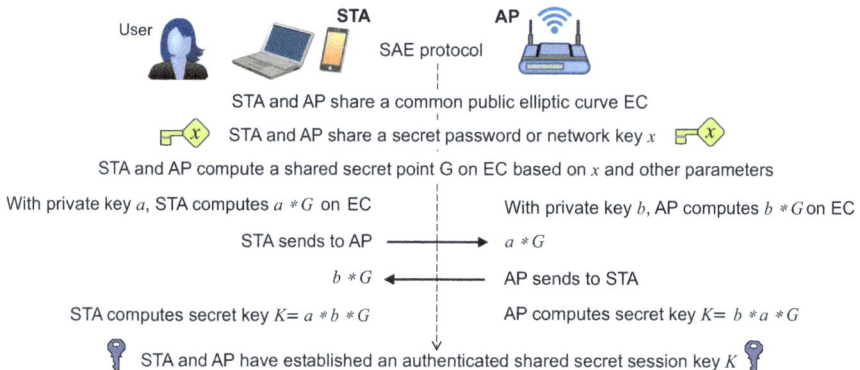

Fig. 7.6 SAE protocol to establish an authenticated session key with the WPA-Personal

An important feature of the SAE protocol is *forward secrecy* (also called *perfect forward secrecy*), which means that an attacker cannot decrypt historical data encrypted with previous session keys K even if the attacker obtains the secret long-term parameter x. The SAE protocol is also called *The Dragonfly Protocol*.

7.3 Bluetooth Security

Bluetooth is a technology of short-range radio communication. A network of Bluetooth devices is called a *piconet*, or a *scatternet*, when several piconets are connected by a Bluetooth device acting as a bridge between them. Bluetooth is primarily used to set up personal piconets, but can also be used in an industrial context. Many types of digital devices support Bluetooth, including mobile phones, laptops, cars, medical devices, printers, keyboards, mice, speakers, headsets, smartwatches, fitness monitors, smart light bulbs, and smart appliances at home and in the office.

7.3 Bluetooth Security

Bluetooth is the English name for King Harald Blåtann, who in the 900 s united the various Danish tribes into one kingdom. In a metaphorical sense, the name Bluetooth symbolizes a technology that unites different devices into one network.

7.3.1 Bluetooth Technologies

The first specification for Bluetooth was published in 1998. Since then, the technology has evolved with several improvements in performance, energy savings and security, among others. Figure 7.7 shows the different versions of Bluetooth standards and technology.

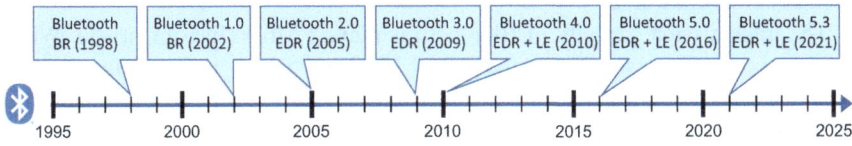

Fig. 7.7 The development of Bluetooth technology

There are two categories of Bluetooth technology. The first is called BR/EDR (Basic Rate & Enhanced Data Rate), also called Bluetooth Classic. This book abbreviates BR/EDR to just EDR. The second technology is called Bluetooth LE (Low Energy).

EDR is mainly used by Bluetooth devices that require relatively large continuous transmission capacity (up to 2 Mbps), and the device has a relatively good power supply (e.g. from the power connector or with laptop/mobile battery). LE is mainly used by Bluetooth devices that have only a small battery (e.g. a coin-sized battery) and/or have only limited transmission capacity requirements, and not continuous data transfer. Many Bluetooth devices can use both EDR and LE in dual mode, such as mobile phones and laptops. Other devices such as smart light bulbs and sensors are typically designed to only use LE. Figure 7.8 shows typical architectures with EDR and LE technology, respectively.

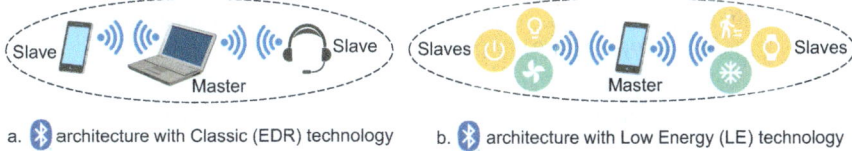

Fig. 7.8 Bluetooth architectures with EDR (Enhanced Data Rate) and BLE (Bluetooth Low Energy

In a Bluetooth connection, there is always one master device and one or more slave units. With EDR, a master device can have a maximum of 7 slave devices connected, while LE allows an unlimited number of connected slave devices. LE is intended to be used in smart environments to control and communicate with an arbitrary number of different sensors and smart appliances.

Bluetooth radio frequencies are between 2.4000 and 2.4835 GHz, which is in the globally unlicensed (but regulated) 2.4 GHz frequency range for industrial, scientific, and medical (ISM) applications, where devices can only transmit at low power (between 1 and 100 milliwatts) providing short range. Bluetooth uses a radio technology called frequency hopping, which means that each Bluetooth connection jumps between 79 different frequencies for EDR, and between 40 different frequencies for LE, with 1600 jumps per second. The advantage of frequency hopping is that different devices do not interfere with each other even when there are many devices in the same area.

The transmission capacity of Bluetooth is max. 2 Mbps. In comparison, Wi-Fi has a maximum transmission capacity of 14 Gbps, which is 7000 times more. Bluetooth has its usefulness for easy communication in small local networks, and is not intended for high-speed access to the Internet. However, Bluetooth devices can be connected to the Internet through a hub, thus being used for IoT (Internet of Things). The transmitting power of Bluetooth devices is determined by the Bluetooth class, where there are 3 main classes, as shown in Table 7.1. Class 1.5 applies to LE only.

Table 7.1 Power classes for Bluetooth devices

Effect class	Max power	Max range	Typical devices
1 EDR, LE	100 mW	100 m	Internet access points, long-range USB adapters
1.5 LE	10 mW	30 m	Digital beacons, portable sensors, smart light bulbs and smarthome devices
2 EDR, LE	2.5 mW	10 m	Mobile phones, short-range USB adapters, speakers, headphones, media in cars
3 EDR, LE	1 mW	5 m	Wireless mouse and keyboard, medical implants, long-life sensors

Any Bluetooth device in discoverable mode transmits the following information when another Bluetooth device asks for it:

- Bluetooth device name (e.g. "Galaxy XCover 5", which is a smartphone model)
- Device class (e.g. PHONE_SMART)
- List of services (e.g. TELEPHONY)
- Technical information (e.g.: features, manufacturer, Bluetooth version, clock offset)

Each device has a unique 48-bit MAC address. However, these addresses generally do not appear in requests. Instead, user-friendly Bluetooth names that are set by the manufacturer are displayed. The device name appears when another user scans for devices and in lists of paired devices.

Bluetooth devices have the default Bluetooth name set by the manufacturer, which is typically the device's model name. This can be confusing, since, for example, there may be several mobiles of the same model within range. On a device with a display and keyboard, the user can change the name of another device in the list of paired Bluetooth devices. This does not mean changing the name that the other device itself emits, but simply the Bluetooth name in the list of Bluetooth devices.

Any Bluetooth device can perform a request to find other Bluetooth devices to connect to, and any Bluetooth device can be configured to respond to such inquiries. However, if the device trying to connect knows the address of the other device, the other device always responds to direct connection requests, and transmits the information shown in the above list if prompted. Use of a Bluetooth device's services may require pairing or acceptance by the owner, but the connection itself can be initiated by any device and maintained until it moves out of range. Some devices can only be connected to one device at a time.

7.3.2 Pairing and Connection

Without security mechanisms, Bluetooth data transmission would leak sensitive data, or devices could be remotely controlled and misused. For security reasons, it is therefore necessary to be able to recognize specific Bluetooth devices, and thus decide which ones you want to connect to. On the other hand, it is also important to have good usability by being able to connect Bluetooth devices sometimes even without user interaction (for example, as soon as the device is within range).

To solve this dilemma, Bluetooth uses a process called device pairing that starts either at a specific request from a user to pair (for example, with the "Add a Bluetooth device" feature), or happens automatically when connecting to a service.

Pairing often involves a certain level of user interaction for authentication between devices. When pairing is complete, a permanent binding is formed between the two devices, which allows the two devices to connect automatically without repeating the pairing process. The user can delete bindings whenever desired (for example, with the function "Remove Bluetooth device").

During pairing, the two devices establish a common cryptographic link key that allows the devices to recognize each other upon subsequent connection, and is used to encrypt data that is transmitted. Removing a Bluetooth device from the list of paired devices means deleting the link key.

The simplest Bluetooth connections, such as Bluetooth *Object Push*, does not require authentication or encryption. The purpose of Object Push is to make the connection as easy as possible for users, but the downside is that it does not provide any security.

7.3.3 SSP and Authentication Between Devices

SSP (Secure Simple Pairing) is the name of the method of creating a common cryptographic link key between two Bluetooth devices. SSP uses the ECDH protocol described in Sect. 4.6.2. SSP specifies four different methods of authentication between devices during the pairing process, as described below. Authentication protects against man-in-the-middle (MITM) attacks during pairing.

- **Numeric comparison.** This method can be used when both Bluetooth devices can display a six-digit number and at least one of the devices allows the user to select "yes" or "no". During pairing, the same six-digit number is displayed on both screens, and the user(s) must answer "yes" on the device(s) if the numbers match.
- **Passkey Entry.** This method can be used when one Bluetooth device can take input from the user (e.g. with a keyboard), while the other device has a display. The screen-only device displays a six-digit number that the user enters on the device with a keyboard.
- **Just Works.** This method is used when at least one of the devices does not have a display or keyboard (e.g. headset). The user accepts pairing without explicit authentication, so Just Works in theory does not provide protection against MITM attacks. However, physical proximity between devices during pairing will give the user an indication of authenticity, making MITM attacks difficult in practice.
- **Out of Band (OOB).** This method means that authentication takes place through a secondary channel, i.e. not through Bluetooth. Such a secondary channel can be, for example, NFC (Near Field Communication) or cable. NFC is a technology for radio communication over a very short distance that is used, for example, to read smart cards at payment terminals or to read passports electronically. Pairing with NFC is easy by placing the Bluetooth devices against each other, followed by the user accepting the connection with a single keystroke. Authentication with OOB is based on direct physical proximity or cable connection.

Once the authenticated link key is created with SSP, data transmission will go encrypted between connected devices.

7.3.4 Bluetooth Security Recommendations

Attacks against a Bluetooth device assume that Bluetooth is enabled on the device, that it is set to discoverable (visible) mode, and within the attacker's Bluetooth radio coverage area. By turning off Bluetooth, it is protected from attacks via Bluetooth. Putting the device in non-discoverable mode (hidden) provides protection against simple attacks, but not against more advanced attacks [31]. When no applications use Bluetooth on a device, it is advisable to turn off Bluetooth, both for security reasons and to save battery. If Bluetooth is used, it is advisable to put the device in non-discoverable mode, which in some cases is done automatically and in other cases will have to be done manually. Devices with display and input typically have menu options for visible and hidden mode. Often, visible mode only lasts a short time before the device automatically goes into hidden mode. Devices without input often have a button that puts the device into discoverable mode for a short time.

Given that Bluetooth has relatively weak security, it is good practice to avoid sending sensitive information via Bluetooth. Passwords and other login information should never be sent via Bluetooth.

Do not accept pairing requests from unknown sources. In areas with lots of people, hackers can send out these requests in hopes that someone accepts pairing and connection.

Pairing between devices without authentication (e.g. with "Just Works") should be done in a secure place, that is, where there are not many people nearby that you do not know. Resist the temptation to immediately pair your new Bluetooth gadget outside the store in the shopping centre where you bought it. Wait until you get home or are back to the office.

Get in the habit of deleting any old Bluetooth pairings you no longer need or use. While most pairings are likely harmless, like that speaker you connected to last year when renting a vacation apartment, unused pairings expose your devices to unnecessary threats, even if the risk is small.

7.4 Mobile Network Security

Mobile telephony has mostly replaced landlines in Norway. It is much cheaper and easier for telecom operators to allow subscribers to connect to mobile networks and the Internet with radio signals than with cables that must be stretched to each building, and on to different rooms within a building. For users, it is also much easier to use a mobile terminal than a terminal connected to the wall with a wire.

However, wired subscriber lines provide a form of physical security that is lost with radio communication. Because mobile networks use radio signals that can be picked up by anyone in the vicinity, the connection and radio signals must be protected with security controls typically based on cryptography.

7.4.1 Mobile Network Technologies

The technology for mobile networks has undergone tremendous innovation ever since 1980, when NMT (Nordic mobile phone) and Advanced Mobile Phone System (AMPS) were introduced. Figure 7.9 shows this development, with each new generation arriving regularly at intervals of about 10 years. Also note that the radio cell (i.e. distance between base stations) decreases as the transmission capacity increases.

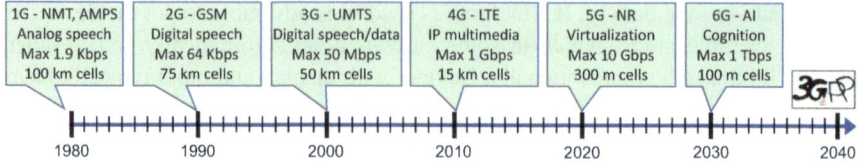

Fig. 7.9 The development of technologies for mobile networks

In NMT (1980) and AMPS(1983), speech was transmitted as analog radio signals in the 450 MHz and 900 MHz bands. The signals were sent unencrypted, which meant that phone calls could be easily intercepted, and mobile subscriptions (but not the mobile subscriber, of course) could be stolen and misused. The subsequent generation was 2G (1990), called GSM, which was originally the name of the working group with the French name *Groupe Spécial Mobile*. GSM was later given the name and status of *Global System for Mobile Communications*. With 2G, speech is sent as a continuous stream (not as data packets) of digital radio signals that are usually encrypted. 2G was developed as a European standard, but was quickly adopted worldwide. As of 2024, 2G is largely still supported by telecom operators around the world. Then came 3G (2000), which was given the name UMTS (Universal Mobile Telecommunications System) to emphasize that it was not only European. What was new with 3G was the transmission of data packets in addition to digital speech. 3G's capacity of transmitting data packets was for example sufficient to support Internet access with the first iPhone launched in 2007. Nevertheless, 3G technology suffers from inefficient use of the frequency spectrum and has been phased out by most telecom operators. The next generation is 4G (2010), which was named LTE (Long Term Evolution) because both voice, video and all types of data are sent as data packets, which is here to stay, and which makes much better use of the radio frequency spectrum. 4G made it possible to significantly increase the capacity of mobile networks, and is by 2024 the most widely used mobile technology worldwide. The latest generation is 5G (2020), which is slowly being introduced by telecom operators during the 2020s. An innovation with 5G is of course higher transmission capacity which is enabled by a new radio interface and access network called NR (New Radio) with an open modularized architecture. 5G also specifies an architecture for the core network and infrastructure of telecom operators that is modularized, open and highly flexible. This means that a telecom operator is no longer forced to purchase a full stack of mobile network equipment (hardware and software) from a specific manufacturer, but that the telecom operator can purchase various network functions in the form of cloud services, thus representing the virtualization of 5G mobile networks. There is already talk of 6G, where the standards will only be finalized closer to 2030. In addition to increasing the capacity of the radio connection with frequencies above 1 THz, it is suggested that AI will play an important role in the 6G network infrastructure, and that users will benefit from a computation fabric with built-in cognition capacities.

While the transmission capacity increases with each new generation, each mobile cell also becomes smaller, that is, the range of a mobile connection becomes shorter so that devices need to be closer to the base stations. The reduction in range is due to

7.4 Mobile Network Security

the use of higher frequencies and bit rates, which means that radio signals are attenuated faster through air and other materials. The bit rates and cell sizes for each generation shown in Fig. 7.9 are only approximate, as there may be significant variations.

The standards for new generations of mobile networks are being developed by 3GPP (The Third Generation Partnership Project), which is a collaboration between seven different organizations around the world working on telecommunications standards. They established 3GPP for 3G, but continue to cooperate within 3GPP for 4G, 5G, 6G, etc.

7.4.2 2G Security Architecture

The GSM network, often called 2G, was designed by ETSI (European Telecommunications Standartisation Institute) in the 1980s when the Cold War was still going on. Due to political pressure from European authorities, the security of GSM was deliberately made weak to allow easy interception of mobile calls. Mobile operators outside Europe were forced to use ridiculously weak encryption, while European operators were able to use somewhat stronger encryption. For this reason, a "strong" and a "weak" set of cryptographic algorithms were designed. Parts of the GSM 2G standards were kept confidential and only distributed to industry partners under non-disclosure agreements. However, the specification for security in 2G was quickly leaked, so researchers quickly found that cracking 2G cryptography was easy. ETSI should have followed Kerckhoffs's principle, which states that system security should not be based on security by obscurity, i.e. on the secrecy of the system's design (see Sect. 4.9.1).

The security architecture of 2G is so simple that it is relatively easy to explain, as shown in Fig. 7.10. Besides the abbreviations explained in the figure, A3, A5 and A8 represent cryptographic algorithms. A5 is a stream cipher as described in Sect. 4.3.

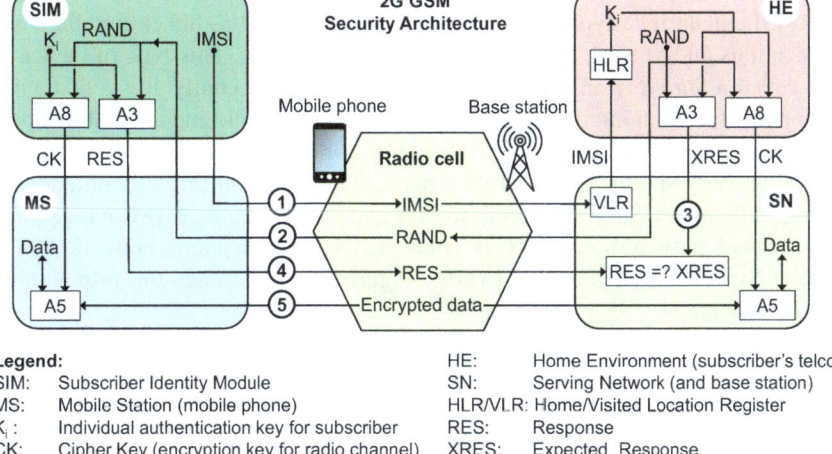

Fig. 7.10 Security architecture in 2G GSM

K_i is a unique authentication key for each subscriber. K_i is stored both in the subscriber's SIM chip and in the HLR (Home Location Register) of the subscriber's telecom operator, called HE (Home Environment). It is interesting to note that SIM and HE have the same cryptographic functions, and that MS and SN also have the same cryptographic functions, performing the same functions, thereby reflecting each other. When a subscriber wants to set up a call, the following messages are exchanged between the entities as shown in Fig. 7.10:

1. The subscriber initiates a call, the SIM chip sends IMSI (International Mobile Subscriber Identity) to HE (via MS and SN) of the subscriber's telecom operator, which looks up and finds K_i in HLR.
2. HE sends a random number RAND to the SIM chip (via SN and the mobile phone MS).
3. HE encrypts RAND with K_i providing XRES (expected response) sent to SN.
4. The SIM chip encrypts RAND with K_i and sends the result as RES (response) to SN (via the mobile phone MS). If SN finds that RES = XRES, it means that the subscriber, i.e. the SIM chip, is authenticated, and the connection can be set up.
5. Encrypted radio transfer of data (voice) between the mobile phone MS and SN can begin.

Because 2G does not allow the subscriber to authenticate the mobile network, the subscriber does not know if it is connected to a real 2G operator's base station, or to a fake base station set up by an attacker. This vulnerability makes it easy for an attacker to set up a rogue mobile network with an IMSI catcher, which is described in the next section.

7.4.3 IMSI Catchers

An IMSI catcher is a type of eavesdropping device used to hijack mobile communications and to track the movements of mobile users. This type of device is also called a *StingRay* or *Cell-Site Simulator (CSS)*. Essentially, it is a fake base station that forms an intermediary between the user's mobile and the mobile operator's real base station. The user is tricked into connecting to the fake base station, which typically spoofs a real network name. This is a man-in-the-middle attack (MitM) which allows the attacker to receive and read the traffic between the phone and the real base station which is connected to the telephone network PSTN (Public Switched Telephone Network). Figure 7.11 illustrates the principle of IMSI catchers.

7.4 Mobile Network Security

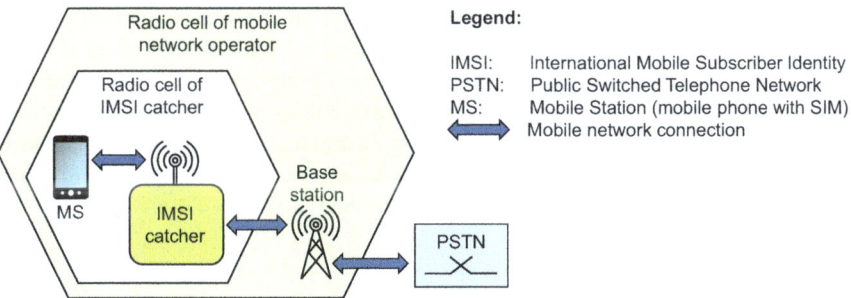

Fig. 7.11 Principle of IMSI catchers

The behaviour of an IMSI catcher is to act like a mini-base station that tricks nearby cell phones into connecting. At the same time, the IMSI catcher is also a mobile terminal that can connect to a real mobile network so that a call from a mobile phone to the IMSI catcher can be forwarded to the real telephone network. Thus, the subscriber will not notice that the call is going through the IMSI catcher and that the call is being intercepted.

An IMSI catcher is a device the size of an airplane-cabin case, or may even be so small that it is handheld. There is a large selection of IMSI catchers with prices ranging from approx. $1000 to $100,000. IMSI catchers also work for 3G and 4G, but not directly on 5G.

The term IMSI (International Mobile Subscriber Identity) is the name given to the unique identifier of a mobile subscriber, which is considered to be a confidential parameter because it can be used for tracking. To reduce IMSI exposure, TMSI (Temporary Mobile Subscriber Identity) is used, which is a temporary non-sensitive parameter because it is constantly changing and cannot be used for tracking. However, IMSI is still broadcast in plain text the first time a subscriber connects to a new network, or if TMSI is not recognised by SN. Thus, it is easy for an IMSI catcher to capture the IMSI of a subscriber to track the subscriber. At the same time, the IMSI catcher can eavesdrop on the conversation, or read data that is being sent. IMSI catchers are typically used by law enforcement agencies to investigate crime and terrorism, but they can also be used by criminal organizations and foreign state intelligence agencies. Primarily, this raises the question of whether the right balance exists today between legal use and criminal use of IMSI catchers. Second, it also raises the question of whether governments and citizens can trust their law enforcement agencies to IMSI catchers lawfully. In countries ruled by totalitarian regimes, IMSI catchers are likely to be used indiscriminately by law enforcement agencies. In countries governed by democratic principles, the use of IMSI catchers is restricted by law. However, even in democratic societies there are cases where law enforcement agencies have been suspected of circumventing the legal restrictions for using IMSI catchers, which is of course problematic.[1]

In addition to the vulnerabilities exploited by IMSI catchers, the Signalling System 7 (SS7) used in the traditional PSTN telecommunications network, is

---

[1] https://therecord.media/secret-service-ice-carried-out-illegal-stingray-surveillance-government-watchdog-says

infested with serious security vulnerabilities, which can easily be exploited.[2] SS7 is used by telecom operators for the management and configuration of telecommunications networks, and for setting up connections through the telecommunications network. SS7 attacks and abuse allow tracking of subscribers as well as eavesdropping and hijacking SMS messages. Hijacking SMS messages is precisely an attack vector against 2FA schemes in which SMS is used as a secondary channel. User authentication with secondary channels is described in Sect. 8.4.4.

7.4.4 Security in 4G

For authentication of subscribers in 4G, the individual authentication key K_i is still used, which is stored both in the SIM and at the subscriber's telecom operator. This is the same key that appears in the 2G security architecture in Fig. 7.10.

For the mobile phone MS to authenticate the mobile network, MS receives from SN a network identifier called SN-Id which is used to set up the connection between MS and SN. While this could in theory block IMSI catchers, it does not work in practice. The reason this solution fails is that the mobile phone MS, or SIM for that matter, is not able to see if a base station is genuine just by receiving the SN-Id. Put another way, there is no semantics in SN-Id that says whether it is real or not. In 4G, the parameter GUTI (Globally Unique Temporary ID) is used to represent the subscriber temporarily instead of TMSI, to reduce the exposure of IMSI. Nevertheless, IMSI is still broadcast in plain text occasionally as in 2G or 3G, which IMSI catchers can still intercept.

7.4.5 Security in 5G

As in 2G, 3G and 4G, authentication of the subscriber in 5G is still done with the individual authentication key K_i stored in the SIM and in HLR at the subscriber's telecom operator HE. This is the same key that appears in the 2G security architecture in Fig. 7.10.

As IMSI catchers have been a threat to all previous generations (2G/3G/4G) of mobile networks, 3GPP decided to finally find a solution that will stop IMSI catchers in 5G.

In 5G, SUPI (Subscription Permanent Identifier) is the permanent subscriber identifier (instead of IMSI), and GUTI (Globally Unique Temporary ID) is the temporary identifier. To avoid sending SUPI in cleartext (as IMSI is sent in cleartext in 2G, 3G and 4G), SUPI is encrypted, where the encrypted version is called SUCI (Subscription Concealed Identifier). Encryption of SUPI is done with the ECIES

---

[2] https://pages.nist.gov/mobile-threat-catalogue/cellular.html

7.4 Mobile Network Security

protocol (Elliptic Curve Integrated Encryption Scheme), which is similar to ECDH described in Sect. 4.6.2.

5G networks are designed with stronger security technology than previous generations, but there is still a general vulnerability because the weak security of 2G, 3G and 4G remains in the network. An attack vector against 5G is that the IMSI catcher forces a mobile phone to use 2G, 3G or 4G so that tracking and eavesdropping is still possible. This is called a *downgrade attack*. Some mobile phones can be configured by the user to not accept networks with 2G, 3G and perhaps 4G, but most models do not have that capability. Paradoxically, it helps little in practice that 5G technology protects against IMSI catchers, when mobile phones can be forced to use 2G, 3G or 4G, which are vulnerable to IMSI catchers. Figure 7.12 shows this paradox metaphorically as chain of links.

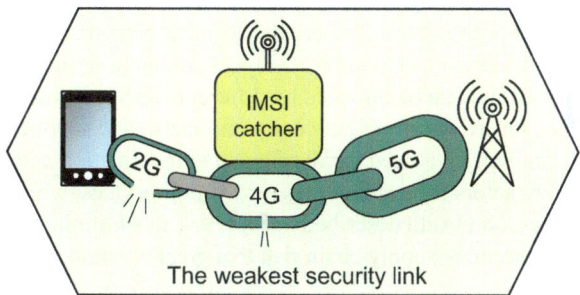

Fig. 7.12 The weakest security link in mobile networks

In fact, there is little telecom operators can do to stop IMSI catchers, because the vulnerability is systematic and lies in the international standards by which all mobile networks and mobile phones are built. Systematic vulnerabilities cannot be fixed like technical vulnerabilities can, as mentioned in Sect. 1.2. When thousands of mobile networks and billions of mobile devices are already built with systematic security flaws, it is almost impossible to fix these flaws.

To detect and possibly stop the use of IMSI catchers, telecom operators could in theory install an IMSI-catcher detector at each base station. However, this would be very expensive and probably inefficient, and is clearly outside what telecom operators see as their responsibility. This realization raises the question of how users can protect themselves from IMSI catchers, and this is where risk assessment comes into play. One threat is that the phone may be tracked with IMSI catchers, which happens passively and is difficult to detect. Therefore, we must always assume that we can be tracked. The risk will be small if the user's movements are not considered sensitive, but will be significant if they are. To avoid tracking with IMSI catchers, the mobile device can be turned off. However, a mobile phone is still active even in the off state with basic operating system functions and other background functions running. Advanced and targeted threat actors will be able to attack and exploit vulnerabilities in these features so that a switched-off mobile device can be tracked. That type of advanced tracking can be prevented by leaving your mobile device at home, or by keeping it in a radiation-protective mobile case with the technical name *Faraday cage*.

Another threat is that data could be intercepted, requiring an active MitM attack with an IMSI catcher. Phone calls and SMS messages can be intercepted that way. The risk is small if calls or SMS messages are considered non-sensitive, but will be significant if they are sensitive. However, attackers will not be able to eavesdrop on connections encrypted at the application layer with HTTPS, or through an app. Therefore, if calls and messages are considered sensitive, HTTPS or a secure communication app should always be used.

The modularity and virtualization of 5G networks bring other types of security challenges. Mobile networks based on 2G, 3G and 4G have largely been built with integrated components supplied by a single manufacturer, where telecom operator typically manage all the components in their respective networks. In such architectures there is no need to secure interfaces between components e.g. with authentication and encryption, other than from a zero-trust perspective (see Sect. 6.6). In mobile networks based on 5G however, functions and components are highly modularized and can be operated by different subcontractors. This creates a need to introduce secure interfaces between components and suppliers, which creates complexity. There will be a need for distributed authentication and access control between different actors for access to resources and services, which can be based, among other things, on OAuth described in Sect. 9.4. In addition, mobile operators need to assess cybersecurity supply chain risk to a greater extent, as described in Sect. 2.1.6.

7.4.6 SIM, eSIM and iSIM

The application is officially called USIM (Universal Subscriber Identity Module), while the terms SIM, eSIM and iSIM are terms for the various physical formats where data and software for the USIM application are stored and executed. Figure 7.13 shows the different SIM formats that exist.

SIM	eSIM	iSIM
• Separate chip received from mobile network operator. • Manually installed and replaced. • Takes up considerable space in the mobile device.	• Coprocessor embedded in mobile device. • Can be configured online. • Takes only little space in the mobile device.	• Integrated into the mobile device's microprocessor. • Can be configured online. • Takes no space in the mobile device.

Fig. 7.13 Different SIM formats

The USIM represents the user's subscription and contains the three important parameters IMSI/SUPI, the phone number, and the authentication key K_i.

Traditionally, we have received SIM chips from our telecom operator, either by picking it up in person or by mail, and then inserted the SIM chip into the mobile device. Physical presence or postal delivery creates a certain level of security because it is not exposed to cyberattacks from the Internet. However, it is relatively cumbersome. In addition, the slot and contacts for the SIM chip take up quite a lot of space in a mobile device.

Embedded SIM (eSIM) is built into the mobile device during manufacture and is intended to simplify deployment of the USIM to subscribers and to save space in the mobile device. The mechanism for transferring SIM data (IMSI/SUPI, K_i and phone number) to the eSIM is done online facilitated by the telecom operator. If permitted by the telecom operator, a subscriber may transfer SIM data from a SIM chip to the eSIM, or transfer SIM data from the eSIM in one mobile phone to the eSIM in another. With eSIM, the user can have many subscriptions in the same device at once.

An iSIM (integrated SIM) is directly integrated into the microprocessor of your mobile device. Because USIM is no longer hosted in separate hardware, iSIM is smaller, cheaper, and perhaps more reliable than eSIM. For mass-produced and low-cost IoT devices that connect to the 5G network, iSIM is a good solution.

Identity theft by taking over someone else's mobile number is relatively widespread, and is the starting point for various types of cybercrime. The scenarios start with a *SIM swap* attack where the scammer pretends to be the owner of your mobile number and tricks the mobile network operator into activating a SIM controlled by the scammer. If this is successful, the scammer has control over your phone number. Anyone calling or texting your number will contact the scammer's device, not your smartphone. This opens up for a whole range of cybercrime which a priori will be blamed on you because your mobile number has been used. Regardless of SIM format, it is therefore important that USIM and telephone subscriptions are managed in a secure manner by the mobile network operators.

7.5 Tasks

1. Wi-Fi
   Assume WPA-Personal with with SAE (Simultaneous Authentication of Equals).
   a. Does it provide user authentication and/or access control? Justify your answer.
   b. Does it provide authentication of AP (Access Point) to STA/user? Justify your answer.

   Assume WPA-Enterprise with user authentication to AS (Authentication Server).

   c. Does it provide user authentication and/or access control? Justify your answer.
   d. Does it provide authentication of AP (Access Point) to STA/user? Justify your answer.

2. Bluetooth
   a. Why should *Just Works* pairing between devices not be done in areas with many people?
   b. What could be the impact if your phone is paired with the device of a criminal hacker?
   c. Bluetooth devices have the standard name given by the manufacturer. What can you do to more easily recognize specific Bluetooth devices?

3. Mobile networks
   a. What is meant by "IMSI catcher", and what does it mean that the "IMSI gets caught"?
   b. What can/should telecom operators do to stop IMSI catchers?
   c. What can/should users do to protect themselves from IMSI catchers?
   d. What new security challenge is particularly significant for 5G networks?

Chapter 8
User Authentication

> *Treat your password like your toothbrush. Don't let anybody else use it, and get a new one every six months.*
>
> Clifford Stoll, American astronomer, author and teacher

User authentication is the process of verifying an entity's claim of holding a given identity. User authentication provides a level of assurance that the correct user is attempting to log in and access a system, application, or resource. An authenticator, such as a password or token, is the thing used to prove identity. An authentication solution can be based on a single authenticator (single-factor authentication) or multiple authenticators (multi-factor authentication). The combined strength of all authenticators used in a particular solution determines the level of authentication assurance, i.e. the certainty about correct identity of the user logging in. Paradoxically, a theoretically strong authentication solution can be bypassed if a user is tricked to send authenticators to the attackers, which can happen through social engineering and phishing attacks. Therefore, an authentication solution should be "phishing resistant" in order to give a high level of authentication assurance from a holistic perspective.

In this chapter, you learn about the methods of user authentication, which mainly fall into three categories: (1) something you know, (2) something you have, and (3) something you are. To strengthen authentication assurance multiple methods can be combined, which then becomes MFA (Multi-Factor Authentication). Finally, you will learn about authentication frameworks, which typically form the basis for specifying user authentication requirements for e-Government and in the government sector in general.

8.1 What Is User Authentication?

User authentication means that a user must prove ownership of a digital identity to log into a system or to access applications and services protected by access control. User authentication is intended to prevent unauthorized entities from gaining access.

A user in this context is not necessarily limited to being a person, but can also be a physical system, or even a logical process implemented in software, which can be called a *bot* (from robot).

Because natural persons are legal entities, user authentication makes it possible to hold the individual accountable for actions performed with this specific identity. That means user authentication supports accountability of activities in systems and networks.

To link natural persons to a digital identity, it is necessary that identity management includes requirements and processes for sufficient documentation and proofing of the natural person's identity. Examples of such requirements are described in *NIST SP 800-63A Enrollment and Identity Proofing* [32]. The verification of a person's identity has traditionally required the person's physical presence combined with documentation. However, due to cost end efficiency considerations, online and remote processes are increasingly being implemented for proofing a person's identity during registration. When considering the emergence AI with its potential for generating deepfake voice and video it is obvious that remote identity proofing is challenging. French and German security authorities published a common report on the topic in 2023 [33].

An *authenticator* is an element that proves the user's ownership of an identity upon login. An authenticator can be a password, a physical token or software app, or a biometric sample. Demonstrating knowledge of the correct password, using the correct token/app or giving the correct biometric sample is interpreted by the system as proof that the user is the rightful owner of the identity. Of course, mistakes can happen, for example when an unauthorized person has stolen a password or an authenticator has been forged.

8.2 Authentication Methods

Autenticators can be divided into three main categories:

- Something you *know*—knowledge-based authenticators, which are typically passwords.
- Something you *have*—ownership-based authenticators, which are typically tokens or apps.
- Something you *are*—physiologic feature-based authenticators, which are typically biometrics.

By using several methods at the same time, authentication assurance can be significantly strengthened, i.e. the certainty of correct authentication increases. *Strong authentication* is typically based on two or more methods being used simultaneously. This is called two-factor (2FA) or multi-factor (MFA) authentication. The three categories of authenticators mentioned above are described in the sections below.

8.3 Knowledge-Based Authenticators: Passwords and Learned Patterns

Knowledge-based authenticators refer to passwords that are remembered, or patterns that a user is able to reproduce. The most common are passwords, but there may also be other things that can be learned and remembered, such as reproducing a movement pattern, a sequence of actions or elements on a picture. Passwords were the first form of authenticator implemented for logging into systems, and are still widely used. This section focuses on passwords with their many interesting aspects.

Disadvantages of passwords include that they

- are easy to share with others (intentionally or not)
- are easy to forget (can become a burden for the help desk)
- are often easy to guess (e.g. with cracking tools)
- can be stolen by malware or a keylogger connected between keyboard and computer
- can be written down (which is both good and bad)

The advantage of password authentication is that it is easy and cheap to implement. The principle of user authentication with passwords is that the user enters a password and that the system or application checks whether the password is correct. Although it sounds extremely simple, it is challenging to create secure solutions for this.

8.3.1 Password Protection Against Cracking

Figure 8.1 shows the basic principle of checking passwords stored in cleartext on the authentication server. Naturally, it is assumed that the password is encrypted if it is sent over the Internet. A serious and persistent security problem is that password files and databases are leaked or stolen. If the passwords are stored in cleartext, the attackers who obtain a leaked database have direct access to all the passwords. The figure indicates that this provides weak security in case the database is leaked.

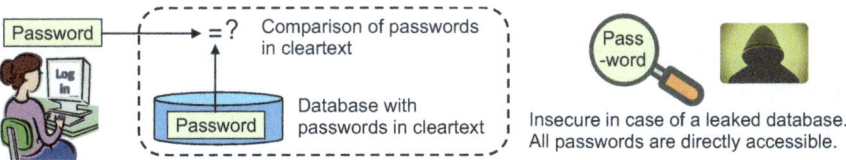

Fig. 8.1 Checking passwords stored in cleartext, with weak security in case of leaked database

Although it still occurs, implementing user authentication with passwords stored in cleartext on the authentication server is extremely bad practice.

An improved method is to let all passwords be hashed before they are stored in the password database. When the user enters a password at login, it must of course also be hashed before it is compared with the stored hash value in the database, as shown in Fig. 8.2. Here, too, it is assumed that the password is encrypted if sent through the Internet. The figure indicates that it provides moderate security in case the database is leaked, because it is relatively difficult for the attacker to recover passwords based on hash values.

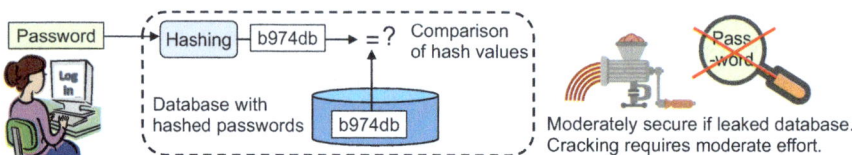

Fig. 8.2 Checking passwords stored as hashes, with moderate security in case of leaked database

The practice of recovering a password from a hash value is called "password cracking". This method involves hashing all possible words until there is a hash value equal to the stored hash value of the password. Then this very word is the password, or at least a password that works. Although this method of saving passwords is still used in many applications, it is not recommended. Attackers quickly understood that cracking could be done efficiently by generating giant tables of hash values and corresponding words for all possible words up to a certain size. Thus, a password could be easily cracked with a simple lookup in the hash table.

The best method for storing passwords is for passwords to be "salted" before they are hashed and stored in the password database. *Salting* means that a random bit string (e.g. 8 bytes) is added to the password before hashing it. The salt value is stored together with the hash value in the database. Of course, when the user enters a password upon login, it must also be hashed together with the salt value before it is compared with the stored salted hash value in the database, as shown in Fig. 8.3. It is assumed that the password is encrypted if it is sent through the Internet. The figure indicates that salting provides good security in case the database is leaked, because it is very resource-demanding for the attacker to recover passwords based on salted hash values.

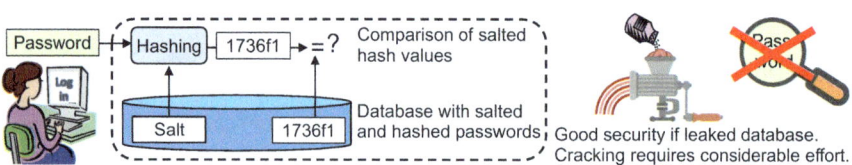

Fig. 8.3 Checking passwords stored as salted hash value, which provides good security in case of leaked database

8.3 Knowledge-Based Authenticators: Passwords and Learned Patterns

When passwords are stored as salted hashes, attackers must use vast resources to crack passwords from a leaked database. It is no longer possible to use ready-made hash tables, because that would require a separate hash table for each possible salt value. For example, if the salt value is 8 bytes, 2^{64} different hash tables are needed, which is not practically possible to produce.

However, attackers can still try to crack certain passwords they consider valuable by investing in high-computational equipment and purchasing large amounts of power. Naturally, how much time and power is needed to crack a single password depends on the quality of the password. A random 8-character password can probably be cracked in a few hours with powerful equipment. However, a 13-character random password would be about a million times harder to crack than an 8-character one, and will therefore be out of reach of that type of cracking.

Unfortunately, it is still a problem that passwords are usually not random, but relatively predictable. This is because humans are human, and often create simple passwords that are easy to remember. Crackers have found that this can be exploited by hashing words that are typical passwords instead of hashing random words. That way, a 13-character password can often be easily cracked if it has a typical regularity to be a password created by a human. The challenge for us humans is to create passwords that humans typically do not create, and at the same time manage to remember them. Then we immediately understand that it is a challenge to create passwords that are both strong and user-friendly.

Password cracking is also done by system owners to check the quality of user passwords in their own organization. If they succeed in cracking a password, the user is typically prompted to change it to a better password.

Password files and databases are stored in different locations depending on your system and network configuration. Some examples are the following:

- Linux: Passwords are stored locally in the/etc/shadow file, which has plain text format. Only Root users have read/edit access.
- Windows: Passwords are stored locally in the Security Account Manager (SAM) database \windows\system32\config\sam, which has NTLM format. Only admin has access.
- Computer networks: In network environments, as a rule, there is a separate application that stores passwords. The Windows application AD (Active Directory) is very widespread. The Linux application LDAP (Lightweight Directory Access Protocol) is used in some Linux-based networks.

Password databases on the Internet have been leaked and stolen ever since the invention of the Internet. Researchers collect and analyse such password databases to understand how we can create better passwords and also to let people find out if their passwords have been stolen.

The *Have-I-Been-Pwned?* website (HIBP)[1] offers a free service to check if a personal user account and/or password has been compromised. The site collects and

---

[1] https://haveibeenpwned.com/

analyses thousands of database dumps of information on billions of leaked user accounts, and allows users to search for their own information by entering their username or email address. It is also possible to check if a specific password is on the list of compromised passwords. Created by security expert Troy Hunt in 2013, the site is a valuable resource for combating identity theft and privacy breaches on the Internet.

As of 2024 the HIBP site has collected over 12 billion compromised accounts containing over 900 million unique passwords.[2] Figure 8.4 lists the 10 most frequently used passwords used among the compromised accounts, where "123456" is at the top of the list occurring 4.5 million times.

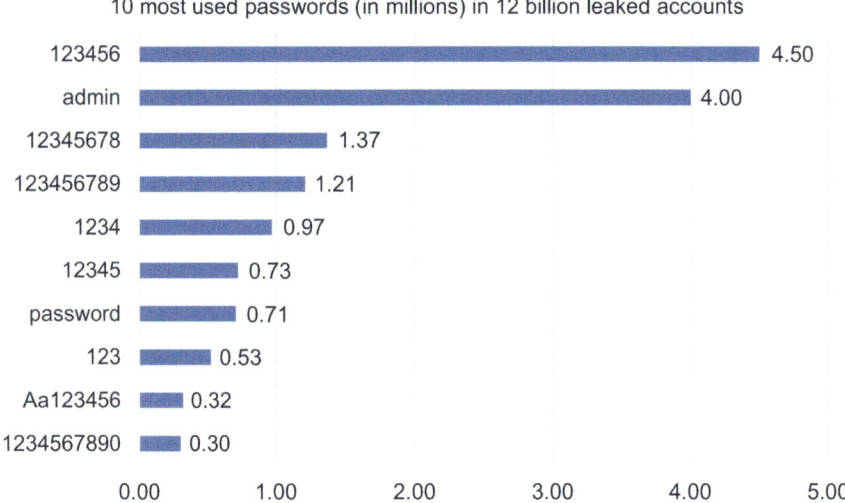

Fig. 8.4 Most used passwords out of a total of 900 million different passwords

It can be assumed that many of the 900 million passwords are relatively old, because they originate from old leaks. Most new applications require a certain length and quality of passwords, so many of the passwords in Fig. 8.4 probably would not have been accepted.

8.3.2 Strong Password Advice

Below is some general advice on strong passwords for cracking and identity theft protection.

---

[2] https://nordpass.com/most-common-passwords-list/

- Never reuse passwords. If a password is reused, and the password is leaked and cracked from one of your services, attackers can steal your identity on all your other services where the same password is used.
- Minimum 13-character passwords for important accounts.
- Never use common words, names, or dates of birth as part of a password.
- Use several categories of characters: uppercase/lowercase, numbers, and special characters.
- Changing your passwords:
  - immediately upon suspicion that an account has been compromised,
  - at regular intervals, e.g. every 17 months, if the account is considered highly sensitive, or the password is used frequently in an environment with potentially high exposure,
  - otherwise, rarely or never.
- Storing your passwords:
  - in your memory,
  - in plain text on paper stored securely,
  - in an online digital device, in which case encryption is required,
  - in an offline digital device (air-gapped), in which case plaintext is acceptable,
  - in an online password manager (see next section).

We need to be aware of the sensitivity of our various user accounts and how we handle the corresponding passwords. Most people have one particular main email address where other accounts and web applications can send emails to restore accounts and recover passwords. This particular email account is an anchor of our digital identity, and hence the password and other authenticators for this account must be handled securely.

Systems and websites for logging in are also responsible for protecting against theft of credentials, for example by blocking the account for 15 min if the wrong password is entered three times. That way, an attacker will not be able to try many passwords manually or automatically in a short period of time.

8.3.3 Password Managers

The task of managing all our different passwords becomes daunting as we accumulate more and more online accounts, combined the requirement of never reusing passwords. This problem is called *password fatigue*, with the meaning that we are overwhelmed from remembering passwords and remembering for which services they are used This is why many people have started to use a *password manager* (PM) which can store your passwords and generate strong passwords for you. It can automatically fill in online login forms with user ID and password because it remembers for which service a specific pair of user ID and password is being used. The three main categories of password managers are briefly described next.

- **Operating System built-in PMs**. These password managers are integrated with the operating system of a client platform. For example, on Apple devices, Keychain is a built-in password manager, so that users do not need to download or install any separate PM app.
- **Browser built-in PMs**. These password managers are integrated with the browser. For example, Firefox and Chrome have built-in PMs that allow users to save passwords for various sites, and to generate strong passwords if the users so desires.
- **Third-party PMs**. These password managers come as dedicated software that typically require users to install an app or a browser extension. For example, third-party PMs include Bitwarden and 1Password.

Hence, password managers have a local component on the client platform, as part of the OS, as part of the browser, or as a separate app installed on the platform. In addition, most password managers also have an online password vault which backs up your passwords and allows you to synch your passwords across your client devices. The principle of a password manager is illustrated in Fig. 8.5.

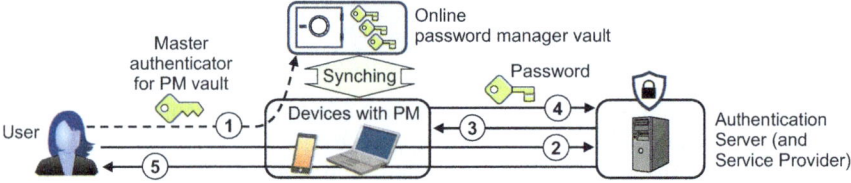

Fig. 8.5 Password manager basic principle

The numbered arrows in Fig. 8.5 are explained below.
1. The user opens the PM vault on the client device, and whenever needed synchronizes passwords from the online PM vault to other client devices.
2. The user requests a service from a service provider (SP).
3. The SP prompts for user authentication.
4. The user fetches the password from the PM, or the PM automatically retrieves the right passwords for the site, which is sent to and verified by the authentication server.
5. The user gets access to the requested service.

Many password managers support passkeys described in Sect. 8.4.2. Hence, the PM vault contains private cryptographic passkeys instead of passwords for websites that support passkeys. One advantage of passkeys is that they are phishing resistant, meaning that a passkey can only be used for authenticating to the genuine website. If the user is tricked e.g. with a phishing email and attempts to log in to a fake website, then passkey authentication will not work.

Once you have started using one specific password manager there is typically no simple and secure solution for migrating to a different password manager. In practice, the user will have to do some manual copying of passwords into the new PM [34]. This shows that currently there is no standardization in the PM market, which shows that this is still an immature technology.

A fundamental requirement for a password manager is of course that your passwords and credentials are protected with strong encryption in the password vault. To unlock the PM vault, you use your master authenticator which of course should be strong. For convenience, biometrics are often used to unlock the PM vault on the client device, but this can be attacked as explained in Sect. 8.5.4. To access the online PM vault, multi-factor authentication of high strength should be used, but oftentimes the minimum requirement can be a simple master password. It is important to note that current online PM vaults are not designed to be phishing resistant, given that the master authenticator is not a passkey. In case authentication to the online PM vault uses SMS-based verification as second factor, then SIM-swap attacks could be an attack vector, as described in Sect. 7.4.6. Note that password managers also can be attacked, where the breach of LastPass in 2022 is a prominent example [34]. It was reported that the encrypted vaults were stolen, giving the attackers the opportunity to perform offline attacks on them. Under such attacks the protection of the credentials in the encrypted vault depends totally on the strength of the master passwords/authenticators of users.

8.4 Ownership-Based Authenticators: Devices

Tokens and other devices that the user carries or controls can form the basis for strong user authentication. These devices may contain cryptographic components with variable parameters that makes the authenticator dynamic and thereby impossible to reuse. The strength of dynamic authenticators is that each authentication is unique, making it impossible to forge authentication by copying and repeating data transmission from a previous authenticator. Tokens are often designed to be tamper-resistant, making them very difficult to clone or counterfeit. An authentication device can also be a secondary channel that the user controls. E-mail or SMS are common secondary channels used for authentication, even if these channels can have weak security. Communication with the SIM card in the user's mobile phone can also be a secure secondary channel. The main types of authentication devices are displayed in Fig. 8.6. These are described separately in the sections below.

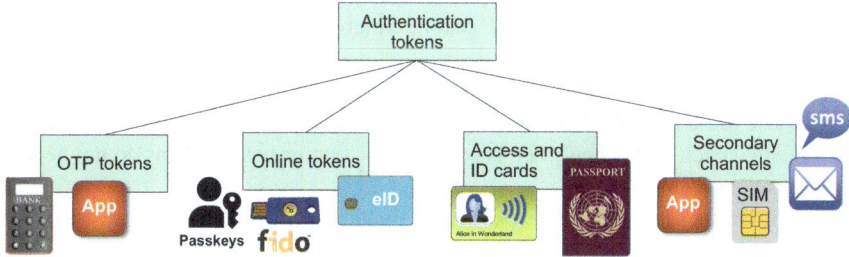

Fig. 8.6 The main types of authentication tokens

8.4.1 OTP Tokens

OTP (One-Time Password) tokens are synchronized with the server for cryptographic calculation of one-time passwords/codes. There are two principles for synchronization: time-based and counter-based.

Time-based OTP tokens contain a clock that is used to calculate a time-dependent OTP code (one-time passcode) which for example is shown on a small display. The user copies the code to the authentication login window, and the server can verify whether the user has sent the correct code. The clocks must always be synchronous, and each code is valid for a time window of around 1 min. The token needs a battery for the clock to run and stay synchronous.

Some tokens require a PIN (Personal Identification Number) for activation so the token cannot be used by unauthorized persons. Figure 8.7 shows the principle of user authentication with a clock-controlled synchronous token that shares a secret key with the authentication server. The token and the server use the same symmetric encryption algorithm to calculate the time-dependent OTP.

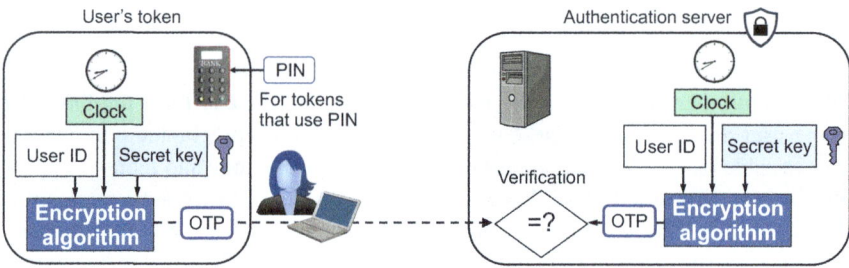

Fig. 8.7 User authentication with synchronous token and symmetric algorithm

In Internet jargon, an authentication server is often called an IdP (Identity Provider), especially in the context of federated identity management described in the next chapter.

Verification is done by the authentication server. If the OTP received from the user is equal to the OTP computed by the authentication server, this is evidence that the user is in possession of the OTP token, which in turn indicates that the user is authentic.

As an alternative to clock, a counter can be used for calculating the OTP, in which case these are called counter-based OTP tokens. This means that a counter is incremented each time a new OTP is computed. Counter-based tokens do not necessarily need a battery.

Synchronous authentication tokens are dynamic because computing the OTP depends on variable time or counter values.

8.4.2 Passkeys and Online Tokens

User authentication with an online token means that the token authenticates on the user's behalf to an authentication server. The online token can be a separate device such as a USB key or smart card to plug in, or a device that communicates with the client terminal via NFC or Bluetooth. Alternatively, the online token can be embedded in a client terminal, e.g. in a smartphone or a computer, either as hardware or software. Regardless of the format, the online token will exchange cryptographic messages with an authentication server through the Internet. This exchange of cryptographic messages between the token and the authentication server uses a cryptographic authentication protocol based on digital signature, as shown in Fig. 8.8.

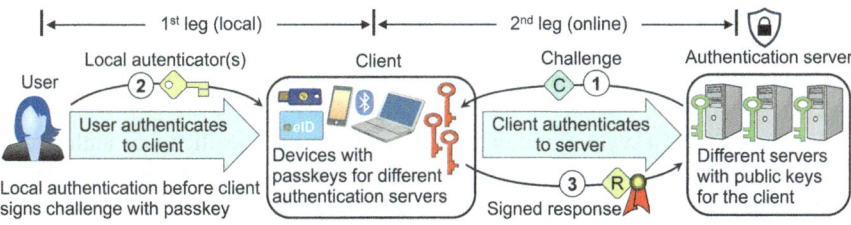

Fig. 8.8 User authentication with passkeys

Typically, the online token will first authenticate the user locally before it in turn performs online authentication, so this is a two-step authentication. The first authentication leg is local user authentication, which is typically based on biometrics or PIN. For biometrics, multimodal biometric authentication can be used, for example with recognition of fingerprints, face and voice. The second authentication leg consists of online authentication against the authentication server.

Passkey tokens contain *passkeys* that are not to be remembered by the user, because these are cryptographic keys. The terms FIDO and WebAuthn are other names for the same type of passkey authentication. The passkey token stores a separate passkey for each different authentication server (service provider) that the user accesses. User authentication with passkeys is considered *phishing resistant*, as explained below. The principle of user authentication with passkeys is shown in Fig. 8.8.

Remember that it is the token that authenticates itself to the authentication server on the user's behalf. The message exchange with numbered arrows in Fig. 8.8 illustrates this principle, as described below.

1. During login, a *challenge* in the form of a random number is sent from the authentication server to the token, via the client device. The authentication server expects a response in the form of a digital signature on the challenge for the user to be authenticated and logged in.
2. The user authenticates locally to the passkey token. Depending on authentication assurance level requirements, different forms of local user authentication may be used.

3. After the passkey token has authenticated the user locally, a signed response is generated and sent back to the authentication server. The user is authenticated by correctly verifying the digital signature of the response.

Suppose your passkey token is connected to or integrated with your client device, which could be a smartphone or laptop, for example. On the client device, the passkey token can be implemented in various ways, such as

- a hardware token (USB chip or smart card) which is connected to the client device, or which communicates with the client via NFC or Bluetooth,
- an app on the client device such as in a smartphone or a laptop computer,
- a function of the operating system on the client device or a function in the browser on such platforms.

When biometrics are part of local authentication on the client, such schemes are typically called authentication with biometrics, which can be misleading because it gives the false impression that biometric samples are sent online to the authentication server.

Private keys (passkeys) stored in the passkey token are specific to the authentication server URL, which means that the platform/device will not sign a challenge from a fake website because it will have the wrong URL. This is the reason why authentication with passkeys is considered to provide phishing-resistant authentication.

Current solutions for synchronization and migration of passkeys to other devices are typically based on cloud storage, e.g. based on a password manager (see Sect. 8.3.3) or a passkey bank controlled by the manufacturer of the passkey token. Access to passkeys in the cloud ultimately requires traditional authentication with a master password or other master authenticator, which ironically is not phishing resistant.

Passwords are simple to understand and are something we are familiar with. Passkeys on the other hand are difficult to understand, and when many users wrongly believe that passkeys involve sending biometric samples through the Internet, it can be challenging t get widespread acceptance of passkeys.

The downside of biometrics is that it can be forged or coerced relatively easily, which could expose users to physical danger. Threat actors can steal user devices and authenticate with forged biometrics, or can force biometric samples from users e.g. by doping them down. For this reason, local user authentication on the client should not be based on biometrics alone, but should include a memorized PIN or pattern code because it is more difficult to steal or force.

8.4.3 Access and ID Cards

An access card is typically a smart card that uses NFC for communication with a card reader. NFC (Near Field Communication) is a technology for transmitting radio signals between devices over distances of a few centimetres. The card reader also generates magnetic waves which are picked up by the antenna on the card to induce electric current for the microprocessor. This and similar technologies are used not only for access cards, but also for payment cards, passports and ID cards. Alternatively, the card can be inserted into a card reader with an electrical contact. Along with the card, a PIN code can also be used. Access cards are used for physical access control described in Sect. 12.3.

An NFC card can receive and send data. The example in Fig. 8.9 shows that the card reader sends a random number as a challenge to the NFC card, which encrypts the challenge and returns a response verified by the card reader. The verification consists of comparing the response received with the computed expected response.

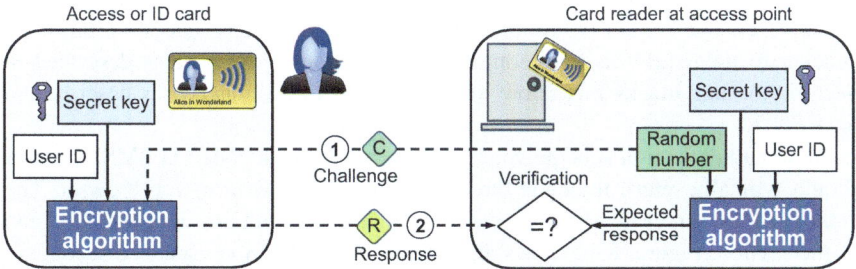

Fig. 8.9 Challenge-response authentication with NFC card against card reader

In the example in Fig. 8.9 a symmetric algorithm with a secret key is used. Another possibility is to use an asymmetric algorithm with a private/public key pair and a challenge-response protocol as in Fig. 8.8. Passports and national ID cards use challenge-response based on private/public key. The card reader can be advanced terminals used for passport inspection at airports, or can be a smartphone with NFC. The private key is in the passport/ID card and the public key in the card reader/terminal. Here, a separate PKI is used to guarantee authenticity of the public key in the card reader/terminal. The challenge-response protocol allows card reader/terminal to authenticate passport/ID cards. Authentication with this type of access and ID card is dynamic because the random number in the challenge is unpredictable and varies every time.

Electronic passports and ID cards are also designed so that they can authenticate the card reader or inspection terminal. The purpose is that passport/ID cards shall not give out sensitive information such as fingerprint templates to an unauthorized terminal. This authentication uses a separate challenge-response protocol that goes in the opposite direction, i.e. with private key stored in card reader/terminal and public key in passport/ID card. The PKI for passports is managed by ICAO (International Civil Aviation Organization).

Passports and national ID cards are mainly used to authenticate us when traveling and crossing national borders, while an eID is used when we authenticate ourselves for access to online eGoverment services. In order for national ID cards to simultaneously be an eID, the card has to be logically connected to an accredited (federated) IdP provider. Some countries have implemented combined national ID cards and eID, e.g. Germany, but many authorities consider it to be too complicated and therefore keep ID cards and eID separate.

8.4.4 Secondary Channels

If web communication is considered a *primary channel*, SMS can be considered a *secondary channel*. Such secondary channels can support authentication for setting up a session through the primary channel. For example, the authentication server can send a code as an SMS to the user's phone. Phone numbers have moderately strong authenticity in that subscribers must prove identity to get a phone number.

However, one should be aware of possible vulnerabilities that could allow an attacker to unlawfully assume control of another phone number as described in Sect. 7.4.6, and attacks that allow SMS messages to be hijacked as described in Sect. 7.4.3.

In an authentication scheme where the user receives a code via SMS, the user should typically return the code through a login window in a web browser. The authentication server interprets this as proof that the correct user is at the other end of the session. Figure. 8.10 shows an example where SMS is used as a secondary channel for authentication.

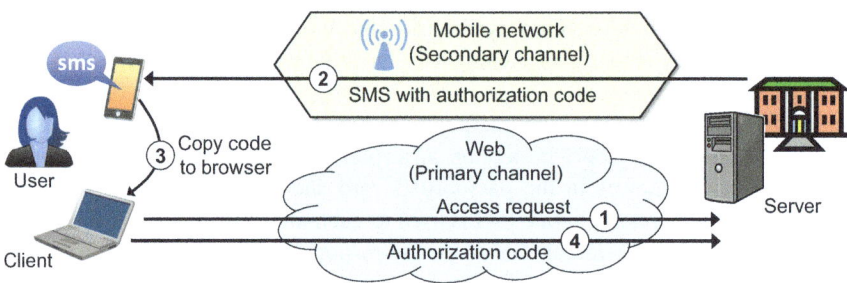

Fig. 8.10 Authentication through secondary channel, with SMS in this example

Any channel that is independent of the primary channel can be used as a secondary channel. Email and web are considered to be independent channels, although both go through the Internet. Therefore, e-mail can be used as a secondary channel for authentication through the web.

A smartphone can support several channels, e.g. SMS, email, web, NFC, Bluetooth and USB. With so many different channels, there are many possibilities for designing authentication schemes with secondary channels. However, these schemes often have vulnerabilities. In order for authentication with secondary channels to provide high authentication assurance, the secondary channel must be secure against eavesdropping, hijacking and man-in-the-middle attacks. Authentication with secondary channels is dynamic because the authentication server sends different and unpredictable codes for each new authentication.

8.5 Biometrics

Biometric authentication is to verify a person's identity based on physiological characteristics. There are a number of different methods called *biometric modalities*, where well-known modalities include

- fingerprint
- face recognition
- eye/retina/iris scan
- hand geometry
- hand signature
- voice/speech characteristics
- keystroke dynamics
- gait

8.5.1 Requirements for Biometric Systems

There are virtually no limits to which physiological characteristics can be a biometric modality, but for a modality to be applicable in practical implementations, seven main requirements apply:

1. **Universality:** The vast majority of people should be able to express the characteristics of the modality.
2. **Uniqueness:** Different people should generally have distinctly different characteristics of the modality.
3. **Permanence:** Characteristics of the modality for a person should remain largely unchanged for a long time.
4. **Collectability:** Characteristics of the modality should be easy to collect and measure in a quantitative way.
5. **Performance:** The speed at which characteristics are collected, the speed of analysis of the measurements, and the resource usage sufficient to achieve the desired speed.
6. **Acceptability:** The extent to which people are willing to accept the use of a particular biometric modality.

7. **Spoofing resistance:** The ability to protect against circumvention, i.e. to detect, attempts to trick the biometric system with falsified biometric samples. This is called *presentation attack detection* (PAD), which is described in Sect. 8.5.3.

Biometric templates and samples represent highly sensitive data where confidentiality is important. In most cases, biometric data is considered personal data, so privacy regulation such as GDPR must be considered when adopting a system of biometric authentication. If biometric data is processed *"for the purpose of unambiguously identifying a natural person"*, which is the usual case, it should be considered special categories according to GDPR. In such cases, the data controller must satisfy one of the processing bases described in GDPR Art. 9 for the processing to be lawful. If biometric templates and samples for a user are only collected and processed locally on the user's device, it is relatively easy to protect biometric data. If biometric data from many users is stored and processed centrally, strict security controls are required to protect such data.

8.5.2 Mode of Operation and Components of Biometric Systems

A biometric template is a set of stored biometric features. A capture device, e.g., a camera or fingerprint scanner collects biometric samples which are further converted into a mathematical file. Thus, the template is a digital representation of biometric samples stored in a biometric database. The formats of biometric templates and samples are very diverse and depend on the modality. For voice recognition, sound recordings are made in which sequences of sound frequencies are analysed and compared, typically with ML-based classifiers. For fingerprints, a two-dimensional image of the lines in the skin of a finger is created, where the geometry of the lines is analysed and compared. Measurements for a specific modality can be described by a set of features.

For example, fingerprint processing is based on detecting and mapping so-called *Minutia*, where there are two different types, as shown in Fig. 8.11.

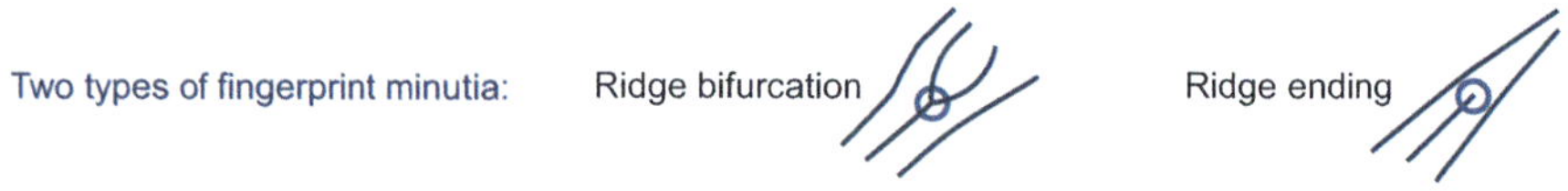

Fig. 8.11 Minutia for fingerprints

In traditional fingerprint processing, the image is analysed to identify lines and to detect and locate minutia. Each minutia is recorded with the geometric location relative to other minutia. This information is eventually encoded and expressed in binary, as shown in Fig. 8.12.

8.5 Biometrics

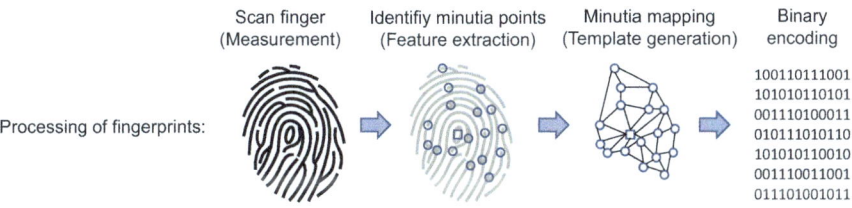

Fig. 8.12 Processing of fingerprints from biometric measurement to binary encoding

For fingerprint authentication, each user must first give samples of their fingerprint to create a fingerprint template. The quality of the collected fingerprint template is crucial to the accuracy of biometric authentication. The system will therefore verify whether the quality is satisfactory before entering a fingerprint template with characteristics into the database. Upon authentication, the user will place their finger on a sensor that records an image of the finger with lines in the skin. The sensor sends the image to the next component, which extracts characteristics from the sample. These are eventually sent to the comparator, which compares characteristics from the obtained sample with characteristics from the fingerprint template in the database. Figure 8.13 shows the interrelation of these general components.

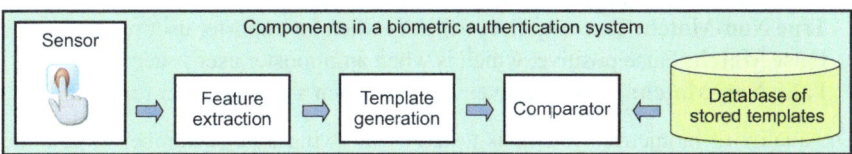

Fig. 8.13 General components of a biometric system

A system as shown in Fig. 8.13 can be used to compare with the "one-to-one" or "one-to-many" methods. If the application has many users, it is common for the user to provide identity, so that the biometrics system can retrieve the template for a single user from the database for comparison with the obtained biometric measurement, which is the method of one-to-one comparison. If the user does not provide identity, it is possible for the biometric system to compare the obtained biometric sample with all the templates in the entire database to identify the template with the greatest similarity, which is the one-to-many comparison method.

It is very resource-intensive to use a biometric system with one-to-many as a method. However, this is precisely the method used in forensic investigation, where a biometric sample is compared with templates from all possible suspects to establish evidence.

Traditionally, specific measurement parameters were used for comparison of biometric samples. In newer systems, machine learning is mostly used to compare biometric samples and calculate scoring to make classification decisions about user authentication.

8.5.3 Quality Aspects of Biometric Systems

The accuracy and quality of biometric systems depends on many factors. It is desirable that a biometric system always accepts genuine users and always rejects imposters. Unfortunately, this is unattainable. Below we describe how the system makes the decision to accept or reject users, and how the rate of incorrect classification determines the quality of the system.

Characteristics of a collected sample are compared with characteristics from a saved template. The comparison gives scoring S. The greater the similarity, the higher the score S.

Threshold T determines whether the sample is classified as *match*, meaning that the sample is sufficiently similar to the template to assumed that the user is genuine, or alternatively classified as *non-match, meaning that the* sample does not resemble the template and it is assumed that the user is an imposter.

Match (presumed genuine user) is when $S \geq T$
Non-match (presumed imposter user) is when $S < T$

However, it is possible that the system incorrectly classifies a sample as match for an imposter user and as non-match for a genuine user. There are four options:

- **True Match:** True positive, which is when a genuine user is accepted.
- **True Non-Match:** True negative, which is when an imposter user rejected.
- **False Match:** False positive, which is when an imposter user is accepted.
- **False Non-Match:** False negative, which is when a genuine user rejected.

The quality of biometric systems is tested by allowing a large number of genuine and imposter users attempt to authenticate themselves. A genuine user trying to authenticate means that the user is attempting to log into their own account. An imposter user trying to authenticate means that the user is not registered in the system, or that the user is registered, but is attempting to log in to another user's account. That way, statistics are collected on how often each of the four options above occurs. Two important statistical measures are *FMR (False Match Rate)*, also called FAR (False Accept Rate), which is the rate of false positives, and *FNMR (False Non-Match Rate)*, also called FRR (False Reject Rate), which is the rate of false negatives. These can be expressed as follows:

$$FMR = \frac{\#\text{False Match}}{\#\text{Imposter users}}$$

$$FNMR = \frac{\#\text{False Non} - \text{Match}}{\#\text{Genuine users}}$$

The threshold T determines the proportions of FMR and FNMR.

If the threshold value T is raised, it becomes harder for imposter users to get a match, so the number of false positives goes down and FMR is reduced, which is

8.5 Biometrics

good. However, it also becomes harder for genuine users to get a match, so the number of false negatives goes up and FNMR increases, which is bad.

If the threshold value T is lowered, it becomes easier for imposter users to get a match, so that the number of false positives goes up and FMR increases, which is bad. At the same time, it also becomes easier for genuine users to get a match, so that the number of false negatives goes down and FNMR is reduced, which is good.

The optimal threshold value T for a biometric system, then, is a trade-off between FMR and FNMR. Figure 8.14 shows how the threshold value T affects FMR and FNMR.

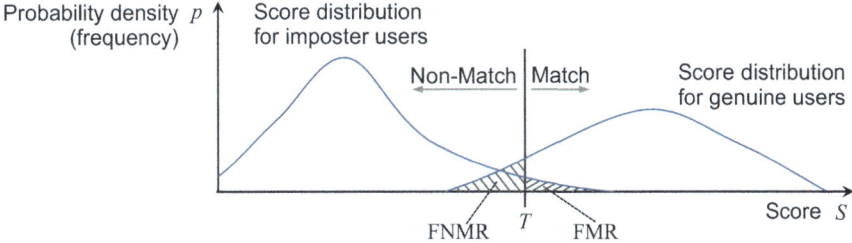

Fig. 8.14 Significance of the threshold value T for FMR and FNMR

The curves represent statistical distributions of scores for genuine and imposter users, respectively, after testing with many users. The height of the curves represents (density of) probability p for a genuine or imposter user to achieve a given score S. The size of the shaded areas of Fig. 8.14 represents FMR and FNMR.

EER (Equal Error Rate) is when FMR = FNMR, i.e. EER = FMR = FNMR. The threshold value T which gives EER is typically seen as the optimal threshold value, but the system owner can still decide to adjust the threshold value T to prioritise reducing either FMR or FNMR, depending on the cost of a false match and false non-match respectively. EER is in general a measure of the quality of a biometric system—i.e. the lower EER, the better the quality.

However, low means EER does not mean that the system cannot be fooled. The distribution of scores for imposter users in Fig. 8.14 does not indicate that they are trying to trick the system—they are just imposter users.

If attackers really try to trick a biometric system, and have the necessary resources and expertise, they will usually always succeed. The term PAD (Presentation Attack Detection) expresses the extent to which attacks against biometric systems with falsified input can be detected and prevented. Unfortunately, robust PAD is difficult to achieve, especially when attackers use AI (Artificial Intelligence) to mimic faces and voices.

In general, biometric authentication systems are relatively vulnerable to attack and should not be used alone for sensitive applications. Together with other methods of authentication, biometrics can provide good security and at the same time provide good user-friendliness.

8.5.4 Safety Aspects of Biometric Systems

It may be appropriate to set requirements for personal safety when using biometric systems, since the user is physically involved. One possible attack on biometric systems could be to force a person to provide a biometric sample, or to steal body parts used for biometric authentication. In Kuala Lumpur in 2005, car thieves chopped off part of the driver's left index finger to start his S-Class Mercedes Benz, equipped with a fingerprint key.[3] In the United States in 2023, several cases have been reported where criminals have drugged victims to unlock their smartphones and access their bank accounts with facial recognition.[4] In such cases, the authentication involves body-based biometric modalities, i.e. a body part, such as the face or finger, is used. Such modalities can be used even if the user is unconscious, so that it is relatively easy to force a biometric sample. For example, it is easy to get a fingerprint from a sleeping person. In behaviour-based biometric modalities, on the other hand, such as voice recognition and keyboard dynamics, the user must be conscious so that it is more difficult for an attacker to force a biometric sample. These examples and assessments show that user safety can be compromised if body-based biometrics are used outside controlled environments. This is especially critical if biometric authentication provides access to significant assets or sensitive information, because it gives attackers strong incentives to physically attack users. It is therefore relevant to require practical use of biometrics to be safe and not expose the user to disproportionate physical danger.

8.6 Multi-factor Authentication

Basing authentication on only one of the methods described in Sect. 8.2 will never be able to provide strong authentication. Put another way, one authentication method alone cannot create a high degree of assurance that an authenticated user is actually the right one. To achieve strong authentication, two or more methods must be used simultaneously. This is called *multi-factor authentication* (MFA), where the different factors are the different authentication methods used in a scheme.

2FA (two-factor authentication) has long been used for highly sensitive services, and is increasingly also used for services with moderate sensitivity. A typical 2FA combination is to use a dynamic authentication device (token or secondary channel) together with a static password.

The biggest obstacles to adopting 2FA or MFA are increased costs and usability challenges. However, a lot of innovation is being done in the development of affordable and at the same time user-friendly methods of authentication. Authentication apps are considered more user-friendly than traditional hardware tokens, while still

---

[3] http://news.bbc.co.uk/2/hi/asia-pacific/4396831.stm
[4] https://www.nytimes.com/2023/03/29/nyregion/indictments-nyc-gay-bars-homicide.html

providing high authentication assurance. This is an example of the successful implementation of two-factor authentication.

For providers of online services, it can be difficult to choose an appropriate authentication scheme. In the next section, we describe various frameworks for authentication that are be helpful in adapting an authentication scheme to the sensitivity of the service.

8.7 e-Authentication Frameworks

Authorities in all countries are increasingly switching to e-Goverment, which means digitisation of public services. Good authentication is important when citizens need to log in to access public services. A system for user authentication in e-Government is called eID. It is natural that eID schemes are harmonised between all public services, state-wide and nationally within each country and regionally. An overview of e-Authentication services around the world can be found in [35].

An e-Authentication framework typically defines levels of authentication assurance, meaning the certainty with which it can be assumed that the correct person or entity has been authenticated. Three general components which contribute to authentication assurance are (a) the initial identity proofing and registration process as well as the provisioning and recovery of authenticators and credentials, (b) the strength of the authentication mechanisms and processes, and (c) the security of the federation process (see Sect. 9.2.3). These components are illustrated in Fig. 8.15 with example requirements from each. The steel chain metaphorically means that the overall authentication assurance cannot be stronger than the weakest link, i.e. the weakest set of fulfilled requirements in each component.

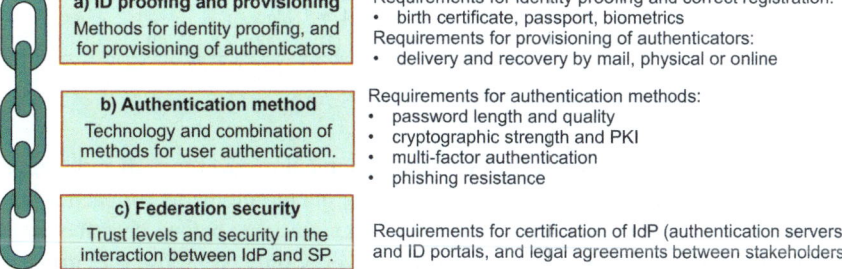

Fig. 8.15 Components of authentication with example requirements

Below is an example of three assurance levels with a simplified description of requirements for each level, which approximately follow the European eIDAS regulation described in Sec. 8.7.2 below.

- **High assurance:** 2-factor (2FA) or multi-factor (MFA) authentication, e.g. static password, as well as a physical token received in person, alternative app on a mobile phone.
- **Substantial assurance:** 2-factor authentication (2FA), e.g. static password and in addition one-time codes received by post to the user's official postal address, alternatively via SMS.
- **Low assurance:** 1-factor authentication, e.g. static password received in person or by mail to the user's official postal address.

Providers of public digital services can use authentication frameworks as a basis for choosing the authentication level and methods of authentication. When a service is assessed as having high, substantial or low sensitivity, the frameworks specify what is required of an appropriate user authentication scheme. Providers of e-Government services may be tempted to always choose the highest assurance level of authentication, but they must also remember that higher assurance levels provide greater complexity and costs in deployment and management. Therefore, it is important that the assurance level is adapted to the sensitivity level of the service. Assessing a service's sensitivity and selecting the appropriate authentication level is an important part of risk management for the service.

Figure 8.16 shows a set of prominent frameworks for e-Authentication and how they define authentication assurance levels.

Frameworks for e-Authentication		Authentication assurance levels		
SP 800-63-4 Digital Identity Guidelines NIST, USA 2024		Assurance Level 1	Assurance Level 2	Assurance Level 3
eIDAS-2 Electronic identification and trust services EU 2024		Low	Substantial	High
e-Pramaan: Framework for e-Authentication India, 2014		1FA	2FA	MFA
IS 29115 Entity authentication assurance framework ISO/IEC 2013	Low (1)	Medium (2)	High (3)	Very High (4)

Fig. 8.16 Frameworks for e-Authentication

Note that the frameworks in Fig. 8.16 give divergent names for each authentication assurance level. On a general level, however, the requirements for each assurance level are relatively well harmonised. The sections below describe each framework in more detail. Among these frameworks only SP 800-63-4 currently describes component c) of Fig. 8.15 with requirements for federation security. Since identity federation is a fundamental component of most identity schemes it is important to take into account this component for defining requirements for authentication assurance levels.

8.7.1 USA: NIST SP 800-63-4 Digital Identity Guidelines

In the US, NIST published the first version of its guideline on digital identity in 2004, called *SP 800-63 Electronic Authentication Guideline* (version 1), to provide "technical guidance to federal agencies implementing electronic authentication". As of 2024, NIST has published the second draft version of *SP 800-63-4 Digital Identity Guidelines* (version 4). The NIST guidelines define the same authentication components as shown in Fig. 8.15, each with three assurance levels. These components are:

(a) **IAL** (Identity Assurance Level) which refers to the identity proofing process.
(b) **AAL** (Authentication Assurance Level) which refers to the authentication process.
(c) **FAL** (Federation Assurance Level) which refers to the identity federation process, see Sect.9.2.3.

NIST SP 800-63-4 specifies requirement that must be met to obtain a certain assurance level for each component. For a given service, the overall authentication assurance level **xAL** is determined by the lowest assurance level of the three components.

Phishing attacks are a threat against which many existing authentication methods provide little protection. A man-in-the-middle attack means, for example, that the user is tricked into sending authenticators to a fake service provider or authentication server, which forwards the authenticators to the genuine service provider/authentication server. To stop such attacks, research is being conducted on developing phishing-resistant authentication. This can be done, for example, by making authenticators a dynamic function of the identity and/or IP address of the service provider/authentication server, so that authenticators cannot be forwarded from the fake to the real service provider/authentication server. In the work on a new revised version of the American *NIST SP 800-63-3 Digital Identity Guidelines*, it is mentioned that phishing-resistant authentication methods are becoming a requirement for the highest level of authentication.

8.7.2 EU: eIDAS

eIDAS is the European regulation for electronic identification and trust services. The term "trust service" is EU jargon for PKI-based certification and authentication services. On 29 February 2024 the European parliament adopted eIDAS 2.0 with many new requirements for the European eID, including the *Digital Identity Wallet*, which European countries must implement by August 2026. The eIDAS act defines three LoA (Levels of Assurance) for authentication.[5] The first version of eIDAS

---

[5] https://ec.europa.eu/digital-building-blocks/sites/display/DIGITAL/eIDAS+Levels+of+Assurance

came into effect on June 15, 2018, but did not have the expected impact. eIDAS 2.0 is therefore much more ambitious.

The revised regulation mandates that all EU countries establish a national eID scheme. The Nordic countries introduced eID already around 2004 based on their national BankID schemes. It is interesting to note that the Nordic BankID providers are private companies established by the banking sector, which now provide authentication as-a-service for e-Government. In other words, the governments in the Nordics opted to rely on the banking sector's authentication schemes instead of establishing state owned authentication services. In contrast, many other EU countries have opted for government owned authentication services. Hence the other Nordic countries and a number of other EU countries have had good eID schemes for many years, but this does not apply to all countries in the EU. The goal of eIDAS, among other things, is to force all EU countries to introduce national eID schemes for their citizens, and for these schemes to work across national borders within the EU.

The digital identity wallet scheme will enable each citizen to use an ID app which makes it easy to present documentation in the form of: *verifiable credentials* in various contexts. A verifiable credential is a digitally signed document that third-party recipients can verify. Verifiable credentials are for example certifications, diplomas, transcripts, references, certificates, licenses, permits, vaccinations, and medical test results. A well-functioning ID wallet scheme has the potential to streamline the management of this type of document and to prevent document forgery.

8.7.3 India: e-Pramaan

e-Pramaan [36] is the Indian national e-Authentication scheme for providing federated authentication for numerous services accessible through mobile and fixed platforms. e-Pramaan is also a standard which specifies requirements for the authentication processes. The Single Sign-On feature provides registered users with single window access to all services that are integrated with e-Pramaan.

e-Pramaan is designed to work in tandem with India's Aadhaar identity scheme which consolidates disparate identity documents across government departments under a single digital profile for each citizen. The assurance levels in e-Pramaan are defined in terms of how many authentication factors are being used, but implicitly also considers the strength of each factor. Aadhaar and e-Pramaan together represent the largest eID scheme in the world, with more than 1.3 billion registered users.

8.7.4 ISO/IEC 29115 Entity Authentication Assurance Framework

Older e-Authentication frameworks typically defined four assurance levels of authentication, as is the case with *ISO/IEC 29115:2013 Entity authentication assurance framework*. However, it turned out that the lowest level, meaning *"little or no assurance of correct identity"*, had no practical application for e-Government. Therefore, modern authentication frameworks have only three authentication assurance levels. Given the enormous innovation in identity management during the last 10 years, it is obvious that the international standard ISO/IEC 29115 from 2013 is outdated. Therefore, it should either be revised or withdrawn.

8.8 Tasks

1. Password

    (a) Why should passwords not be stored in cleartext on the authentication server?
    (b) What is meant by password salting, and what is its purpose?
    (c) Which known service on the Internet is used to check if a user account data (and password) have been stolen? Check if one of your own user accounts has been stolen, and change the password if it is.
    (d) Name a reason why passwords should never be reused.

2. Biometrics

    (a) Explain how the quality of a biometric authentication system can be expressed with EER.
    (b) If a biometric system is configured with a very low FMR (False Match Rate), what is the consequence for FNMR (False Non-Match Rate)?
    (c) Imagine a system configured with very low FMR. To what extent does this make the system robust against intentional presentation attack?

3. Passkeys

    (a) How are phishing attacks carried out against traditional user authentication?
    (b) What does it mean that an authentication method is phishing resistant?
    (c) Why are passkeys (FIDO/WebAuthn) considered phishing resistant?
    (d) Assume that passkey authentication becomes widespread, so that classic phishing can no longer be used to steal identities. However, threat actors will still find ways to attack. Imagine that you are a threat actor, and propose ways to attack and phish passkey-based user authentication.

Chapter 9
IAM—Identity and Access Management

> *Identity is the new security perimeter.*
>
> Chuck Robbins, CEO, Cisco

In the cybersecurity community, the statement that *identity is the new security perimeter* reflects that data is increasingly stored, accessed and processed online, and that the companies' own physical data networks to a lesser extent represent a security perimeter for data. Digital security for data and applications online is therefore dependent on robust online identity and access management, which means that identity has become an important security perimeter. In general, identity and access management (IAM) is the security and business discipline which ensures that the right entities can access the right assets at the right time for the right reasons.

In this chapter, you learn how to apply user authentication in a wider context. By integrating the authentication methods described in the previous chapter with security protocols, it is possible to design identity federations and eID solutions that are user-friendly and scalable across the Internet. *Identity management* is the term that covers such advanced schemes of user authentication. You will also learn about *access management*, which consists of methods that ensure that authenticated users can only access data and services for which they are authorized.

9.1 Definition of IAM

The concept of IAM (Identity and Access Management) covers the two general functions of identity management and access management (see Fig. 1.13 in Sect. 1.11). Identity management focuses on the registration and authentication of user identities. Even if a user has been authenticated, the resources that the user can access are limited. Access management focuses precisely on access authorization for registered users, and then on access control for authenticated users to enforce the requirement that users only shall be able to access data and resources for which they have been authorized. There are several models of access control, traditionally called *security models,* which are methods of controlling authenticated users' access to resources. Identity management and access management are closely linked in practical implementations. A good definition of IAM is articulated by Gartner:[1]

> Identity and Access Management (IAM) is a security and business discipline that includes multiple technologies and business processes to help the right people or machines to access the right assets at the right time for the right reasons, while keeping unauthorized access and fraud at bay.

While Fig. 1.13 in Sect. 1.11 presents IAM in a relatively abstract form consisting of a configuration phase and a usage phase, Fig. 9.1 shows an alternative representation of IAM in the form of a typical IAM scenario. Each stage of IAM in Fig. 9.1 is described below.

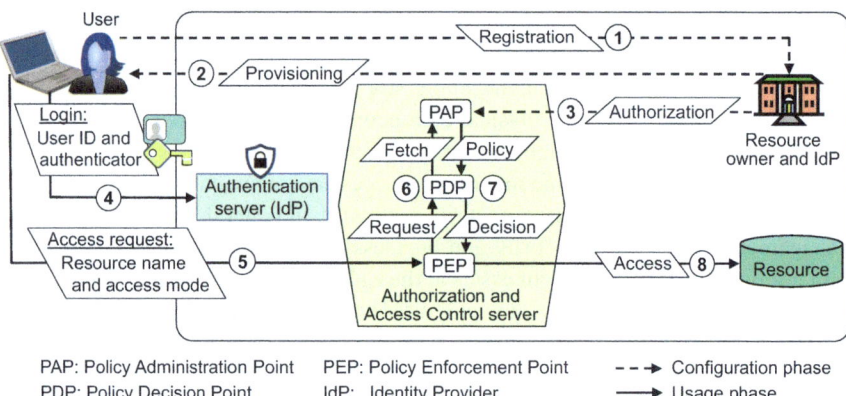

Fig. 9.1 IAM scenario

Configuration phase
1. The user's identity and user ID are registered by the authentication server (IdP).
2. The IdP prepares authenticator(s) that are provisioned to the user.
3. The resource owner authorizes the user by specifying access policies and rules in the Policy Administration Point (PAP). An authorization policy can change over time and is in general determined based on the user's current job role.

Usage phase
4. The user logs in with user ID and authenticator.
5. The user requests access to a resource.
6. The access control server fetches access policy from PAP.
7. PDP (Policy Decision Point) makes an access decision.
8. PEP (Policy Enforcement Point) enforces the decision and grants or denies access accordingly.

---

[1] https://www.gartner.com/en/information-technology/glossary/identity-and-access-management-iam

The configuration phase and the usage phases take place in parallel. Typically, there will be a need to change access authorizations and to provision new authenticators for a user while the user account is in active use.

The configuration and usage phases of IAM are shown in Fig. 9.1 with dashed and solid arrows, respectively, so that steps 1–3 are part of the configuration phase, while steps 4–8 are part of the usage phase. From another perspective, identity management consists of steps 1, 2 and 4, while access management consists of steps 3, 5, 6, 7 and 8.

The fact that the user makes an access request in step 5 does not necessarily mean that the user explicitly specifies the resource name and access mode. In practice, it depends on applications how the user makes an access request. The following outlines three different ways to make an access request:

- Access can be granted automatically after authentication without the user having to make a request. Access control and authentication are thus combined in one step.
- The application can be configured to display only resources for which the user is authorized. In this way, access control consists of the application showing resources for which the user is authorized and hiding resources for which the user is not authorized.
- The application may display, or allow the user to select, all resources that the application may offer, regardless of whether the user is authorized to access or not. Access control then happens when the user actually makes a request, as shown in Fig. 9.1.

The terms PAP, PDP and PEP represent abstract functions, and not necessarily specific software modules. These features are specified in the eXtensible Access Control Markup Language (XACML) standard published by OASIS. This standard can be used for the implementation of ABAC (Attribute-Based Access Control), which is an access control model described in Sect. 9.3 below.

The scenario for IAM in Fig. 9.1 is relatively simple, and scales poorly for accessing services on the Internet. Identity federation has become a central component of modern IAM architectures, as described in the next section. Advanced models for access management are described in Sect. 9.3.

9.2 Identity Management

Identity management consists of registering identities and authenticating users based on their identities. There are different models for identity management, where the silo model is a classic and old-fashioned model, while federated models are modern in terms of increased usability and flexibility. Since it's helpful to have a clear understanding of what identity means, this is described in the next section.

9.2.1 Identity

In everyday speech, the term "identity" can have multiple meanings. We would like to have a precise definition of identity in order to describe IAM in a consistent manner. We must start from entities that have a physical, legal or procedural existence. *Entities* can be people, organizations, systems, or even system processes in the memory of a system. Such an entity may be given an identity that makes it possible to refer to the entity. Etymologically, the term "identity" means "the same one as last time", which precisely means that we can recognize the entity based on an identity that is already known. More specifically, an identity is just an amalgamation of attributes that describe or are assigned to the entity. A typical attribute is a name, but it can also be a date of birth, address, or almost any piece of information related to an entity. Digital identity is simply the representation of an entity's identity attributes in electronic form. The digital identity of an entity (e.g. person or organization) makes it possible to reliably recognise the entity during online interactions. Concepts related to identity are illustrated in Fig. 9.2.

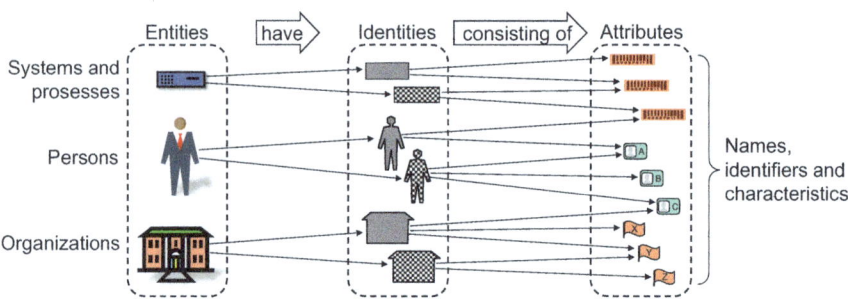

Fig. 9.2 Correlation between entity, identity, and attributes

Different entities and different identities can share common attributes. For example, two people or a company and a person may have the same address. However, two persons each have their own national identity number, while a company has a unique enterprise number for tax purposes, which makes it possible to clearly distinguish the entities from each other.

It is common for each identity to have a unique attribute that represents an identifier or database key within a namespace. A social security identity number is an unambiguous identifier within the namespace of country with a nationwide social security scheme. A common name, on the other hand, is not necessarily unambiguous within the namespace of personal names, since, for example, there are many people named John Smith. User accounts always have a unique name within the domain where an account is registered. In everyday speech, it is common to call such an unambiguous name a user ID. Along with the user ID, additional attributes can be recorded, such as common name, address, etc., and the identity itself consists of the entire set of attributes. Hence, an identity domain consists of identities registered with a unique user ID and other attributes.

An entity can have multiple identities, with each identity consisting of a separate set of attributes. It is possible to have multiple identities within the same domain if each identity has a separate user ID, but this is unusual and is often considered to be problematic from a security perspective. On the other hand, it is very common to have several different identities spread over different domains, such as when using different online services.

A namespace is said to be centralized when managed by a single entity, such as the namespace of user IDs of all employees of a company or the namespace of social security numbers managed by a country. A namespace is said to be distributed when managed by different entities in common. Phone number and e-mail namespaces are distributed because they are managed jointly by different entities.

An Identity Provider (IdP) is an authentication server that maintains a register of user IDs for all users in a domain and performs user authentication. Note that an IdP does not necessarily manage the namespace of user IDs. The term "IdP" can be confusing because it sounds like "identity generator", whereas in reality an IdP only performs user authentication. For example, in the Nordics where the national identity number is the user ID for access to e-Government services, the governments generate user IDs, while the commercial BankID companies perform user authentication. In this case, the BankID companies are the IdPs, while the governments generate the user IDs (national identity numbers) and manage the user ID namespaces. So even if the governments provide identities, they are not IdPs according to the IAM jargon.

The combination of different namespaces and IdPs forms the basis for designing siloed and federated identity management models, which are described in the following sections.

9.2.2 The Silo Model for Identity Management

> The silo model for identity management means that the service provider performs authentication of its users.

In the classic silo identity management model, each SP (Service Provider) maintains a register of its own user identities and performs authentication of these. In other words, the SP and the IdP (Identity Provider, aka. The authentication server) are part of the same organization. In this way, each separate organization creates its own silo for registering and authenticating its users, where different silos are isolated from each other. A user who wants to access services from multiple organizations must create a user account with a separate pair of user ID and authenticator in each silo domain, as shown in Fig. 9.3.

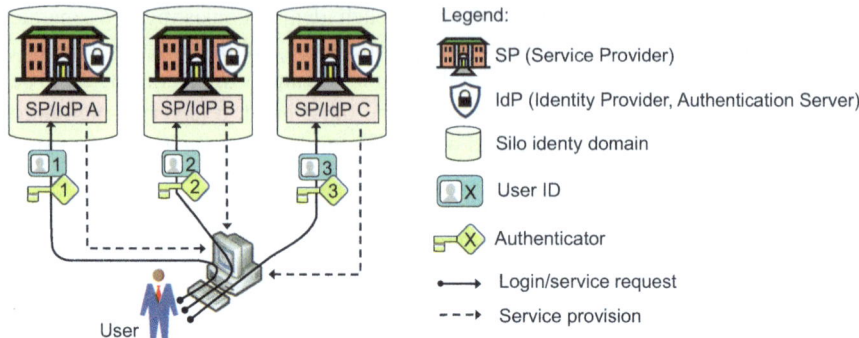

Fig. 9.3 Use-case with silo model for identity management

An advantage of the silo model from the user's perspective is that it provides good privacy protection as a result of silo domains being isolated. In this way, the exchange of personal information between domains is nut supported by the architecture. From the perspective of service providers, this is a disadvantage because they often want to exchange information about users, which of course must always be done in a legal manner with consent and other legal basis, etc. according to the GDPR.

The silo model is also easy to implement for service providers, but provides poor usability. Before the Internet era– in a world without online services—the silo model was the only sensible thing. It eventually turned out that the silo model scaled poorly to the global Internet. In the 2000s, many new online services were established that required login. Initially, these services were implemented with the silo model, which required users to create a separate pair of identity and password for each service, resulting in identity overload and password fatigue for users.

Research showed that users who were actually interested in using an online service often gave up because they simply could not tolerate creating another identity with yet another password.

While the silo model is still justified and sensible in many contexts, federated identity management has become prevalent on the Internet during the last 10 years.

9.2.3 Federated Identity Management

> The federated model for identity management means that the same user ID is recognised by multiple SPs and that user authentication is performed by separate IdPs.

Federated identity management is highly flexible in the sense that modular components can be used to build different architectures that are adapted to the application and business process. In federated identity management, identity domain

9.2 Identity Management

management and user authentication are distributed functions that multiple SPs can utilize. Note that the term "relying party" is sometimes used with the meaning of SP (service provider), in the sense that the SP "relies" on the IdP performing correct user authentication. Of course, keeping a register of user IDs and performing user authentication comes at a cost, so IdPs are often commercial services. For example, it costs money to have a phone number even if you never call, text, or have a data plan. You simply pay to have an identity in the identity domain of phone numbers provided by a telecom operator, which with country codes is a global identity domain. On the other hand, it is free for citizens to have a national identity number, except that you pay tax. SPs that get their users authenticated with commercial IdPs might pay between a few cents up to a dollar for each authentication instance. Internet services that authenticate users using federated authentication from Facebook, Google, or X also pay every time a user is authenticated—not in money, but by collecting information about us users. Identity federation not only improves usability, there are also major business interests behind the development of such services.

For example, a relatively simple architecture is a centralized federation where an IdP authenticates all users for multiple SPs. After authentication, the IdP will send an identity token to other service providers telling them that a user with a certain user ID has been authenticated so that they can allow the user to access the service. Figure 9.4 shows a simple use case where a user authenticates to a centralized IdP and then accesses services from both SP A and SP B.

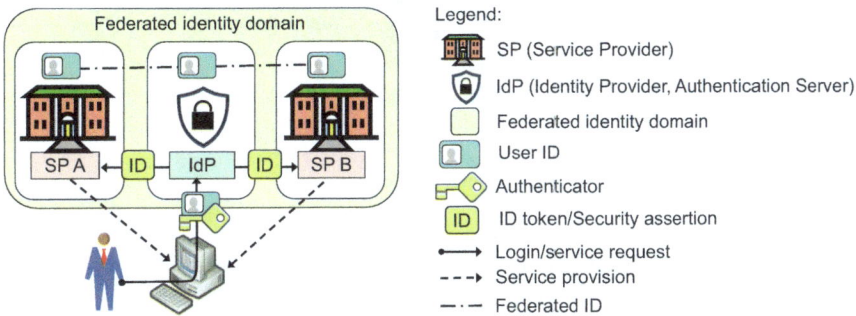

Fig. 9.4 Simple use-case with federated identity management

For SP A and SP B to accept and verify the ID token from the IdP in Fig. 9.4, They must have established a technical and business relationship in advance, such as the exchange of cryptographic key certificates and the establishment of legal agreements. This is a prerequisite for all identity federations, which creates a certain complexity.

ID federation is akin to SSO (Single Sign-On), which means that a single login is sufficient for access to many different services from different providers. Kerberos is the name of a technology for SSO developed in the 1980s at MIT in the United States that is still in widespread use. Kerberos is based on a centralized AS (Authentication Service), a KDC (Key Distribution Centre) and a TGS (Ticket Granting Service). When the user authenticates to AS, the user will then implicitly be authenticated for other services internally in the computer network using cryptographic keys and "tickets" that the services receive and verify automatically. However, Kerberos is limited to a single computer network, and does not scale to the Internet. By considering the federated domain in Fig. 9.4 as a computer network at an organization, the scenario is similar to a use-case based on Kerberos, where IdP corresponds to AS/TGS in Kerberos and an ID token corresponds to a Kerberos ticket.

9.2.4 ID Federation Protocols

Entities that interact in identity federation are user/client, service provider (SP), authentication server (IdP), and sometimes also an ID portal as an intermediary.

Communication between these entities is based on a cryptographic security protocol where the main functions are that the IdP authenticates the user and then sends an ID token to SP, which verifies the token before the user receives the service. A simplified scenario which includes a portal is shown in Fig. 9.5.

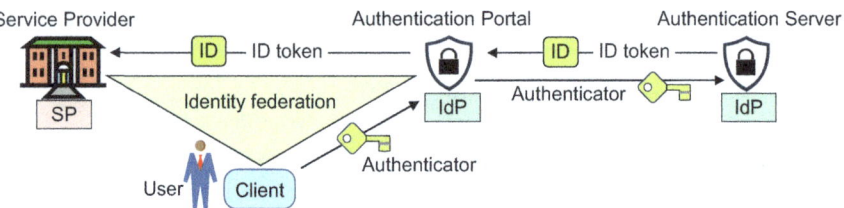

Fig. 9.5 Roles and message exchange in identity federation

Both the authentication portal and the authentication server in Fig. 9.5 are considered IdPs. An example of an authentication portal is *Eduroam* which is an international identity federation in the education sector for providing Wi-Fi access. Note that an authentication portal is not always used.

Until about 2018, SAML (Security Assertion Markup Language) was the most widely used standard/protocol for identity federation worldwide. SAML is increasingly being replaced by the *OpenId Connect (OIDC)* standard, described below. SAML and OIDC are not compatible, so all players within a federation must change technology at the same time to change the federation standard from SAML to OIDC, which can be challenging when many players are involved. Another standard from the early 2000s is WS-Federation, which gained little traction despite being developed in conjunction by the prominent companies Microsoft and IBM.

The different standards for identity federation specify different message formats and methods (profiles) for message exchange and application, which means that

9.2 Identity Management

they are not compatible. All entities in a federation must therefore have installed application modules based on the same standard, otherwise they cannot talk to each other. Different standards form a technical obstacle to integrating different identity federations. Although two identity federations may be based on the same standard, many political, legal and economic constraints also come into play, which means that different identity federations are usually separate.

> eID is a federated authentication scheme for use in e-government.

Many governments around the world have established eID schemes for access to e-Government services. Note that these schemes are quite different from the identity federations on the Internet. Access to e-Government services through a national eID provider and access to online services through the Facebook IdP, for example, are two separate identity federations, although both can be built with OIDC. You cannot submit your tax declaration by logging in with Facebook, and we cannot rent an apartment on Airbnb by logging in through the national IdP.

9.2.5 OpenID Connect

OpenID Connect (OIDC) is part of OAuth, described in Sect. 9.4, or more precisely: OpenID is a specific method ("scope") when using OAuth for user authentication. Identity federations based on OIDC are, for example, online social networks such as Facebook, Apple and Google, where these have the role of IdPs. OAuth uses the general term "authorization server", which in the case of OIDC is an IdP (authentication server). A key element of the OIDC protocol is the *ID token*, which is a cryptographic message and confirmation that the user is authenticated. Message exchange in the OIDC authentication protocol is shown in Fig. 9.6.

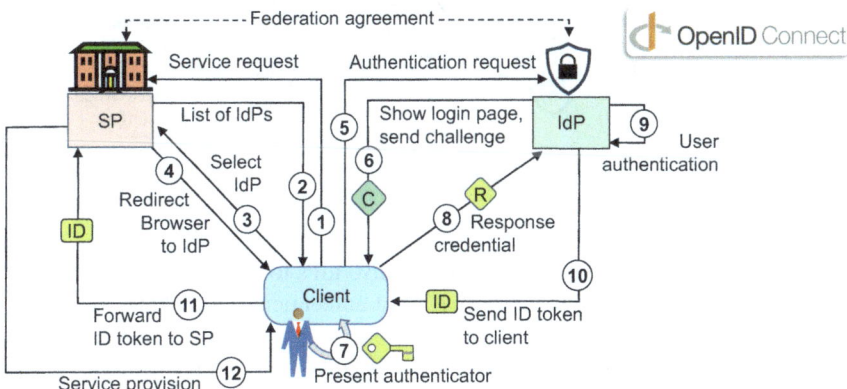

Fig. 9.6 Federated authentication with OpenID connect

Steps in the scenario in Fig. 9.6 is described in the figure, and therefore does not need a separate explanation. OIDC specifies different types of message flows, where Fig. 9.6 shows so-called *Implicit message flow*. Message (5) in Fig. 9.6 contains "scope = OpenID", which causes the IdP to generate an ID token in message (10).

When comparing the scenario for user authentication based on OpenID Connect in Fig. 9.6 With the user authentication scenario based on the silo model in Fig. 9.3, we see that identity federation entails remarkably greater complexity. However, this complexity is transparent for us users, and we only experience increased usability by easily accessing many different services with just one pair of user ID and authenticator.

In addition to user authentication, policies and models for access control are naturally needed for access between different organizations. This is a challenging topic—not only technically, but also legally and politically. Access control is described in Sect. 9.3 below.

9.2.6 Indian eID and e-Pramaan Federation

An IdP used in e-Government is called an eID provider. For example, the Indian Aadhaar is the national eID provider operated by the UIDAI (Unique ID Authority of Inda). In addition, the Indian Income Tax Department authenticates users based on their PAN (Permanent Account Number) which is the Indian tax number, and the RTOs (Regional Transport Offices) authenticate users based on their driver license. Figure 9.7 shows a simplified scenarios for e-Pramaan, which lets users get access to e-Government services such as health, education, tax and ordering passports.

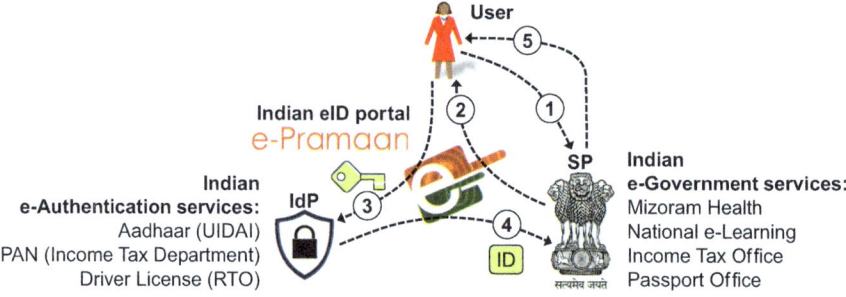

Fig. 9.7 The Indian e-Pramaan identity federation

1. The user requests a service from a SP.
2. The SP displays login window with alternative IdPs to the user.
3. The user selects an IdP, enters user ID and authenticator(s), which are sent to the IdP.
4. The IdP authenticates the user and sends ID token and user attributes via the portal to the SP.
5. The SP checks the ID token and provides the requested service according to policy.

9.2.7 European eID and Identity Federation

eIDAS 2.0 mandates that each member state in the EU/EEA shall establish a national eID scheme, and many EU member states have done so already. Some states have established a government operated authentication service such as in Germany, while other states rely on commercial authentication service providers such as in the Nordics.

eIDAS also mandates that the national eID schemes shall be interoperable in the form of a super federation, in other words a European federation of national identity federations. As of 2024, this super federation still does not exist. Anyway, the idea is that European citizens shall be able to access e-Government services when they are visiting other European countries, such as medical services and for making tax declarations to local tax authorities. Figure 9.8 illustrates a simplified architecture of the EU federation of identity federations, where a citizen from state A is requesting a service while visiting state B.

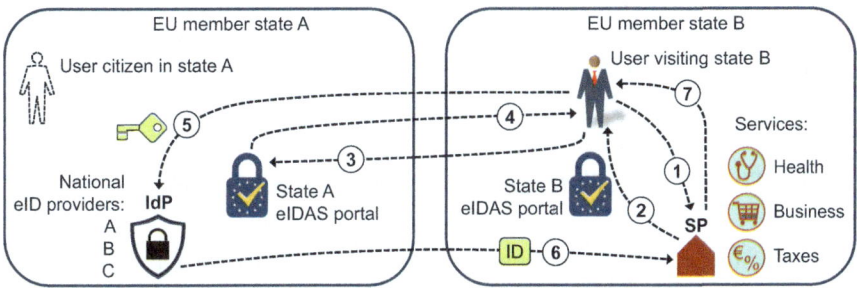

Fig. 9.8 Use-case of the European eIDAS federation of national identity federations

1. The user requests a service from a SP.
2. The SP displays an eIDAS portal window for selecting the user's EU/EEA state of citizenship.
3. The user selects a state, and is forwarded to the national eIDAS portal
4. The national eIDAS portal displays a window for the user to select an IdP.
5. The user selects an IdP, enters user ID and authenticator(s), which are sent to the IdP.
6. The IdP authenticates the user and sends ID token and user attributes via the portal to the SP.
7. The SP checks the ID token and provides the requested service according to policy.

In an ideal world, every person is a citizen in a single state, i.e. no dual citizenships. However, more and more people have multiple citizenships, and can obtain an eID from multiple states. The ability to act with different identities in different situations opens up many possibilities for fraud. Hence, it would be in the interest of European authorities to map every eID belonging to the same person. However, governments do not usually share the list of citizens with each other, and even of they tried, it would be difficult to know which identities correspond to the same person. From a privacy perspective, it would be desirable not to link a person's various eIDs. Thus, there will be a political debate about whether different eIDs for a person should be linked, and whether such linking will even be possible.

9.2.8 eID in the USA

Paradoxically, despite having a first-class e-Authentication framework in the form of NIST SP 800-63-4 Digital Identity Guidelines (see Sect. 8.7.1), the landscape of real implementations of eID and identity federation for e-Government in the USA is immature. As of 2024, the U.S. federal government does not issue general purpose eID and digital credentials for accessing e-Government and other services. Instead, each state must issue their own eID, and in practice, most implementations are simply a state-issued mobile app for the driver license, called mDL. However, many states do not even have mDLs [35]. Unfortunately, the mDLs typically no not provide high authentication assurance levels, resulting in fraud being committed based on abusing the mDLs.[2]

9.2.9 The Identity Federations Facebook, Twitter, Google etc.

More and more Identity federations are based on OAuth and the OpenID Connect standard. On the Internet for example, Google, Apple, facebook and Microsoft have established themselves as IdPs for accessing many types of SP. Figure 9.9 illustrates a simplified identity federation scenario for accessing services on the Internet based on the above mentioned IdPs. It is common for an SP to register users with an email address as their user ID, and allow users to choose which IdP they want to use for authentication. It is of course a prerequisite that the same e-mail address is specified as user ID at both the SP and IdP.

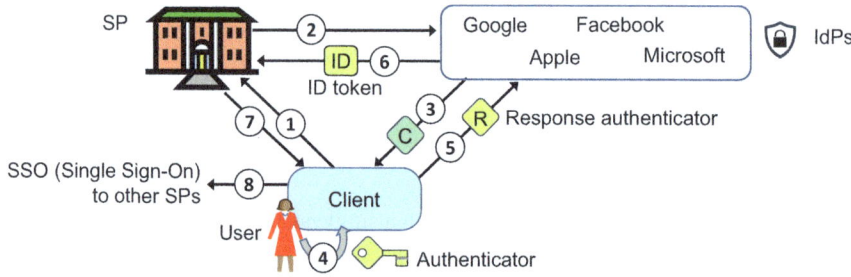

Fig. 9.9 Simplified identity federation scenario based on OpenID connect

---

[2] https://www.biometricupdate.com/202303/banks-hit-with-biometric-fraud-fake-mobile-drivers-licenses

9.2 Identity Management

The message exchange scenario in Fig. 9.9 is described below. This is a simplified version of the scenario in Fig. 9.6 where some messages have been omitted below.

1. The user requests a service from the SP.
2. The SP forwards the browser to the IdP for authentication.
3. The IdP displays a login window (optionally with a passkey challenge).
4. The user presents authenticator(s) to the client (optionally with biometrics).
5. The authenticator(s) is/are sent to the IdP (optionally with passkey response).
6. The IdP verifies the authenticator(s) and sends the ID token to the SP (via the client).
7. The SP checks the ID token and provides the requested service according to policy.

The authentication levels of IdPs on the Internet do not follow the guidelines of, for example, eIDAS or NISTs *SP 800-63-4 Digital Identity Guidelines*. Nor are they considered eID providers because they are not used for authentication to e-Government services. However, these IdPs offer 2-factor authentication (2FA) through secondary channels such as mobile apps or SMS authentication, and they even offer authentication with passkeys, as described in Sect. 8.4.2.

9.2.10 Categorization of Identity Federation

Identity federation is very diverse, as shown in the description of various standards and architectures in the preceding sections. It can therefore be challenging to provide a logical categorisation of identity federations. Nevertheless, below we make an attempt to provide a general categorization of identity federations.

We consider two aspects:

1. Whether the user-ID namespace is managed by a central entity or by many entities in a distributed manner.
2. Whether the federation has only one centralized IdP for each user ID, or whether the IdP role that performs authentication of a user ID is distributed across multiple entities.

We have talked a lot about the role of IdP in previous sections, but little about the role of namespace custodian and its importance, so such a discussion might fit here.

The namespace for user IDs defines the amount of possible user IDs. It is important to be aware of the legal and commercial status of a namespace. States typically define a namespace for national identity numbers with a limited set of different names (national identity number). Such a national identity number is thus part of a centralised namespace, precisely because it is managed by a single national authority.

The e-mail address defines a namespace with an unlimited number of different names (e-mail addresses) that is managed by all the entities that own a domain name on the Internet. The e-mail address is thus a distributed namespace, precisely because it is managed by many different entities.

When establishing an identity federation, a namespace that is convenient for its application must be selected. Sometimes a centralized namespace fits best, while other times a distributed namespace fits best. It must also be decided whether it is appropriate to have a centralized authentication service or whether it is appropriate to have different and thus distributed IdPs for authentication. The combination of centralized or distributed namespace, and centralized or distributed authentication, forms four categories as shown in Fig. 9.10, where examples of prominent identity federations are assigned to the four categories.

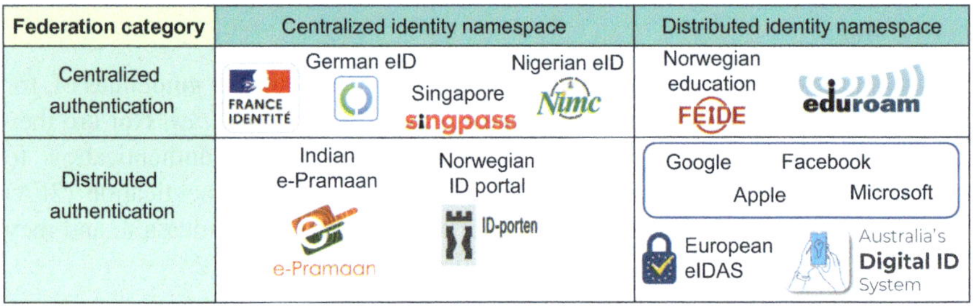

Federation category	Centralized identity namespace		Distributed identity namespace
Centralized authentication	German eID FRANCE IDENTITÉ	Nigerian eID Singapore singpass Nimc	Norwegian education FEIDE eduroam
Distributed authentication	Indian e-Pramaan e-Pramaan	Norwegian ID portal ID-porten	Google Facebook Apple Microsoft European eIDAS Australia's Digital ID System

Fig. 9.10 Categories of identity federation, with examples

Below is an explanation for why each identity federation shown in Fig. 9.10 is assigned to a specific federation category.

France Identité is the French government-issued eID which in practice requires an app on the user's smartphone to be used. The eID namespace and authentication are both centralized. To initialize the app and create an eID, users need to already have obtained a physical national ID card so the eID and ID card can be linked. The German physical national ID card serves as an eID for access to e-government and a set of private services, and hence no app is needed. The ID card connects to the smartphone by holding the card against the phone so that they can communicate via NFC (Near Field Communication). The authentication protocol is similar to that of Passkeys described in Sect. 8.4.2. eID in Germany has an identifier (ID card number) consisting of 9 characters (letters and numbers) in a centralized namespace, and has *Bundesportal.de* as centralized state IdP (eID provider). Singpass in Singapore and the Nigerian eID are similar centralised eID schemes which a government controlled IdP.

FEIDE is the identity federation of the Norwegian education sector. Eduroam is the international education sector federation for letting staff and student get Wi-Fi access while visiting other educational institutions. FIEDE and Eduroam have distributed namespaces with user IDs defined by each educational institution. In Norway, user IDs for students, pupils and employees in Norwegian educational institutions are registered in distributed namespaces. In Eduroam, a user ID is defined in the same way as in FEIDE, i.e. by each educational institution, but globally and not only nationally. Authentication is centralized in both FEIDE and

9.2 Identity Management

Eduroam, because a user can only be authenticated by the educational institution where the user ID is registered. However, it could be possible for educational institutions to outsource the authentication service to various actors, which would thus make the authentication distributed.

e-Pramaan, described in Sect. 9.2.6, is the Indian identity federation, where "pramaan" is the Hindu word for "proof". There are multiple IdPs, hence the authentication is distributed. *Aadhaar* (from Hindu: "basis") is the main IdP in e-Pramaan. The Aadhaar identity is a twelve-digit unique identity number for residents of India, which is stored in a profile along with the person's biometric and demographic data. The data is collected by UIDAI (Unique Identification Authority of India), a statutory authority set up in January 2009 by the Government of India. Aadhaar is the world's largest biometric ID system, with over a billion registered people in a centralized database.

The Norwegian ID portal uses multiple IdP's for user authentication, hence the authentication is distributed. The namespace for the ID portal is the national identity number which is managed by the Norwegian Tax Administration, and hence is a centralised namespaces.

The identity federation of social networks, where companies such as Google, Apple, Facebook, and Microsoft play the role of IdPs, can be categorized as having a distributed namespace because an SP allows users to register with all kinds of email addresses and phone numbers from different domains and countries. Users who use any e-mail address or phone number to access Internet services can choose different IdP's that are thus distributed.

With the eIDAS 2 regulation of 2024, the European Commission is planning to establish a European identity federation by 2026, as described in Sect. 9.2.7. We can assume that such an identity will be based on a distributed namespace of user IDs, where different countries each manage their own subset of the namespace. We can also assume that different authentication services will be able to perform user authentication within each country's namespace. Hence, the eIDAS will be a federation of distributed namespaces and distributed authentication.

The Australian Digital ID System follows the federal Trusted Digital Identity Framework (TDIF), where both government and private IdPs can be accredited to provide authentication services. Authentication is thus distributed in the Australian Digital ID System. The identity namespace is also distributed, where TDIF states that *"No single identifier is issued by the Identity Exchange to Identity Service Providers, Attribute Service Providers or Relying Parties."* [37] However, a user can link their various IDs from different relying parties, so that the distributed name space in practice has characteristics of a centralised name space. The term "relying party" has the same meaning as "SP" (service provider).

In some countries or federations, such as the UK and USA, the political and public opinion is such that a national or federal eID is seen as an unacceptable threat to privacy. In those regions, solutions for eID are typically fragmented across states (USA) and across sectors (UK).

9.3 Access Control

The fact that a user is correctly authenticated within a computer network or by a service provider does not mean that the user therefore should be able to access to all resources managed within the computer network or by the service provider. Access control is the security function that ensures that users can only access resources to which they have been authorized beforehand. The following sections describe methods and models for controlling access to resources.

In access control terminology the *subject* is the entity who requests access, and the *object* is the resource to which the subject wants access. An access request means, for example, that the subject attempts to open a file to read/edit, open a directory to see what it contains, or to launch/execute an application. The subject is typically a user, but can also be a system or a process. It is often relevant that a request for access to a resource also specifies *access mode*, such as read, write or execute, but this is often implicit. It is assumed that access authorization has already been specified during the configuration phase, as shown in Fig. 9.1 in Sect. 9.1 and in Fig. 1.13 in Sect. 1.11. When the subject makes an access request, the access control function in Fig. 9.1 checks relevant access authorization attributes and decides on whether to allow or deny access.

Access control models are traditionally called "security models", which date from the 1970s, when access control was the most important security feature in computer systems. Classic models are DAC (Discretionary Access Control), MAC (Mandatory Access Control) and RBAC (Role-Based Access Control), all of which are still widespread. However, these models scale relatively poorly to modern business processes with identity federation, distributed access control and extensive interaction between different organizations. A more modern and flexible model for access control is therefore ABAC (Attribute-Based Access Control), which is essentially a generalization of all the previously mentioned models. In addition, there are architectures for distributed access control in open environments. These models are described in the following sections.

9.3.1 DAC: Name-Based Access Control

DAC (Discretionary Access Control) is a model of access control based on subject and object names. It generally means that rules of access are expressed on the form *"subject X has access to object Y with access type Z"*. Such access authorizations rules can be called access policies. Figure 9.11 shows a simple example with an ACL (Access Control List) that specifies access authorizations for users. An important question is whether the rule should be stored as part of the subject profile, as part of the object profile, or separately. For example, the rules can be stored as part of the object profile, that is, as metadata for a file or domain. The rules can also be stored in the subject profile for users, e.g. by specifying which domains a user belongs to and should have access to.

9.3 Access Control

	Access Control List	HR department	Service department
Subject names	Alice in Wonderland	read, write	
	Bob the Builder		read, write

Object names span HR department and Service department columns.

Fig. 9.11 DAC: Name-based access control

In Windows and Linux, DAC is implemented as ACL (Access Control List). Each object has an ACL that specifies which subjects are authorized for access and with which access mode. However, it is not possible to specify individual subjects, only three categories: *owner*, *group* and *others*. With this granularity, the owner can specify access rights for themselves, for a group of subjects, and for all subjects. Individual subjects can be added to each group. For each object, access rights can be specified for each category with one or more of the access modes *read*, *write* and *execute*. In a similar way, access rights to directories can be specified.

The term "discretionary" in DAC reflects that the owner of the file can specify at "their discretion" who should have access to the file. The term DAC originated in the 1980s, when DAC access models were first implemented in systems. All operating systems support DAC whcih is the most widely used model for access control in both the civilian and military sectors.

9.3.2 MAC: Label Access Control

MAC (Mandatory Access Control) is a model for access control based on security labels on subjects (users), and on objects (data and services). For users, the label is e.g. a security clearance, and for object the label is e.g. a security classification. For example, the NATO hierarchy of security clearance and classifications are the following:

TOP SECRET
SECRET
CONFIDENTIAL
RESTRICTED

Documents that are not classified are called UNCLASSIFIED, which is not strictly a classification, but is often applied to documents in environments where a high proportion of classified documents are otherwise handled to mark a clear distinction between classified and unclassified information. There are many different grading scales that are specified in other legislation or by other countries.

Access control based on MAC generally means that the access control feature compares the subject label to the object label to make an access decision. The term "mandatory" in MAC reflects that the owner of the file cannot specify at their discretion who should have access to the file—it is the labels that dictate who shall have access. The MAC designation originated in the 1980s, when MAC access models were first implemented in systems.

The most common model, or set of rules, for comparing subject and object labels is called *Bell-LaPadula* after authors David Elliott Bell and Leonard J. LaPadula, who first described this model [38]. Bell-LaPadula describes a set of different rules, but the two most important are called "no-read-up" and "no-write-down", as illustrated in Fig. 9.12. Both rules are illustrated with a subject who has security clearance SECRET and a set of objects with different classifications.

Fig. 9.12 MAC access rules from the Bell-LaPadula model

The "no-read-up" rule, illustrated on the left in Fig. 9.12, is intuitively easy to understand. If a person has clearance SECRET, then that person can read documents with the same or a lower classification, but not documents with a higher classification. This rule prevents the leakage of highly classified information to personnel with lower security clearance.

The "no-write-down" rule is not as intuitive and needs an explanation. The logic is that a person with a high security clearance might have highly classified information in their head. If this person could write the information from memory to documents with a low security classification, then other personnel with a lower security clearance could read it, which would cause the leakage of highly classified information to low-security clearance personnel.

In practice, it is not what information a user has in their head that is relevant to "no-write-down", but what other documents the user has opened simultaneously in an active logged-in session. "No write-down" is intended to prevent the user from copying information from a highly classified document to a low-rated document in a session on a workstation.

If a user wants to both read and write simultaneously on different documents, then the user and all documents must have the same security clearance and classification level. However, a user with the highest security clearance TOP SECRET must be able to work with all documents classified at the same level or lower. This is done by defining a Subject-Max label that is equal to the security clearance, and a Subject-Current label that the user selects at login, where Subject-Current can

9.3 Access Control

have the same level or lower level than Subject-Max. This principle is illustrated in Fig. 9.13, where the user has security clearance (and Subject-Max label) SECRET, but is logged in with Subject-Current label CONFIDENTIAL. This allows the user to work with (read and write) documents classified CONFIDENTIAL during that session.

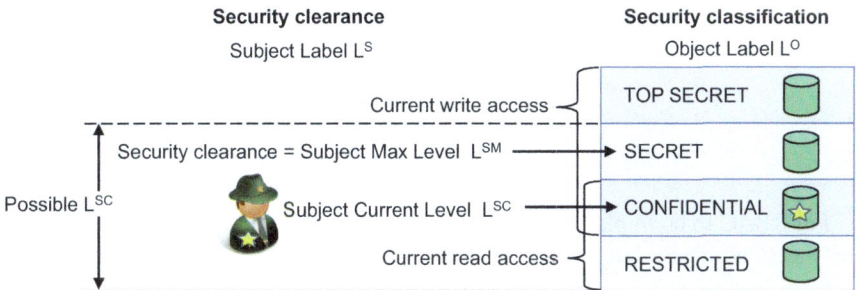

Fig. 9.13 Practical use of Bell-LaPadula by having the user select subject-current-label for a session

Access control based on MAC/Bell-LaPadula is typically used in military systems where it is common to process classified documents.

Another important principle of access control is "need-to-know", i.e. that a user should only be authorized to access documents they need for their work, regardless of the user's security clearance. DAC can be used to implement need-to-know. In that way employees can be authorized to access different domains of data and services in the business according to DAC, while at the same time being restricted by MAC. The next section describes RBAC, which is another principle that can also be used together with MAC and DAC.

9.3.3 RBAC: Role-Based Access Control

Managing access rights requires a lot of work in organizations. When employees change their job function, old access rights must be removed, and new ones added. A natural approach for managing this is that access rights are defined for roles and not for individual users. It also specifies possible roles that a user is allowed to take at logon, which then allows the user during that session to automatically have all access rights defined for the role. This access control model is called *RBAC (Role-Based Access Control)*. Figure 9.14 illustrates a simple example with access rights based on RBAC.

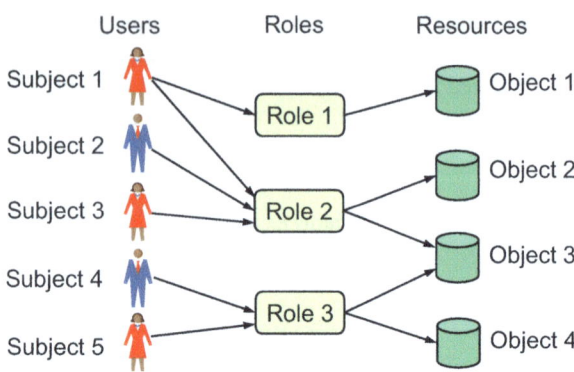

Fig. 9.14 RBAC—role-based access control

For RBAC to be implemented efficiently, roles must be defined to cover typical job functions. Thus, defining a user's right to assume has the effect of defining the user's access rights to resources. RBAC is one possible way to implement need-to-know in organizations. Restrictions can be placed on whether a user can only take on a single role in a session, or whether it should be possible to take on multiple roles at the same time.

RBAC has its advantages, but in practice it often turns out that it is necessary to define a large number of roles and that maintaining access rights for roles requires considerable work.

It is interesting to note that MAC can be implemented with RBAC in the sense that a role is defined for each security level of TOP SECRET, SECRET, CONFIDENTIAL, and RESTRICTED, and that a user can select a role that corresponds to the user's clearance level or lower. The selected role for a session is thus specified by the subject-current-label according to the MAC/Bell-LaPadula mode described in Sec. 9.3.2.

9.4 OAuth and Distributed Access Management

OAuth is an open standard that describes how websites or applications can access and exchange resources among themselves by the resource owner giving consent and delegating access rights in a secure manner. The first version of the OAuth protocol was published by the IETF in 2010.

In OAuth, the security token can represent different things (have different scopes), which for access management is an access token. When OAuth is used for authentication, the meaning ("scope") of the security token is defined as the ID token, in which case the protocol is OIDC (OpenID Connect), as described in Sect. 9.2.5. In that sense, the OIDC protocol is just a specific application of the OAuth protocol.

9.4 OAuth and Distributed Access Management

There are various scenarios where OAuth supports distributed access control. The next sections first describe a scenario with online social networks, and then a scenario of interaction for access to patient records between healthcare institutions.

9.4.1 ABAC: Attribute-Based Access Control

Many different aspects may be relevant for specifying access rights. For example, the DAC, MAC, and RBAC models described above focus on names, labels, and roles. In addition, it may be relevant to include context such as geographical location, time of day, weekdays/weekends, emergency level and operational status. By considering all these things as *attributes*, access control rules can be defined to take into account many different types of attributes, which thus become ABAC (Attribute-Based Access Control).

ABAC allows organizations to specify access authorizations based on all kinds of attributes related to subject, object, access mode, context, and the policy (rules) itself that apply to an access request. Thus, the main categories of attributes are request attributes, subject attributes, object attributes, context attributes. The AC policies can be thought of as the rules for interpreting the different attributes to make a decision about access. Systems can automatically make and enforce access decisions based on these attributes. Figure 9.15 illustrates a simple ABAC example where access is based on the attributes and policies mentioned above.

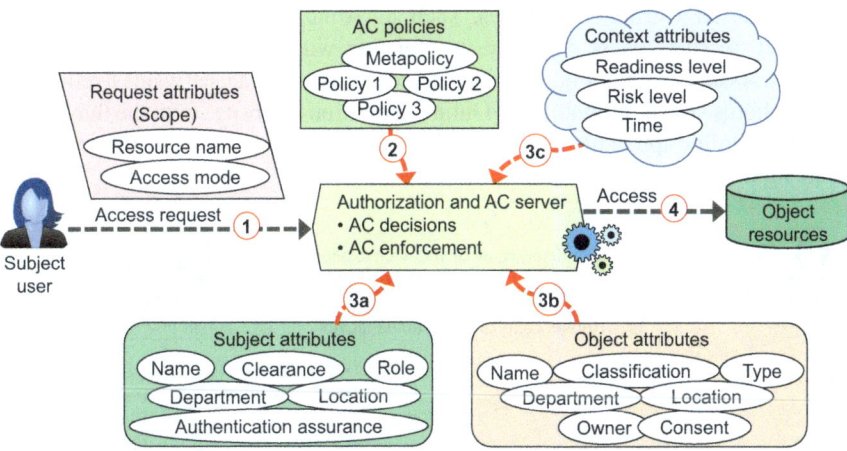

Fig. 9.15 ABAC: attribute-based access control—simple scenario

Note that ABAC can be used to implement DAC, MAC and RBAC, which means that ABAC is a general and flexible model for access control. DAC can be implemented with ABAC in that both subject and object attributes can be names, and the AC policy is simply an ACL (Access Control List). MAC can be implemented with

ABAC in that subject and object attributes can be labels that specify clearance and classification levels, and the policy is simply Bell-LaPadula. RBAC can be implemented by defining a set of roles as intermediaries consisting of both a role object that user subjects have access to and a role subject that has access to object resources, so that subject users' access to object resources goes indirectly via the intermediary role.

Practical solutions based on ABAC have increased rapidly since 2010. Together, ABAC and federated identity management form the foundation for flexible and scalable solutions for access management and collaboration in distributed environments.

In a distributed infrastructure, it may be necessary to obtain attributes from different organizations. A typical example is NHN (Norwegian Health Network), where an employee in a healthcare institution can make requests for access to patient data from another healthcare institution. First, the healthcare worker must be authenticated through the HealthID federation and then make a request for access. Next, an authorization server must assemble relevant attributes and policies as a basis for deciding to grant or deny access to the resource. A possible architecture for distributed access control in this type of environment is typically based on OAuth, which is described in the next section.

9.4.2 OAuth for Online Services

OAuth is used by major websites, such as Amazon, Google, Facebook, and Microsoft, to share users' resources/information with third party applications or websites. As an example, Fig. 9.16 shows a scenario where a user wants to edit their private photos stored in PhotoBank Online (a fictitious website) with the third-party PhotoEdit Online app (a fictitious app). The numbered arrows indicate the sequence, so the figure is self-explanatory.

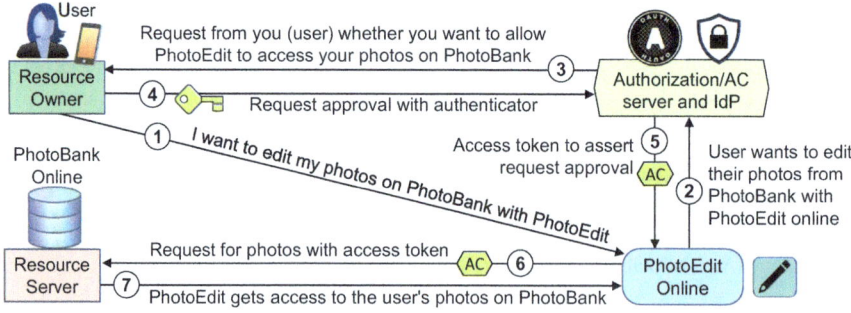

Fig. 9.16 OAuth based distributed access control for sharing information between online websites

9.4 OAuth and Distributed Access Management

The motivation behind the specification of OAuth was precisely the type of scenarios shown in Fig. 9.16, because it is a flexible protocol that forms the basis for business processes and business models on the Internet.

At the same time, OAuth can be very invasive with respect to our personal data. Although we technically give consent through OAuth for Internet service providers to exchange data about us, most of us fail to grasp the extent of what we are giving consent for, and we typically fail to understand that this can also be a way for Big Tech companies to collect personal data. Tracking with personal data is described in Sect. 10.3.

9.4.3 OAuth for Access and Interaction Between Healthcare Institutions

OAuth is applicable in all kinds of sectors, not just commercial Internet services. When a patients is being treated by a healthcare provider, for example, a challenge can be that critical patient data may be stored in different healthcare institutions, e.g. GPs, hospitals or pathology laboratories. The exchange of patient data is subject to strict control, and often requires consent. OAuth and OIDC can be used for secure exchange of patient data.

Figure 9.17 shows a possible way in which OAuth can be used in the Norwegian Health Network (NHN) for distributed access control between healthcare institutions. NHN is a Norwegian state-owned enterprise established and owned by the Ministry of Health with the task of developing and operating national ICT infrastructure for interaction between institutions in the healthcare sector. The numbered arrows show the sequence of messages, so the figure is self-explanatory.

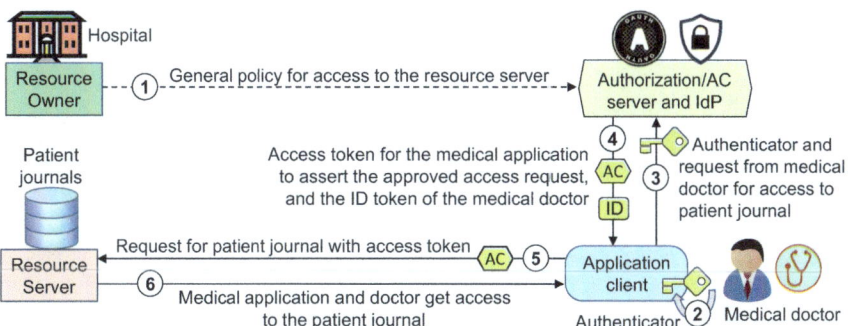

Fig. 9.17 OAuth for distributed access control and collaboration in the healthcare sector

A variation of the scenario in Fig. 9.17 is if the resource owner wants to make fine-grained access decisions for healthcare workers at each access request, rather than access decisions based on a general policy sent in message 1. In this case, the authorization server and the resource owner will exchange messages corresponding to messages 3 and 4 in Fig. 9.16.

9.5 Tasks

1. IAM concepts

    (a) Briefly explain the following concepts related to identity management:

    1. Entity
    2. Identity
    3. User ID (identifier)
    4. Digital identity
    5. Autenticator

    (b) Explain what is meant by the term IAM (Identity and Access Management).

2. Silo model and federated model

    (a) Describe the silo model for identity management.
    (b) Describe the advantages and disadvantages of the silo model.
    (c) In general, describe the federated model of identity management.
    (d) Describe the advantages and disadvantages of the federated model.

3. ABAC: Attribute-based access control

    Attribute-based Access Control (ABAC) is a flexible model for access control.
    (a) Name categories of attributes in ABAC.
    (b) Explain how DAC can be implemented with ABAC.
    (c) Explain how MAC can be implemented with ABAC.
    (d) Explain how RBAC can be implemented with ABAC.

Chapter 10
Information Privacy

> *If you're not paying for it, you're not the customer; you're the product being sold.*
>
> Andrew Lewis, technology and privacy advocate

Information privacy, aka. data privacy or data protection, is the protection of personal information from illegal and unethical collection and processing. Privacy is eroded as personal information is increasingly being shared across social networks, websites, and mobile apps. The social web does not provide free services, in fact we pay for it by sharing our personal information, or said differently, through the loss of our privacy. In turn, Big Tech social networks leverage this information for profit. As individuals, we have little power to protect our privacy in this arena, only regulation accompanied with strong enforcement can bring a degree of fairness. In fact, information privacy is regulated by various laws and regulations around the world, such as GDPR in the EU. The extraterritorial scope of GDPR means that it applies not only to organizations based in the EU but also to any organization that processes the personal information of EU citizens, regardless of the organization's location. This has led many non-EU countries to align their information privacy laws with GDPR principles to ensure compliance when processing information of EU citizens.

In this chapter, you first learn the need to strike a balance between digital surveillance and privacy. Next, you get an insight into the industry behind online tracking, with some advice on how to protect yourself. Furthermore, you learn about the legal protection of personal data through the European General Data Protection Regulation (GDPR). Finally, you learn when and how to perform a Data Protection Impact Assessment (DPIA), which is a requirement that organizations must follow in order to process personal data legally under the GDPR.

10.1 What Is the Difference Between Privacy and Information Privacy?

Information privacy is a subclass of the general concept of privacy. While there is no generally accepted definition of "privacy", it can be understood as the gist of Article 12 in the UN Universal Declaration of Human Rights and Article 8-1 of the European Convention for the Protection of Human Rights and Fundamental Freedoms:

> *No one shall be subjected to arbitrary interference with his privacy, family, home or correspondence, nor to attacks upon his honour and reputation. Everyone has the right to the protection of the law against such interference or attacks.*
> United Nations Universal Declaration of Human Rights, Article 12

> *Everyone has the right to respect for his private and family life, his home and his correspondence.*
> European Convention on Human Rights, Article 8-1

It is reasonable to interpret the word "correspondence" in both articles above as referring to "information privacy". In the Internet age, "correspondence" is how we live our digital lives by digitally representing and communicating personal information. According to the declaration of human rights thus, our digital lives have the right of protection from arbitrary interference. Information privacy covers aspects such as the ability to know and control when, how and how much information about one's own person is disseminated to other parties.

Other terms are being used for what we here refer to as information privacy. The term *data protection* is typically used in the legal context in the EU, and is for example the meaning of the letters "DP" in the abbreviation GDPR. *Digital privacy* and *data privacy* are other terms having more or less the same meaning.

10.2 Privacy in the Digital Age

The digitization of society leads to rapidly increasing generation and sharing of personal data. Previously, personal data was usually explicitly defined and relatively static categories such as address, school, workplace, health information and membership in associations. Now when almost all of our activities have a digital component, personal data is generated and stored dynamically and continuously, offering massive potential for monitoring.

First, let us simply assume that massive collection of personal data takes place. On the one hand, it can support new business processes, streamline public governance, and make our private and working lives much more practical. It also provides law enforcement with tremendous opportunities to investigate crime and enforce law and order. On the other hand, it can have serious negative consequences if private companies and government authorities decide to abuse the massive amounts of personal data. Possible negative consequences are that sensitive information about us gets collected and shared against our will, that we get unwanted attention, that we

10.2 Privacy in the Digital Age

are subject to discrimination and harassment, and that it gives authorities the opportunity for abuse of power and oppression that can threaten our freedom.

Let us then consider a situation where there is very limited collection of personal data. On the one hand, citizens may be left alone, avoid unwanted attention and to a lesser extent fear abuse of power and oppression by the authorities. On the other hand, it would also have serious negative consequences. For example, it would hinder the development of smart digital business processes, it would make public governance inefficient, and it would hinder effective criminal investigations. Criminals obviously have a great interest in protecting their personal data by acting anonymously and leaving no trace when they commit crimes, so that they cannot be traced. Strong information privacy for criminals and terrorists would prevent law enforcement, which in turn could threaten the stability of our society.

Based on the considerations above, it is obvious that there needs to be a balance, as illustrated by the surveillance barometer in Fig. 10.1. Different cultures, countries and regions have different perceptions of where this balance should lie, and a distinction should be made between authorities and private enterprises as potential surveillance actors. China is an example where the government's policy is to practice a high surveillance over the population. At the opposite end of the scale, the EU/EEA is a region that through GDPR has decided that we should have a relatively low surveillance both by government authorities and private companies. To be more precise, GDPR allows the use of personal data whenever relevant for business models, but requires that it is done in a controlled and transparent way which *"safeguards the data subject's rights and freedoms"*. When disagreements arise about the legality of processing personal data, it should of course be settled according to the rule of law. Numerous fines have been issued for the violation of GDPR, where the website GDPR Enforcement Tracker reports known cases.[1] For example, the Irish DPA (DPC) has fined Meta Platforms Ireland Limited EUR 390 million for violations regarding the processing of personal data for the purpose of personalized advertising.

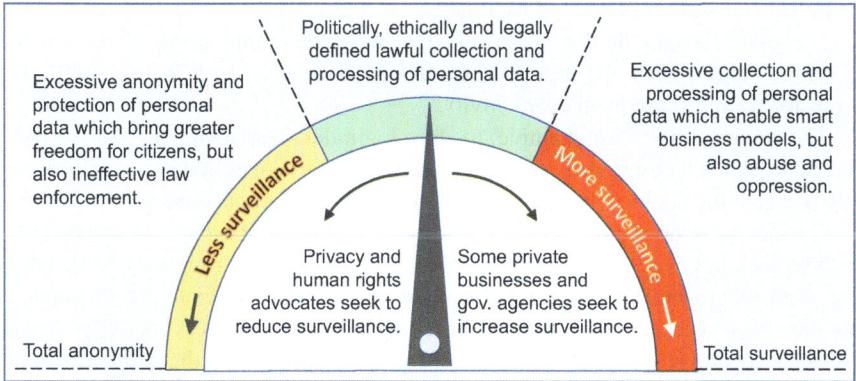

Fig. 10.1 The surveillance barometer

---

[1] https://www.enforcementtracker.com/

Although the GDPR defines what is acceptable processing of personal data, the law is relatively difficult to enforce. This is because the collection and processing of personal data takes place in companies' private systems that third parties have little insight into. Since the processing of personal data forms the basis for highly profitable business processes, it is tempting for businesses to engage in more data collection and processing than is permitted by law. With thousands of companies and organizations in every country, it is difficult for a national data protection authority to detect breaches of GDPR, except in cases where a serious incident has occurred, or the unlawful use of personal data can be deduced from the way an organization conducts business.

10.3 Privacy-Invasive Technologies

There are many methods and technologies for collecting personal data, some of which are clearly invasive by collecting information about us and tracking our activities without our knowledge or consent. The sections below describe some privacy invasive technologies as well as some privacy-enhancing technologies and methods.

10.3.1 Tracking with Cookies

The advertising industry generates profits based on personal data, because such data enables targeted marketing that has a higher impact than non-targeted marketing has. The web industry and the advertising industry have therefore developed highly advanced methods for collecting data about our online activities Cookies are a key technical element for collecting user data.

A cookie is a data file that is downloaded and stored on the user's client (smartphone or laptop) when the user accesses a web page, and which is returned to the website on each subsequent user activity.

The cookie is used, for example, to store login details, remember what the user has done on the website, or keep track of shopping carts in online stores. The website that sent the cookie can customize its services based on its content.

Cookies were originally intended to only be sent from first-party websites' own web servers, but over time it became common to use cookies from third parties that e.g. send advertisements or offer API services to websites. For example, third-party cookies on a website allow advertising companies to map users' activity across websites, as shown in Fig. 10.2. There are also third-party actors that only track user data without having any function on the website and without sending advertising. This is done by the browser loading dummy elements from third-party companies that then receive information about the client platform and can track the user by storing cookies in the user's browser. Websites that send a web page to the browser

10.3 Privacy-Invasive Technologies

insert URL links to advertising and other things that the browser itself loads from third parties. The fact that elements of a web page come directly from third parties is not clear on Fig. 10.2, which is an oversimplification. In deciding which items to load from third parties, first-party websites have some power with regard to controlling the scope of tracking.

When designing webpages, websites typically include additional features through third-party application programming interfaces (APIs). Such API services may be, for example, Google Analytics, which monitors traffic and activity on websites, services for accepting payments, or services that customize content. Such API services may also serve cookies and thus collect user data in the same way as advertising.

Many people are puzzled by how it is possible that they first search or read about a thing through a specific channel or platform, and then receive advertising about the same thing through another channel or client platform. The explanation lies in the fact that data about the same user is tracked across websites and client platforms. It is usually easy to match cookies about the same user based on similar content. Figure 10.2 shows how information about the same user flows through different websites and third parties, and how this information is ultimately aggregated by specialized tracking companies that compile data about the same user.

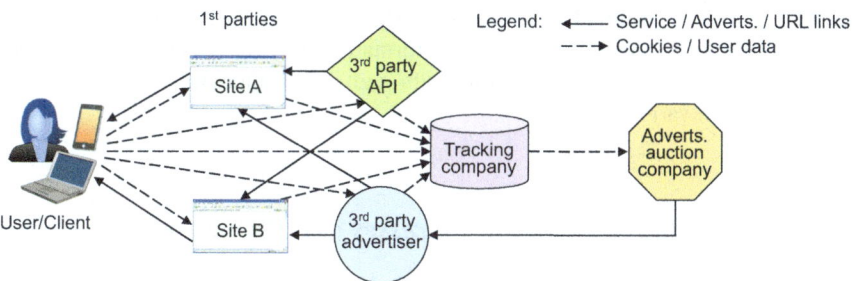

Fig. 10.2 Cookies to collect user data for ad auction

In order to streamline advertising on the web, programmatic advertising and real-time ad auction were introduced in 2007 [39]. Websites may offer advertising space to advertising providers who auction these ad spots. Each time a user clicks on a web page, ad space is offered, and advertisers automatically bid to purchase that view. Thanks to tracking of user data and auctioning of ad space, advertisers can now pay to show ads to users who have a specific profile. Advertisers' bid strategies are pre-programmed, which means that each ad auction takes only milliseconds. Within this small window of time, a surprisingly large number of players are involved in a complex, automatically choreographed bidding round. The auction process takes place in three stages, as explained below.

1. The auction opens when advertisers receive a bid announcement for a webpage that a user has clicked on. The bid announcement specifies the URL of the page,

the website category, the user's type of client and browser, one or more user identifiers that bidders can use to recognise a user from a different context, and then cookies about the user which may include assumed gender, interests, physical location (city or GPS coordinates) and more.
2. The auction itself takes place automatically based on competing advertisers' programmatic bidding strategies, which are specified based on the bid announcement and additional data about the user that the bidder has been able to obtain from other tracking sources.
3. In the end, the highest bidder wins the auction and automatically sends the advertisement embedded in the web page presented to the user, hoping that the user sees and perhaps clicks on the advertisement.

This type of collection and processing of user data is very invasive, and at the same time obscure for most people. It is interesting to consider this in relation to GDPR. Even if a website obtains consent, it would not be able to meet transparency requirements regarding the advertising auctioning process and would not be able to ensure that all parties involved have an adequate level of security for the protection of user data. On that basis, one could argue that this industry is probably violating the principles of GDPR.

An even more grim scenario is if a similar mechanism of targeted influence is used to manipulate political elections, as happened during the 2016 US presidential election. Entities located in the Internet information flow ecosystem can accurately map users' profiles and political attitudes through aggregation, tracking, and analysis of data from various sources.

10.3.2 Tracking with Email Addresses and Phone Numbers

There is a trend among browsers and operating systems to block the use of cookies and app tracking. This makes it harder to track based on cookies. Tracking operators and advertising agencies are therefore working to find alternative methods of tracking. Tracking based on email addresses is one such option that is increasingly used on the Internet. Tracking based on email address has the advantage (for advertising companies) that it can provide an easy unambiguous identification of users.

Tracking based on email address does not need to use the email address itself to match the same user on different websites. Instead, a kind of address fingerprint is generated by the email address, which thus constitutes a pseudonym. In this way, user data can be exchanged together with the email pseudonym so that tracking actors can match user data from different sources. To protect their business model, providers of such pseudonymous identifiers may diversify each user's identifier for different recipients (e.g., advertising agencies) so that recipients cannot cooperate among themselves by exchanging identifiers with associated user data.

Phone numbers are also a unique address that can be used for tracking in the same way as email addresses. However, tracking with a telephone number is less common than with an e-mail address, because we use telephone numbers as an identifier on the Internet to a lesser extent. However, tracking with telephone numbers is more common with mobile apps.

10.3.3 Cross-Platform Fingerprint Tracking

Web browsers send information about settings and characteristics about the client platform to the web server to customize how web pages can best be presented. For example, the browser sends information about graphics resolution on the client platform so that the size and resolution of images can be customized. Other examples of such attributes are which plugins are installed, which letter fonts are installed, and which time zone is set. By combining these and other attributes, it is possible to uniquely identify a browser and platform with high probability. As an analogy, it is possible to track a single car without using the car number, but only based on the make of car, model, color, tire type, optionally installed equipment, approximate mileage, and perhaps by mapping scratches in the paint.

The number and types of attributes determine how unique each platform fingerprint is. Presumably, a large proportion of "fingerprints" are completely unique, so that in practice it can be used to unambiguously identify a user. There are online services to check how unique a platform is, such as EFF's *Cover Your Tracks*.[2]

Unlike cookie-based tracking, which involves influencing the way a browser works and leaving traces by placing a cookie, cross-platform fingerprint tracking occurs hidden in the background on the server side, which is therefore harder to detect and prevent.

Cross-platform fingerprinting tracking is done at scale, complementing tracking with cookies and email addresses. Both first-party websites and third-party API and advertising providers engage in cross-platform fingerprinting tracking. If a user deletes cookies from a browser, platform fingerprinting can be used to regenerate cookies. This is done by companies that track user data storing cookies indexed with platform fingerprints. When a specific platform is detected, it is a simple matter of retrieving associated cookies that are again stored on the user's browser.

10.3.4 Mobile App Tracking

Mobile apps usually require access to various resources on your mobile device, such as contacts, photos, microphone, camera, and location. We often consent to everything a mobile app asks for without giving it much thought, because we want to start using the app as soon as possible and assume that these are resources that the app

---

[2] https://coveryourtracks.eff.org/

needs to work. We are also asked if the app can use these resources only when it is used or all the time, i.e. also when it is not being used. The recommendation is clearly to only allow an app to use resources when the app is in use, but even then, many apps collect significant amounts of information that is sent to the company behind the app and to third parties. This in itself is both a privacy issue and a security issue.

In 2023, many state governments in the US, Europe and Africa started banning the apps TikTok and Telegram on devices owned by the government. Technically, TikTok and Telegram are like any other app, they collect data for the features they offer, and they get the accesses they request, and that the user accepts. The difference is that TikTok is Chinese-owned, and Telegram is assumed to be Russian-controlled, and that the states that issued the ban consider both China and Russia as threat actors within the cyber domain. Legislation in China requires Chinese companies to hand over any information that the authorities may request, as described in Sect. 17.6.4. Similar legislation or practice presumably applies in Russia.

From a sinister perspective, one can imagine that apps could turn on a microphone or camera to spy, which is technically possible. Apps can access both users' personal data and business information stored on their mobile device or accessed by their mobile device.

An app like TikTok does not just send media files from the TikTok app to TikTok's data center. Apps have common access to data stored in shared areas, such as contacts, photos, messages, and documents. TikTok can also copy that type of information from users' devices, and upload it to TikTok's data centre. If the Chinese authorities request it, they can have that kind of information handed over.

But it would be a mistake to only be sceptical of TikTok and Telegram; We must be sceptical of most apps. There exists an extensive tracking industry for both mobile apps and regular websites with cookies that collect large amounts of information for compilation, analysis and resale.

Just as websites have third-party cookies that collect information about users, most apps also contain API features from third parties such as Alphabet (Google), Meta (Facebook), Twitter, Verizon, Microsoft, Amazon and other Big Tech companies. When third-party API features are embedded in an app, they can send user data back to the third parties. Such API features make it easier to develop apps with advanced features, which is attractive for app developers. The API features can also be used to send advertisements to the user so that app providers can also monetize it.

The large amounts of user data collected in this way are primarily used for commercial targeted advertising similarly to the process shown in Figure 10.2. However, many such collected datasets can be purchased by anyone, including criminals or state actors, and thus used as a source of reconnaissance and OSINT (open-source intelligence) as part of an APTattack (see Sect. 16.1.2). In this way, the advertising industry is not only a threat to privacy, but also a resource that can be exploited by criminal and state threat actors. The best advice is to have as few apps installed as possible on mobile devices. Apps that are no longer used should be deleted.

10.4 Anti-tracking

In response to the increasingly invasive technologies of the tracking and advertising industry, technologies and methods are also being developed that strengthen privacy by protecting against and preventing invasive tracking.

Explicit notification and freedom of choice about consenting to tracking cookies depends on first-party websites acting against third party tracking requests. In this context, we should consider that first-party websites often rely on third-party services for functionality and revenue, so there may be a lack of incentives for first-party websites to adopt such controls.

Within the EU/EEA, the GDPR requires obtaining user consent to track with cookies. Unfortunately, the principle of consent can fail in practice when we tire of clicking to change cookie settings, or the website collects more than we actually have, or think we have, consented to. To get away from this frustrating situation, many users install client-side technologies that effectively block tracking anyway. Anti-tracking in the browser is the most commonly used. Anti-tracking exists as browser plugins such as Ghostery, Disconnect, Adblock, Adblock Plus, Privacy, Badger and others, or it can be an integral part of common browsers such as Safari, Firefox, Chrome and Edge. The default setting does not always provide maximum blocking, but, for example, "balanced blocking" as in Microsoft Edge. Therefore, it may be necessary for users to adjust the setting for blocking third-party cookies. For example, in Microsoft Edge, select "Settings", then select "Privacy, search, and services". Under "Tracking prevention" there are three options: Basic, Balanced and Strict, where Strict blocks most tracking. Tracking is typically based on lists of IP addresses of third-party trackers. From 2024, Google's Chrome browser can be set to actively block tracking with third-party cookies. While it sounds encouraging from a privacy perspective, it still leaves Google itself as a third-party with the ability to track. This is because the Chrome browser sends telemetry data to Google about the user's browsing habits. This puts Google in a monopoly situation for collecting Chrome users' data, which regulators are investigating as anti-competitive practice.[3] The third parties which have been locked out of tracking with cookies are not necessarily willing to rely on Google for obtaining user profiles for targeted advertising. Instead, they switch to other tracking methods, such as described in Sects. 10.3.2 and 10.3.3. In the EU/EEA where people have been constantly confronted with cookie consent banners since the introduction of GDPR, they are now starting to be confronted with consent banners for "platform-ID tracking". In the end, it is business as usual for the advertising industry, they will always find a way to track Internet users for targeted advertisements.

Many types of browser plug-ins are also designed to also block advertising from third parties. As a result, advertisers do not get what they pay for, namely ad views.

---

[3] https://www.forbes.com/sites/forrester/2024/01/16/google-commits-to-third-party-cookies-deprecation-in-2024/

In response, advertisers may, for example, alert first-party websites and ask them to block web services or reduce functionality for users who block advertisements.

One way to prevent tracking with email addresses is to use different email addresses for different services. However, this is very inconvenient because we then had to remember which e-mail addresses we use for different services, and also remember different passwords related to each e-mail address. However, there are services that automatically generate different alias email addresses, such as Apple's *Hide My Email* and *Firefox Relay*. Using different phone numbers to block phone number tracking is even more inconvenient, and will also be expensive because it costs to have a phone plan for each number.

To block cross-platform fingerprint tracking, one must take drastic steps, such as constantly changing client platforms (replacing PCs and mobile phones) and changing browsers, and constantly changing IP address using VPN solutions. For most people, this is too impractical.

The battle between actors who want to track, and users who try to avoid tracking on the web, will always be like a cat-and-mouse game. Business models based on tracking users are hugely profitable, so strong market forces will drive innovation for the development of new advanced methods of tracking [40]. Unfortunately, there are few financial incentives for developing technologies that prevent tracking and strengthen privacy. The GDPR, described in the following sections, is intended to compensate for this bias, and restore the balance between privacy and market interests.

10.5 GDPR: The EU General Data Protection Regulation

For cybersecurity, organizations have strong business incentives to introduce controls to reduce cyber risks. In contrast to cybersecurity, there are less business incentives for organizations to safeguard privacy of personal information. To put it bluntly, without regulation there would be no information privacy. In the EU there is strong political will to protect citizens' information privacy through legislation for data protection, which has resulted in the introduction of GDPR. For people who mostly work with IT and cybersecurity, GDPR can seem relatively complex and difficult to understand because it is articulated in legal terms.

Apart from GDPR, there are a number of technical approaches to strengthening privacy in digital environments. Such technologies are often called *PET (Privacy Enhancing Technologies)*. PETs are typically technologies for anonymisation and pseudonymisation. This book does not focus on PETs.

GDPR (General Data Protection Regulation) is an EU regulation intended to strengthen and harmonize data protection in the EU and the EEA through rules for processing personal data. An EU regulation is a law that has binding effect in all EU member states. Each member state must then immediately adopt the regulation as national law, without the possibility of changing a single article. This is in contrasts to EU directives, which member states have the freedom to implement in their own way over time. The EEA countries can decide together that an EU regulation or directive is relevant to them, in which case the regulation or directive must be incorporate it into their legislation. The EEA countries chose to incorporate GDPR, but this did not happen until a few months after the regulation entered into force in the EU in 2018.

Although GDPR is a regulation, it allows for a number of national adaptations and additions to the GDPR. As an example, Art. 84 GDPR says that each member state shall determine the maximum penalties imposed on individuals in the event of serious violations of the GDPR. As a national adaptation, Norway, through section 48 of the Personal Data Act, have set the maximum penalty to fines and/or imprisonment of up to 1 year (3 years in particularly aggravating circumstances).

GDPR applies primarily within the EU/EEA, but also deals with personal data originating in the EU/EEA that is being processed outside the EU/EEA and with the transfer of personal data out of the EU/EEA. To ensure compliance and "adequacy" for processing personal information of EU citizens, many non-EU countries have aligned their information privacy laws with GDPR. In the EU, GDPR came into force on 25 May 2018, replacing the previous Data Protection Directive. In the EEA, GDPR came into force on 20 July 2018.

10.6 Roles in GDPR

GDPR describes a set of roles that have different rights, responsibilities and accountabilities, as shown in Fig. 10.3.

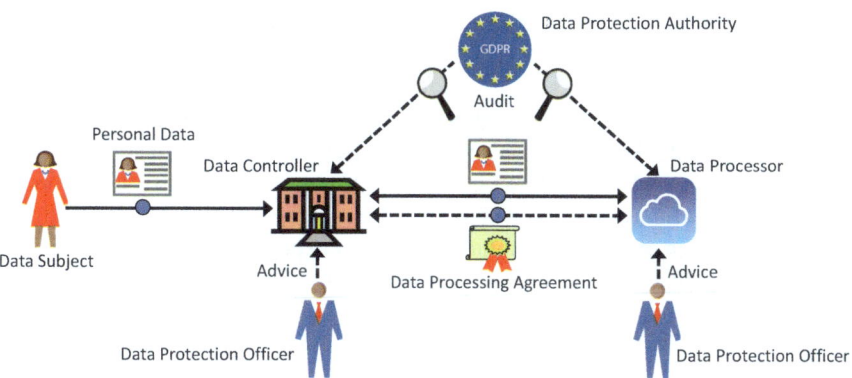

Fig. 10.3 Roles defined in GDPR

The various GDPR roles are described in the sections below.

10.6.1 The Data Subject and the Associated Personal Data

A *data subject* is a natural person who either

1. is a citizen of the EU/EEA,
2. is located in the EU/EEA, or
3. has its personal data stored in the EU/EEA.

Personal data is defined under the GDPR as *"any information relating to an identified or identifiable natural person (data subject)"* (GDPR art. 4, no. 1). The following description of what constitutes personal data is taken from the Irish Data Protection Commission's website:

> *Personal data basically means any information about a living person, where that person either is identified or could be identified. Personal data can cover various types of information, such as name, date of birth, email address, phone number, address, physical characteristics, or location data – once it is clear to whom that information relates, or it is reasonably possible to find out.*
>
> *Personal data doesn't have to be in written form, it can also be information about what a data subject looks or sounds like, for example photos or audio or video recordings, but data protection law only applies where that information is processed by 'automated means' (such as electronically) or as part of some other sort of filing system.*[4]
>
> The Irish Data Protection Commission

The GDPR also defines *special categories of personal data* (often interpreted as sensitive personal data), which, if collected and processed, create a particularly high risk to the *rights and freedoms* of data subjects. Therefore, the processing of such data is generally prohibited. According to Art. 9(1) GDPR, this relates to *"racial or ethnic origin, political opinions, religious or philosophical beliefs, or trade union membership, and the processing of genetic data, biometric data for the purpose of uniquely identifying a natural person, data concerning health or data concerning a natural person's sex life or sexual orientation "*. However, Art. 9 specifies a set of exceptions where, nevertheless, it is permissible to process special categories of personal data.

What is meant by "rights and freedoms" are first and foremost the rights described in GDPR art. 12-22, but also rights enshrined in EU member states' national constitutions, in the UN Universal Declaration of Human Rights and the European Convention on Human Rights, which describe, among other things, the right to privacy, protection of communications, freedom of expression, freedom of religion, the right to organize, and freedom from discrimination.

---

[4] https://www.dataprotection.ie/en/dpc-guidance/what-is-personal-data

10.6.2 The Data Controller

That *data controller* is typically the owner of the service or application that processes personal data, and is accountable for ensuring that the processing of personal information is lawful. The controller determines the purposes for which and in what way the personal data is processed. Conversely, if an organization decides "why" and "how" the personal data should be processed, then it is the data controller.

It is possible that the responsibility for processing is shared between several organizations that jointly decide "why" and "how" personal data should be processed. Joint processing responsibilities must involve an arrangement outlining their respective obligations to comply with the GDPR, such as who performs the DPIA (see below).

The data controller is accountable for the processing of personal data, whether that processing is actually done by the data controller or the data processor. It means that in case of breach of the GDPR, the data controller receives the fine disregarding which party caused the improper handling of personal data.

10.6.3 Data Processor

The data processor is a third party whose task is to process personal data on behalf of the data controller. The data processor's responsibility vis-à-vis the controller must be specified in a *data processing agreement*. For example, the agreement must specify what happens to the personal data when the business relationship has ended. A typical activity for the data processor is to offer IT solutions, including cloud storage. The data processor may only hand over parts of the processing to a subcontractor, who thus also becomes a data processor if there is prior written approval from the data controller.

10.6.4 Data Protection Officer

The data protection officer (DPO) advises the data controller or the data processor on obligations that the company has under the GDPR. The DPO participates in the DPIA (see below). All organizations can have a DPO, but not all organizations are obliged to have one.

If the data controller is a public agency (with certain exceptions) or an organization with more than 250 employees, a DPO must be appointed. The same applies to data processors where the core task is the processing of personal data. Other organizations may decide for themselves whether it is appropriate to appoint a DPO. The role of DPO may be combined with other roles such as CISO, or may be solely as DPO, as required.

10.6.5 The Data Protection Authority and Penalties for Infringement of the GDPR

Data protection authority (DPA) is an independent public authority in each EU/EEA member state that monitors organizations' compliance of GDPR. The DPA advises and handles complaints of breaches of the GDPR.

A DPA can sanction violations of the GDPR with fines or imprisonment. GDPR explicitly states that some violations are more serious than others, and therefore specifies two categories of fines.

The less serious violations can result in a fine of up to €10 million or 2% of the company's worldwide annual revenue, whichever is higher. Violations of this type are typically that the data controller or processor has not followed the correct procedure for establishing a data processing agreement or have not adequately performed a DPIA.

The more serious violations are those that breach the very principles of the right to privacy and the right to be forgotten, which are at the heart of the GDPR. These types of infringements can result in a fine of up to €20 million or 4% of the company's worldwide annual revenue, whichever is higher. Violations of this type are typically that the data controller collects, processes and forwards personal data without a legal basis, without consent, without providing access or without deletion after the processing has ended. Serious intentional violations of the GDPR can also result in prison sentences for accountable managers and directors. Likewise, a prison sentence may be imposed for gross negligence in securing sensitive or large amounts of personal data against theft and unauthorized access.

The purpose of the draconian penalties is to act as a deterrent against violations of the GDPR. The challenge of enforcing GDPR is that violations of GDPR can be very difficult to detect because it can take place hidden in private computer networks. The draconian penalties are precisely meant to compensate for the difficulty of monitoring and enforcing compliance with GDPR.

10.7 Particularly Relevant Articles in GDPR

GDPR consists of a total of 99 articles. This book briefly describes seven of these articles that are particularly relevant from the perspective of designing IT solutions with adequate security and privacy. The following articles are described in the following sections.

- Art. 5: Principles relating to processing of personal data
- Art. 6: Lawfulness of processing
- Art. 25: Data protection by design and by default
- Art. 32: Security of processing
- Art. 35: Data protection impact assessment (DPIA)
- Art. 45 and 46: Transfer of personal data to countries outside the EU/EEA

10.7.1 Article 5: Principles Relating to Processing of Personal Data

Art. 5 can be considered as a summary of the entire GDPR. This article defines what is meant by the term privacy under the GDPR and defines seven concise principles that should be easy to remember. These principles are briefly described below.

- **Art. 5.1.a: Lawfulness, fairness, and transparency**
  Lawfulness means, firstly, that there must be a *legal basis* for the processing, where Art. 6 describes each specific basis for processing. The controller must justify that the processing falls under at least one of the processing bases described in Art. 6. Secondly, legality also depends on the processing complying with all the other articles of the GDPR, and here there is a lot to watch out for. For example, Art. 25 requires that, by default, only a minimum of personal data necessary for the purpose shall be collected and processed. Art. 32 requires that the storage and processing of personal data be subject to adequate security controls. If personal data is to be transferred to countries outside the EU/EEA, a *basis for transfer as* described in Art. 45 and 46 is required. These are just some examples.

  By *fairness* is meant what the population of the EU/EEA generally perceives as fair. This principle is thus somewhat vague, and can legally be used to capture things that are not specifically specified by other articles of the GDPR.

  Transparency means that data subjects shall be informed when personal data is being processed, what information is being processed, and what the purpose is. This is crucial for the data subject to be able to make use of the other rights that the GDPR provides, such as the right to have their data corrected, or to withdraw consent and have their data deleted.

- **Art. 5.1.b: Purpose limitation**
  Purpose limitation means that personal data shall only be processed for specific and legitimate purposes. All processing purposes of personal data shall be described precisely and explained in such a way that all concerned have the same understanding of the purposes. That a purpose is legitimate means that it must have a legal basis and that it must be in accordance with other ethical and legal norms. Personal data collected for a specific purpose may not be reused for purposes that are incompatible with the original purpose.

- **Art. 5.1.c: Data minimisation**
  Data minimisation means that the collection and processing of personal data shall be limited to what is necessary to realize the purpose of the processing. If personal data is not necessary to achieve the purpose, it shall not be collected. Furthermore, each application shall only retrieve and process personal data necessary for the processing, even if the organization may have stored a larger set of personal data.

- **Art. 5.1.d: Accuracy**
  Personal data processed shall be correct in relation to the purpose of the processing. This may, for example, be crucial for healthcare personnel to be able to provide the right treatment. If aspects of a person change, personal data about these aspects shall be updated. This means that the controller must immediately correct personal data that is outdated or incorrect, or must delete personal data that cannot be corrected.

- **Art. 5.1.e: Storage limitation**
  The principle of storage limitation means that personal data shall be deleted or anonymised when they are no longer necessary for the purpose for which they were collected. However, it is not always clear when the purpose ceases. For example, in a hiring process where several candidates have applied for the same position, applications to those who were not hired must normally be deleted when the position is filled. However, in the event that the new employee changes their mind after a short time, it will be convenient for the company to take a closer look at the next applicant on the list and offer the position to this person without having to advertise the position again. In addition, other laws can set requirements for the storage of personal data, including for criminal investigations and for archiving accounting documents.

- **Art. 5.1.f: Integrity and confidentiality**
  This principle focuses on cybersecurity, which means that personal data must be protected with adequate security controls. Although not explicitly mentioned, availability should be included here, so that this principle means CIA for personal data.

- **Art. 5.2: Accountability**
  The principle of accountability means that the data controller is accountable for ensuring that the processing of personal data takes place in accordance with the GDPR. Demonstrating accountability means that the data controller must document all relevant aspects of the processing, such as purpose, basis for processing, types of personal data, DPIA and risk assessments, etc., and that the documentation can be presented during audits. It also means that the data controller cannot avoid accountability in the event of a breach of the GDPR, even if, for example, the direct cause of the breach lies with a data processor. To demonstrate accountability, organizations must act proactively by establishing the necessary organizational and technical controls needed to comply with GDPR.

10.7.2 Article 6: Lawfulness of processing

Art. 6 is central to the GDPR in that it describes a set of categories of purposes for processing personal data. Any processing of personal data must be explained in relation to one of the categories. Art. 6 has 4 paragraphs, of which paragraph 1 containing letters (a) to (f), are reproduced below.

10.7 Particularly Relevant Articles in GDPR

Section 1. Processing shall be lawful only if and to the extent that at least one of the following applies:

a. *the data subject has given consent to the processing of his or her personal data for one or more specific purposes;*
b. *processing is necessary for the performance of a contract to which the data subject is party or in order to take steps at the request of the data subject prior to entering into a contract;*
c. *processing is necessary for compliance with a legal obligation to which the controller is subject;*
d. *processing is necessary in order to protect the vital interests of the data subject or of another natural person;*
e. *processing is necessary for the performance of a task carried out in the public interest or in the exercise of official authority vested in the controller;*
f. *processing is necessary for the purposes of the legitimate interests pursued by the controller or by a third party, except where such interests are overridden by the interests or fundamental rights and freedoms of the data subject which require protection of personal data, in particular where the data subject is a child.*

Point (f) shall not apply to processing carried out by public authorities in the performance of their tasks.
GDPR, Art. 6

To illustrate the application of Art. 6, we look at some cases below.

- *Case 1: A private hospital needs to process personal data about a patient who is unconscious and needs immediate treatment.*
  There are two possible bases here:
  - Pont (d): The patient's vital interests dictate that the hospital may process personal data without consent or other agreement.
  - Point (c): The hospital may have a legal obligation to provide treatment to save lives.

- *Case 2: An insurance company needs to process personal data about customers and other persons of interest to investigate and detect possible insurance fraud.*
  There are two possible bases here:
  - Point (f): Insurance companies have a legitimate interest in preventing fraud and should therefore be able to process relevant personal data for that purpose, without consent or other agreement.
  - Point (e): It may be considered to be in the public interest to prevent insurance fraud.

- *Case 3: The company has a website and wants to track visitors with cookies.*
  There are two possible bases here:
  - Point (a): The company has no contract with the user (the data subject). Therefore, consent is an adequate basis for processing.

– Point (f): The company can argue that it has a legitimate interest in using cookies, but here the purpose is crucial. If it is to support security and reliability in the service, that is OK. However, if it is to send targeted advertising, it is certainly not OK.

10.7.3 Article 25: Data Protection by Design and by Default

Art. 25 deals with data *protection by design,* which generally means that the data protection principles of Art. 5 shall be taken into account in all phases of the life cycle of a service or application that processes personal data. These phases are described in Sect. 11.3 and can be summarized as *training, requirements, design, implementation, test, release, and response*. Art. 25 consists of three paragraphs summarized below, i.e. the text below is not verbatim as it stands in Art. 25, but only a summary.

1. Technical and organizational controls shall be implemented with a view to the effective implementation of the principles of protection of personal data, e.g. data minimisation, and to integrate the necessary safeguards into processing in order to comply with the requirements of this Regulation and to protect the rights of data subjects.
2. The controller shall implement appropriate technical and organizational controls to ensure that, by default, only personal data necessary for each specific purpose of the processing is processed. That obligation applies to the amount of personal data collected, the scope of processing of the data, how long they are stored and their availability.
3. An approved certification mechanism pursuant to Art. 42 Certification may be used as a factor to demonstrate compliance with the requirements laid down in paragraphs 1 and 2 of this Article.

Paragraph no. 2 above requires websites and apps to collect by default only personal data that is necessary for the service requested by the user. This is particularly relevant for websites' use of cookies and for mobile apps' collection of information from sensors and other apps. When we visit websites, we as users are often asked what types of cookies we accept. According to point no. 2 above, the default setting should always be the fewest possible cookies, but it seems that many websites sin against this, so we are forced to click through to the choice of cookies to opt out of the unnecessary ones. Similarly, there are many mobile apps that sin against point no. 2 by collecting many more types of information than is necessary for the app's actual function and service to the user. This is an obvious violation of Art. 25, and it seems that that Data Protection Authorities in the EU/EEA do not adequately enforce this article. It should not be necessary to click on anything on web pages for the collection of personal data to be set to minimum.

10.7.4 Art. 32: Security of Processing

CIA security objectives of confidentiality, integrity and availability are also important for the protection of personal data. Art. 32 precisely defines the requirement for CIA for personal data. Art. 32 consists of four paragraphs summarized below, i.e. the text below is not verbatim as it stands in Art. 32, but only a summary.

1. The controller and processor shall implement appropriate technical and organizational controls to achieve a level of security and to balance security risk.
2. When assessing an appropriate level of security, particular account shall be taken of the risks associated with the processing.
3. Adherence to an approved code of conduct referred to in Art. 40 Codes of Conduct or an approved certification mechanism referred to in Art. 42 Certification may be used to meet the requirements of paragraph 1 of this Article.
4. The controller and processor shall ensure that personnel with authorised access only processes data on instructions from the controller, unless when required by Union or Member State law.

Maintaining CIA security goals for personal data is essential. Personal data are very attractive targets for threat actors. For example, personal data may be valuable to

- state (sponsored) actors engaged in espionage to obtain knowledge about important persons
- criminal actors who engage in blackmailing by threatening to disclose sensitive personal data

These types of security incidents can have very serious impacts. Paragraphs 1 and 2 set requirements for security for the processing of personal data based on risk assessment in relation to the CIA security objectives. The probability and impact of security incidents involving personal data shall be included in such risk assessments. Assessment of cyber risk is described in Chap. 19.

10.7.5 Articles 45 and 46: Transfer of Personal Data to Countries Outside the EU/EEA

This section covers both of the articles below that are closely related:

– Art. 45: Transfers on the basis of an adequacy decision
– Art. 46: Transfers subject to appropriate safeguards

According to Art. 45, the European Commission may consider that third countries, i.e. countries or territories outside the EU/EEA, ensure a level of protection that provides at least as good protection of personal data as the GDPR. This is called an *adequacy decision*, which means that a country's privacy laws have been found adequate. Art. 45 states that an organization may transfer personal data to such states without further considerations. The European Commission maintains a list of such third countries. As of 2024, the European Commission recognises Andorra, Argentina, Canada (commercial organizations), Faroe Islands, Guernsey, Israel, Isle of Man, Japan, Jersey, New Zealand, Republic of Korea, Switzerland, the United Kingdom under the GDPR and the LED, the United States (commercial organizations participating in the EU-US Data Privacy Framework) and Uruguay as providing adequate protection.

For countries not covered by an adequacy decision, Art. 46 may be applied. Art. 46 requires organizations to assess for themselves whether it is safe to transfer personal data to specific data processors in states outside the EU/EEA.

There has been considerable controversy regarding the adequacy decision for the United States. Until 2020 the adequacy decision was based on the Privacy Shield agreement. However, the European Commission was sued by Max Schrems because he believed that Privacy Shield was inadequate. In 2020 Max Schrems won the case which was called the Schrems-II judgement because he had already won another case called Schrems-I against the previous Safe Harbour agreement. The basis for Schrems-II was that the US authorities may, through a set of laws, require access to personal data about residents of the EU/EEA when the personal data is stored under US jurisdiction. These laws are:

- *FISA Section 702 (Foreign Intelligence Surveillance Act)* allows for the collection of intelligence on non-U.S. persons not located in the United States.
- *Executive Order 12333* governs all U.S. foreign intelligence activities, including activities that fall outside FISA, such as conducted overseas against non-U.S. persons. Intelligence activities can be conducted secretly.
- *Presidential Policy Directive 28* allows for the bulk collection of (personal) data for surveillance without the persons concerned being specifically suspected or designated as persons of interest. However, the data collected can only be used for national security, and not, for example, for industrial espionage.

The consequence of Schrems-II was in theory dramatic, but in practice minimal. Every organization that previously relied on Art. 45 now had to consider Art. 46 for transferring personal data to the US. For example, many customers of SaaS services like Microsoft Office 365 would in theory have to stop using the service because it would be illegal. For example, in Sweden it was decided that state authorities could not use O365 because it typically processes highly sensitive personal data which causes high risks for the rights and freedoms of data subjects. This put the EU/EEA Data Protection Authorities in a limbo, because it would be very disruptive if the DPAs were to start banning the use of O365 and other US cloud services for the processing of personal data. In the end, O365 and other US cloud services were not banned, and the DPAs did not heavily investigate whether organizations assessed privacy risk of data transfer according to Art. 46. Instead, they waited. Finally, the new EU-US Data Protection Framework (DPF) entered into force on July 2023, which reinstated the adequacy decision for US companies that subscribe to the EU-US DPF. As expected, Max Schrems is preparing to sue the European Commission once more by claiming that the EU-US DPF does not satisfy the requirements of GDPR. The process might take a couple of years, and the outcome is uncertain.

10.8 Article 35: Data Protection Impact Assessment—DPIA

Art. 35 states that businesses must consider possible negative privacy impacts, and possibly reduce these to an acceptable level, before the enterprise operates a service or application that processes personal data. This is called *the Data Protection Impact Assessment (DPIA)*.

10.8.1 The Process Around DPIA

Figure 10.4 illustrates the process around DPIA which starts at the top left and ends at the bottom right. Before a DPIA is performed, it should first be assessed whether a DPIA is necessary at all, which is done by assessing whether it is likely that the processing will result in a high risk to privacy. This can be considered as preparatory work to the DPIA. Although processing is considered to pose a high risk to privacy, there are certain exceptions, such as processing that is necessary for national security, since it takes priority over privacy.

The DPIA itself is about assessing whether the planned processing may have disproportionately negative privacy impacts for the data subjects, and whether the scope of processing of personal data can be justified by the need to be able to fulfil the purpose of the processing. If negative impacts are disproportionately large (the risk to privacy is great) and the scope of processing is greater than necessary, it shall be considered how the scope can be reduced or how the negative impacts (risk to privacy) can otherwise be reduced to an acceptable level.

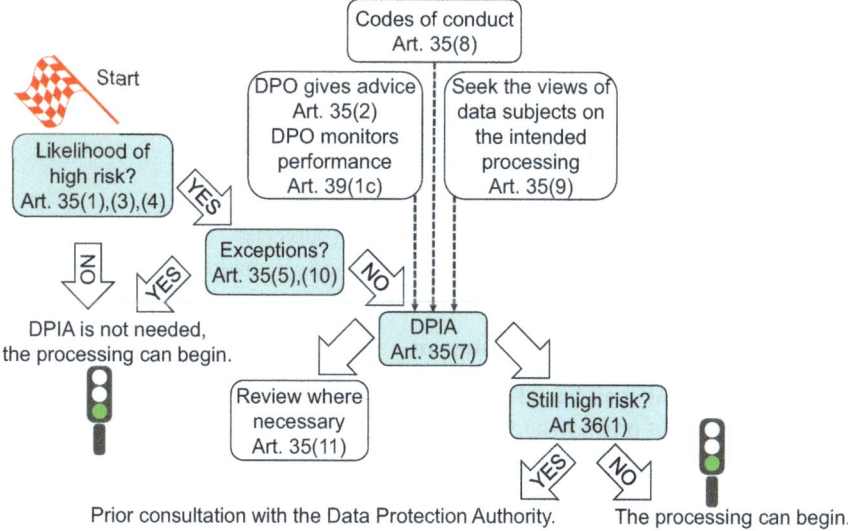

Fig. 10.4 Process around DPIA

As a result of a DPIA, there will be a residual risk assessment report. If the management of the controller considers that the residual risk is disproportionately high, but at the same time does not see how it can be reduced without imposing serious restrictions on the purpose of the processing, the controller may request prior consultation with the DPA. If, on the other hand, management finds that the residual risk is low and acceptable, it may be decided to start processing personal data for the specified purpose.

10.8.2 When Is It Necessary to Perform a DPIA?

According to Art. 35, the controller shall carry out a DPIA if it is likely that the processing will result in a high risk to the rights and freedoms of natural persons, by taking into account the *nature, scope, purpose and context* of the processing.

Examples of the nature of processing are that:

- It is difficult for the data subjects to exercise their rights.
- There is unpredictability, uncertainty, and low transparency about safeguarding principles.
- Systematic processing is planned.
- The processing of special categories of personal data is planned.
- There is an unequal balance of power between the data subjects and the controller.
- Technology is being used that can create new types of impacts.

Examples of the scope of processing are:

- Number of data subjects involved (number or per cent).
- Volume of data (number of variables, details).
- Storage time (short, time-limited, permanent).
- Geographical scope (local, regional, national, international, global).

Examples of the purposes of the processing are:

- Control purposes.
- Processing to make decisions that affect data subjects.
- To make decisions about individuals based on systematic and comprehensive analysis of personal data.

Examples of the context of processing are:

- Expectation of confidentiality (health, welfare and working conditions).
- Expectation of privacy (home, recreation).
- Processing of personal data from different data sets collected for different conditions.

10.8 Article 35: Data Protection Impact Assessment—DPIA

In case of uncertainty as to whether it is necessary to carry out a DPIA, it is recommended to do so anyway as a useful tool to ensure that the data controller is compliant with the GDPR and to have documentation in case of complaint from the data subjects or inspection from the DPA.

It is absolutely necessary to carry out DPIA in the case of:

- systematic and comprehensive assessment of personal circumstances when the data is used for automated decision-making,
- processing special categories of personal data on a large scale, and
- systematic monitoring of public areas on a large scale.

10.8.3 Threat Actors as an Element in Risk Assessment

When assessing privacy risks, it is always useful to identify who the potential threat actors are. For traditional assessment of security risks, it is typically assumed that threat actors are criminal organizations, terrorists or (hostile) state actors, as shown on the left in Fig. 10.5. When assessing privacy risks under the DPIA, we must assume that the data controller and data processor are the threat actors, as shown on the right in Fig. 10.5.

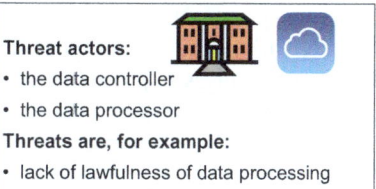

Fig. 10.5 Various threat actors as an element in the assessment of security risk (left) and DPIA (right)

The paradox of DPIA is that data controllers and data processors who carry out DPIA are also potential "threat actors" in the sense that they can be tempted to carry out illegal processing to support their business agenda. This is like "letting the fox guard the hen house" which means that data controller/processor sometimes will have a conflict of interest when performing a DPIA. They need to assess their own planned processing of the data subjects' personal data and make decisions that sometimes go against their own business interests to protect the interests of the data subjects. The data controller/processor have the task of assessing privacy risk on behalf of the data subjects, not on behalf of themselves. This is different from the traditional assessment of an organization's own security risks where the risk is assessed from the perspective of the organization.

10.8.4 Who Should Perform DPIA?

The controller has overall responsibility for the performance of a DPIA, but the work itself can be delegated to others. The Chief Information Security Officer (CISO) and the DPO (Data Protection Officer) can recommend performing a DPIA and can contribute to the process. Team members for performing a DPIA are e.g. the following:

- Data Protection Officer
- Representative(s) of the controller
- Data Processor Representative(s)
- Representative of the data subjects, if relevant, to obtain the views of the data subjects
- General: Relevant actors and, if relevant, experts in the field

The DPIA team should ensure that the processing is in principle lawful and meets basic requirements that follow from other provisions of the GDPR. A DPIA must be performed prior to the start of processing personal data and should be done simultaneously with planning the processing. The DPIA must be repeated in the event of significant changes in processing, for example by including new types of personal data and new types of decisions based on personal data. It is good practice to define and document the following:

- Roles and responsibilities of the processing
- If relevant—why processing should be carried out contrary to the views and interests of data subjects
- If relevant—why views are not sought from data subjects

10.8.5 The Steps of the DPIA

Execution of a DPIA consists mainly of the following five (possibly six) steps:

1. Make a systematic description of the planned treatment and the purpose of the treatment.
2. Assess whether the processing is necessary and proportionate to the purposes.
3. Assess the risks to the rights and freedoms of data subjects.
4. Specify the planned controls to manage the risks and demonstrate GDPR compliance.
5. Get management validation of the DPIA.
6. Prior consultation with the Data Protection Authority.

In practice, the DPIA team will first draft a systematic description, then make an iteration over points 2–4 until it can be concluded that "processing does not affect the rights and freedoms of the data subject", and finally edit the final version of the systematic description, which is validated by management in point 5). Each step is described in detail below.

1. **Systematic description**

This step consists of making a systematic description of the planned treatment and the purpose of treatment. Here you will review, and quality check the overview and description of the treatment with regard to:

- the nature, scope, purpose and context of the processing,
- recipients, data flow and storage,
- functional description of the treatment and all assets, and
- relevant references for documenting the processing.

When describing the processing, the purpose of each type of personal data shall be specified and why the processing cannot be done without it. The objective of this step is that the data controller has a complete overview of the processing and does a quality check that the descriptions made are complete and clear.

2. **Assessment of proportionality**

This step consists of assessing whether the processing activities are necessary and proportionate to the purposes (proportionality). This is done by considering:

- legal basis for processing,
- purpose(s),
- whether data minimisation has been considered,
- whether there is quality control for the correctness of the personal data,
- whether reasonable limitation of storage time is specified,
- whether the rights and freedoms of the data subjects are adequately safeguarded.

If possible, suggest controls to improve the way privacy principles and data subjects' rights and freedoms are safeguarded.

The objective of this step is to obtain quality assurance that the proposed processing is necessary and proportionate to the purposes for complying with the GDPR.

3. **Residual risk assessment**

This step consists of assessing the risks to the rights and freedoms of data subjects. This is based on the following points:

- Consider the potential negative impacts of inadequate privacy. Examples of impacts are that:
  - there is a lack of transparency regarding the processing of personal data,
  - more personal data is collected than necessary for the purpose,
  - personal data processed is not correct,
  - personal data is stored longer than necessary for the purpose,

- data subjects lack the information to make an informed choice,
- data subjects are not able to find out what personal data is stored about them,
- data subjects are unable to get personal data corrected or deleted,
- data portability has not been facilitated,
- data subjects' rights are not protected in case of profiling,
- the processing may result in discrimination,
- the processing restricts freedom of expression.

- Estimate the probability that the negative impact will occur, and assess the resulting severity for each risk.

The objective of this step is to know the risks to the rights and freedoms of data subjects.

4. Proposed controls

This step consists of proposing and specifying controls to manage the risks and to demonstrate compliance with the GDPR. Examples of controls are:

- collect less personal information,
- reduced processing of personal data,
- permissible right of reservation,
- ongoing information, multiple channels,
- specially adapted access portal,
- automatic deletion,
- anonymization,
- explainable AI,
- alternate manual procedure for decision-making.

The objective of this step is to have a treatment that does not affect the rights and freedoms of the data subjects.

5. Management validation and decision

The management must understand in what ways the processing may produce a high risk to the rights and freedoms of the data subject, which thus requires DPIA. Management must also understand the steps of the DPIA, identified risks and proposed controls. Failure to implement the DPIA, performing DPIA incorrectly or not consulting the correct authorities may result in administrative fines from the Data Protection Authority.

Based on the DPIA report, management can typically choose one of three possible decisions:

1. The DPIA is approved and validated, which means that the processing can start.
2. The DPIA is approved on condition of improvements, with an explanation of what improvements are expected. This management will then receive a revised DPIA.
3. The DPIA has been rejected, which means that the processing cannot start.

It is necessary to request a prior consultation with the Data Protection Authority in case the DPIA has been presented to the management team more than once, and satisfactory controls have not been found to reduce the risk adequately, while the motivation to process personal data is still high.

The enterprise must document that the risk cannot be reduced. The decision to request prior consultation shall be made by management.

6. Prior consultation with the Data Protection Authority

In the event of high risk, which cannot be reduced, the Data Protection Authority shall be involved in prior consultation. There are requirements for documentation to be submitted, and that the data controller has followed guidelines.

The maximum processing time for requesting prior consultation is normally eight weeks, but in special cases the processing may take another six weeks.

The objective of prior consultation is to find a good solution for the processing of personal data that does not entail a high risk to the rights and freedoms of the data subjects with regard to privacy.

10.9 Notification of Personal Data Breaches

Art. 4 (12) defines a "personal data breach" as *"a breach of security leading to the accidental or unlawful destruction, loss, alteration, unauthorised disclosure of, or access to, personal data transmitted, stored or otherwise processed"*.

This represents a traditional security breach that involves personal data. Such incidents can be very damaging if, for example, they affect many people, or if sensitive information has been leaked.

In the event of a personal data breach, the controller is obliged to notify the DPA without undue delay, and if possible, within 72 h. Three days may seem like plenty of time, but many things need to be done. Information must be obtained and assessed whether a notification should be sent to the DPA at all, the content of the notification must be articulated and, not least, the notification must be submitted through the official DPA portal by an authorised person.

The 72-h deadline should be complied with—otherwise the reason for the delay must be stated in the notification. Note that notification can be given step by step by the controller providing continuous updates of the incident.

In addition to the duty of notification to the Data Protection Authority, the controller should notify the data subjects and whenever relevant the press in a controlled manner at an appropriate time.

10.10 Tasks

1. Roles

    (a) Can an entity be a data processor and at the same time have a role as data controller for the same processing? Justify your answer.
    (b) Can an entity be the data controller without appointing a data protection officer? Justify your answer.

2. Consent

    (a) Can the data subject consent to the collection and processing of personal data without the controller having defined a purpose for the processing? Justify your answer.

    Imagine that you are making an audio recording of a conversation between your colleagues or friends. Consider the following questions in relation to this:

    (b) How should you inform those present during the audio recording?
    (c) How can you use the audio recording later, with a view to playing it (i) for those who were present during the recording and (ii) for others who were not present?
    (d) Is it meaningful to say that someone "owns" the audio file or the content of the audio file?

3. Outsourcing

    Assume an IT company in a European country that processes personal data. The company decides to let the processing of personal data be outsourced to a cloud service in a foreign country.

    (a) Is this allowed, according to GDPR, and how does GDPR define "foreign"?
    (b) What does the company need to do to outsource the processing of personal data?

Chapter 11
Secure by Design

The details are not the details. They make the design.

Charles Eames, American designer

The concepts of *secure by design* and *privacy by design* were coined around 2010, and have gradually gained popularity. Secure by design and privacy by design are principles of good practice in system and software development that emphasize integrating security and privacy measures and considerations into the design and architecture of systems, applications, and products from the outset. In addition, these principles emphasize that security and privacy considerations are considered during the whole lifecycle of systems, applications and products. This is in contrast to treating security and privacy as an afterthought or as a separate layer, which would be inadequate in most cases.

In this chapter, you learn what is meant by secure by design and privacy by design. You will also learn about secure system development, application security and security in the cloud, which are three essential elements of secure by design.

11.1 Secure by Design

Secure by design generally means explicit consideration of cybersecurity throughout the lifecycle of software products and applications.

One company that discovered the need for secure by design early on was Microsoft. Versions of Microsoft Windows prior to Windows Vista (2006), Windows 7 (2009), Windows 8 (2012) and Windows 10 (2015) had extremely many vulnerabilities, which were exploited on a large scale by threat actors when all computers eventually connected to the Internet. For Microsoft, it quickly became apparent that protecting users from malware would require a whole new approach to security than it had had until then. Microsoft Security Development Lifecycle (SDL) was a result of the need to apply a methodology that is easy for developers to understand and that enabled them to practice secure software development in a consistent manner. The SDL methodology was adopted by Microsoft in 2004 and used internally until

2008, when the SDL methodology was also published publicly. The SDL methodology consists of seven phases illustrated in Fig. 11.1 in Sect. 11.3. Microsoft SDL can be interpreted as a *de facto* standard for secure-by-design system development.

Microsoft SDL has become an integral part of the software development process at Microsoft and is used by many other companies around the world. Development, practical use and constant improvement have allowed SDL to mature into a well-defined and practically applicable methodology.

BSIMM (Building-Security-In Maturity Model) and SAMM (Software Assurance Maturity Model) are from 2009. But the most modern secure-by-design guidelines are published by CISA (cisa.gov/securebydesign).

11.2 Privacy by Design

Privacy by design is defined in *GDPR Art. 25 Data protection by design and by default*, which is described in Sect. 10.7.3. The main principles of data protection by design according to GDPR Art. 25 are to minimize the collection and processing of personal data in relation to the purpose of the application, and that applications by default shall not collect and process personal data without the data subjects' consent.

Since GDPR Art. 25 is articulated in legal jargon that makes it relatively difficult to access for developers of software and applications, the Norwegian Data Protection Authority has published a guideline called *Software development with Data Protection by Design and by Default*[1] which is very readable and at the same time more comprehensive. This guideline follows the seven phases of Microsoft SDL as shown in Fig. 11.1. It makes a lot of sense that the guideline from the DPA is harmonized with Microsoft SDL, because both naturally have the same lifecycle, and because it makes it easy for organizations to integrate secure by design and privacy by design.

11.3 The Seven Phases of Secure by Design

The lifecycle of software systems can be described with a process consisting of seven phases that combines both secure by design and privacy by design. This process is illustrated in Fig. 11.1.

---

[1] https://www.datatilsynet.no/en/about-privacy/virksomhetenes-plikter/data-protection-by-design-and-by-default/

11.3 The Seven Phases of Secure by Design

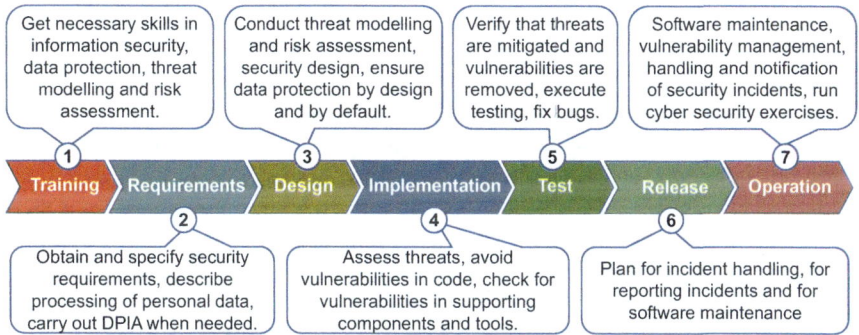

Fig. 11.1 Security Development Lifecycle for security and privacy by design

Each phase of the process in Fig. 11.1 is described in the sections below. Details of secure software development covering the design, implementation and test phases are further described in Sect. 11.4 about secure software development.

11.3.1 Training

The purpose of training is that everyone who participates in the development and operation of digital services and applications should have basic knowledge and understanding of digital security and privacy as well as the risks associated with this. Ideally, employees should get this expertise from their education. However, if employees lack the necessary expertise, the employer is responsible for ensuring that they acquire this competence. The individual employee must know when and why digital security and privacy are important in their work tasks, and what should be done to ensure adequate security and privacy.

It is obvious that everyone working in IT must have security competence. As an analogy, it would be unacceptable to educate building architects and engineers without giving them knowledge of fire safety, because without this knowledge architects and engineers could build firetraps into our buildings. Similarly, it is unacceptable to educate computer scientists and computer engineers without compulsory courses on cybersecurity, because without this knowledge, IT people would necessarily build vulnerable ICT infrastructures.

Unfortunately, there are still many IT designers around the world who lack the necessary expertise in cybersecurity simply because there was no curriculum on security in the study program they took. Universities and university colleges have had study programmes in computer science and IT since the 1980s, but it was only after 2000 that these study programmes gradually began to include electives in information security.

Education on safe use of the Internet should start as soon as children start using the Internet. Children must learn early to look for both left and right before crossing the road, otherwise they will be run over. Metaphorically, children must also learn to look left and right on the Internet, otherwise they will quickly be overrun by identity theft, fraud and privacy violations.

Many young people are very interested in exploring what is possible with ethical hacking, and often become adept at this. It is extremely important that these young people learn early on where to draw the line between legal and illegal hacking. There are many examples of young people who initially have good intentions of hacking a business to show the business that it has vulnerable systems, and then being reported to the police by the same company. This will typically be very traumatic and demotivating for a young person who otherwise could have used their talent for the good and be appreciated. To avoid such tragic cases, young people must learn where to draw the line for legal ethical hacking.

11.3.2 Cybersecurity and Privacy Requirements

Taking cybersecurity and privacy into account is a fundamental aspect of developing systems, applications, and services. Security requirements must be continuously updated to reflect changes in required functionality, changes in the threat landscape, and changes in laws and regulations, as described in Sect. 1.7. The optimal time to define security requirements is in the initial design and planning phase, as this allows the development team to integrate security in ways that minimize how it impacts functionality and user experience. Factors influencing security requirements include legal and industry norms, industry best practices, internal standards and practices, experience from previous incidents and known threats. These requirements should be mapped so that a comprehensive set of requirements can form the basis for system design.

For application security requirements, industry best practices are defined through the OWASP Top 10 and OWASP ASVS (Application Security Verification Standard), which are described in Sect. 11.6.2.

For data protection requirements, the applicable norms in each region must form the basis, where GDPR applies to the EU/EEA region, as described in Chap. 10.

11.3.3 Secure Design

The purpose of secure design is to achieve a final product that is as hardened and robust as possible against existing and future security threats.

An important part of the design phase is to specify "secure function", in that software functions are well designed with security in mind. To achieve this, developers will often rely on security mechanisms, such as cryptography, authentication, logging, etc. In many cases, selecting or implementing security mechanisms has proven to be so complex that poor design or implementation choices are likely to lead to vulnerabilities. Therefore, it is crucial that designers have good expertise in the use of such features, and that these are used consistently and with an understanding of the protection they provide.

11.3 The Seven Phases of Secure by Design

To avoid vulnerabilities in the planned system, it is important to do threat modelling, which means identifying ways in which an attacker can abuse access points to functions and data flow or other aspects of the system. Identified threats with the potential to create security incidents reflect the existence of weak points and vulnerabilities. After threat modelling, the next step is to suggest how the design can be improved to remove or mitigate the vulnerabilities, which is equivalent to stopping identified threat scenarios.

11.3.4 Implementation and Secure Coding

The purpose of secure coding is to avoid vulnerabilities being built into the system during software development, and that all security functions work according to requirements and design.

Modern software development largely uses third-party components (both commercial and open source). When selecting third-party components, it is important to understand the impact a security vulnerability may have on the security of the system into which they are integrated. It is essential to have an overview of all functions, APIs, third-party libraries and modules used in software development. Components deemed insecure must be prohibited, and those that are outdated or contain known vulnerabilities must be updated before use. Code scanning tools can be used to identify vulnerable components. Mapping vulnerabilities in functions and modules can be done, for example, with tools such as OWASP Dependency Check.[2] This tool detects known, publicly disclosed vulnerabilities in software libraries used by an application. SBOM (Software Bill Of Materials) is an other approach which is being integrated into software vulnerability scanning tools, as described in Sect. 3.3.

Static analysis of source code before compilation is a scalable method for verifying that secure coding rules are being followed. Automatic code parsing and code review tools do this job efficiently. In addition, it is recommended to do a manual review of the code, although this method can be time-consuming. Identified prohibited features may optionally be replaced with safe alternatives.

From a privacy perspective, the code should be checked to disable unnecessary tracking, logging and collection of personal data.

11.3.5 Software Security Testing

The purpose of security testing is to uncover vulnerabilities that have remained undiscovered through design and the coding phases so that these can be removed. Security testing provides greater certainty that the code adequately safeguards

---

[2] https://owasp.org/www-project-dependency-check/

security and privacy. There are various forms of testing; Dynamic testing, fuzz testing and penetration testing are described below.

11.3.5.1 Dynamic Testing/Vulnerability Analysis

Dynamic testing of the complete software checks the functionality that only becomes apparent when all components are integrated and run together. Dynamic testing is mainly done by using tools, but also by manual review, which analyses how the software behaves in different situations. For example, testing will verify that users can only access the information and functionality for which they are authorized, and that attempts to acquire unauthorized information are logged as security events. Typical vulnerabilities that should be tested while running the software are Cross-Site Scripting (XSS) and SQL injection, which are described in more detail in Sect. 11.7.

From a privacy perspective, it is relevant to test session management, access control and the use of cookies to verify that personal data does not leak from secure sessions.

11.3.5.2 Penetration Testing

Penetration testing, often called *pentesting*, is to go a step further after vulnerability analysis. A penetration test is an authorized simulated attack performed to evaluate the security of the system. The test is performed to identify vulnerabilities, and then the potential for the attacker to exploit the vulnerabilities to gain access to the system's functions and data. Term *"ethical hacking"* is roughly synonymous with penetration testing. Pentesting is described in more detail in Sect. 14.7.

Security testing and pentesting also require an assessment of whether a system to be tested only contains test data or if it has production data with actual personal data. In other words, security and penetration testing can cause privacy risk in itself.

11.3.5.3 Fuzz Testing

Fuzz testing is to provoke errors in the software by providing corrupt input values in the form of random and malformed data through all possible interfaces to the software. This is done using automatic tools, which also analyse how the software fails. If the software has multiple interfaces, each one should be tested, both individually and, if possible, in combination.

Fuzz testing is also how attackers detect zero-day vulnerabilities that can be used to create potent attack tools. Fuzzing constitutes an entire industry where active participants are both software developers and threat actors, as well as actors whose sole goal is to sell vulnerabilities to software developers or threat actors, depending on who pays the most, and on the actor's ethical attitude.

11.3.6 Release

Setting software in production is an important phase and milestone in a system's life cycle. Two important steps in this phase are to describe a plan for maintenance and incident management, as well as to formally approve that the software is put into production.

11.3.6.1 Plan for Operation, Maintenance and Incident Management

Software must be operated and maintained continuously. Before release, a procedure for operation must be defined that includes updating and patching, both of proprietary software and of third-party code. A procedure shall be established for reporting nonconformities and how nonconformities are followed up.

Before release, the software or system should be considered regarding incident management, e.g. to define the criticality and priority of the system. This topic is also described in Sect. 14.4. The incident management plan should also describe the handling of incidents resulting from vulnerabilities in code inherited from third parties.

Response capacity should be defined in terms of criticality. For example, software related to emergency care in health is likely to require a 24/7 rapid response. Channels for reporting incidents shall be defined in the plan. The security of the reporting channels themselves should also be assessed according to the sensitivity of the content of reports. If the incident involves confidentiality, integrity or availability breaches related to personal data, the organization should notify relevant authorities, and if possible, the data subjects. In the EU/EEA area, GDPR mandates that personal data breaches must be notified to the Data Protection Authority within 72 h and the data subject immediately, as described in Sect. 10.9.

The plan should define criteria for classifying various incidents. Handling an incident can consist of, for example, detecting, analysing, responding, recovering, and reporting. Furthermore, the plan should describe logging of data related to incident management, including the guidelines set by the data protection regulations for this. Finally, the plan should include procedures for evaluating actual incident management (lessons learned), to support continuous improvement of the incident management process.

11.3.6.2 Formal Approval of Production Setting

Prior to formal approval of production setting, it should be verified and documented that all security and privacy requirements are met, and that identified vulnerabilities have been removed or sufficiently mitigated.

The organization must define the authority for releasing software. The CISO or equivalent role is responsible for approving that security requirements are met and function as intended. The Data Protection Officer or equivalent role is responsible for approving that all defined privacy requirements are met.

All relevant data and documentation from the entire development process must be archived. This is important to maintain quality in the work on cybersecurity and privacy, and to have a basis for audit or inspection.

11.3.7 Operation and Incident Management

Operation of IT systems consists of many different activities. These activities can be divided into two main categories, where *operations and maintenance* are one and *incident management* is the other. These are described below.

11.3.7.1 Operation and Maintenance of the Software

Procedures and procedures for operation and maintenance of the software shall of course be followed. This includes, among other things, routines for how privacy and security are to be safeguarded over time. As the software is further developed, it may be necessary to repeat security tests, vulnerability analyses and penetration testing of software, infrastructure and networks. Regular audits should also be carried out to document compliance with relevant cybersecurity regulations that contain security requirements. In general, organizations should have a cyber governance program, also called ISMS (Information Security Management System), that includes the acquisition, management, operation, and maintenance of software. The topics of cyber governance and ISMS are described in Chap. 18.

Errors discovered through bug reporting or other means must be followed up routinely. Logging is important for monitoring systems and networks, but can produce enormous amounts of data and can in itself constitute a breach of privacy if not described and considered in the personal data processing plan. Therefore, it is important to define what can and should be logged, and that the logs are secured. Logging entails an obligation to actually use the logs for threat intelligence and to use the logs to investigate security incidents. If the logs are never used, they should not be collected. Although logging of personal data is in principle legal when used for the right purposes, it is important to ensure that it is not exported to other applications and made available to anyone other than those who have a defined need.

11.3.7.2 Incident Management

The organization must be prepared to handle attacks and incidents that may lead to breaches of confidentiality, integrity and availability related to personal data. There must be a defined SOC function that can detect and handle incidents. The organization must also have a plan for informing users and affected parties. The topic of incident response planning is described in Chap. 14

Deviations and incidents shall be reported via the channels described in the plan for incident management. It should be easy for users of the software to report errors, vulnerabilities and deficiencies, so that the software can be continuously improved and further developed.

In principle, the plan for incident management must be followed, but when an acute incident occurs, the nature of the incident may entail changes in how the plan is followed. The plan itself should be maintained and improved as needed. The response team is scheduled to know who to contact as needed to perform necessary tasks. The plan should also define which priorities apply and how decisions are to be taken when there is a real crisis. In order for this to work smoothly in practice, employees need regular exercises. Evaluation of events should follow the plan. A more detailed description incident management is given in Chap. 14.

11.4 Secure Software Development

There are many different methodologies for software development, where *the waterfall method* is considered traditional and old-fashioned, while *agile software development* is the modern approach. These are briefly described below.

11.4.1 The Waterfall Method

The waterfall method for software development consists of six sequential stages, where each stage depends on the results of the previous stage, as illustrated in Fig. 11.2. Note that the stages of the waterfall method are the same as phases 2–7 in the model for secure by design described in Sect. 11.3.

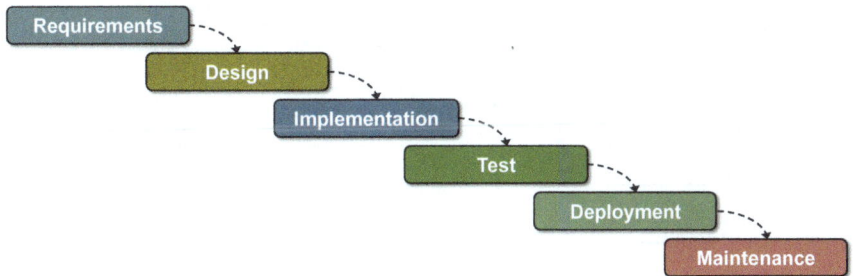

Fig. 11.2 The waterfall method of software development

The term "waterfall" metaphorically means that this is a one-way process, in the same way that gravity forces water to flow downwards. If necessary, it is of course possible to return to an earlier phase to make changes, but this causes complexity and entails relatively large costs.

In software projects, there is almost always a need to make changes along the way. As the waterfall method is not iterative and does not take into account the need to make changes along the way, it tends to be rigid and inflexible. However, it is a well-structured method, which was used as early as the 1970s for software development.

Secure software development according to Microsoft SDL (described in Sect. 11.1) fits well with the waterfall method. In other words, the best practice for secure software development with the waterfall method is to follow Microsoft's SDL.

11.4.2 Agile Software Development

Agile software development is an umbrella term covering different methods for software developers that take into account continuous changes based on specific principles that were outlined in *The Agile Software Development Manifesto*.[3] The key principles consist of early and continuous delivery of software modules, easy handling of changes, frequent delivery of working production software, clients and developers working together, motivated developers and ample resources to succeed. Agile development focuses on face-to-face and shoulder-to-shoulder collaboration and measuring progress through working software. Agile development assumes that a self-organized team is best suited to develop good architectures, requirements, and designs. Agile development encourages self-awareness and the ability to adjust and improve the process accordingly. Figure 11.3 illustrates a general process for agile software development.

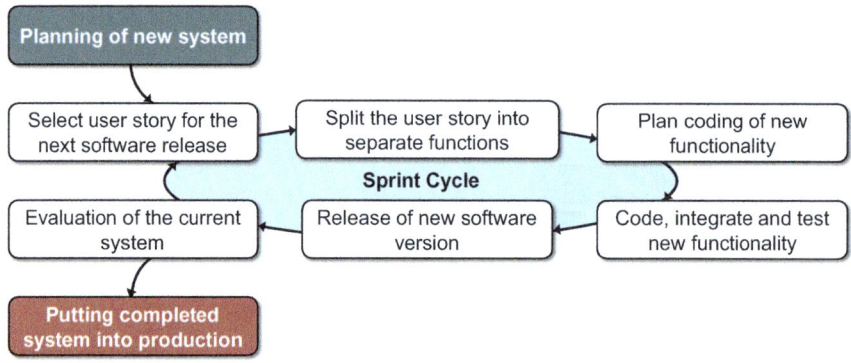

Fig. 11.3 Agile software development process

---

[3] https://www.agilealliance.org/agile101/the-agile-manifesto/

11.4 Secure Software Development

Planning the system and specifying requirements consist of describing user stories that the system will support. User stories can be changed, and new user stories can be added. Development takes place in iterative cycles called *a sprint*, with a typical duration from a week to a month. For each sprint, one or more user stories are selected. During a sprint, design, planning, coding, testing, release, and evaluation of the new modules in the software will be done. When there are no more user stories remaining to be implemented, the system is finished.

11.4.3 Secure, Agile Software Development

Agile secure-by-design software development requires a different approach than to follow the seven-phase process from Fig. 11.1, since that process is not defined as iterative. What is needed is to expand the iterative process for agile software development to include security activities, in the same way that Microsoft SDL extends the waterfall method to include security activities. Such an extended process for secure agile software development is illustrated in Fig. 11.4.

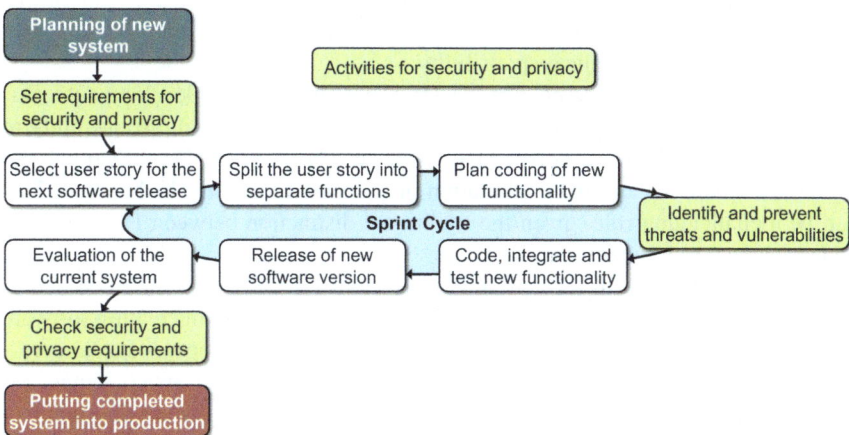

Fig. 11.4 Process for secure agile software development

The yellow boxes represent security related activities in the agile software development process. There are mainly three additional activities that focus on security:

- **Set requirement for security and privacy**. This activity is the same as that in phase 2 of the process for secure by design in Fig. 11.1. This initial activity ensures that basic security and privacy requirements are identified. Under GDPR, performing DPIA is included in this phase (GDPR Art. 35 no. 1).
- **Identify and prevent threats and vulnerabilities**. This activity is part of every sprint cycle. The activity consists of threat modelling and vulnerability analysis for the new functionality being implemented in the sprint. Necessary controls

shall ensure that relevant threats and vulnerabilities are eliminated. This activity is described in more detail in Sect. 11.5.1.
- **Check security and privacy requirements.** This activity happens when the system development is completed, but before the system is put into production. While each sprint contains an activity for testing, a final test and assessment of the system must be done to verify that security and privacy requirements are met.

With additional security activities, it is natural that secure agile software development becomes slightly less agile than traditional agile software development. However, security and privacy are mandatory requirements in an environment of increasing exposure to cyber threats, which means we have to pay that price.

11.4.4 Security Champion

While it would be great if all participants in a (secure) agile software development team have cybersecurity expertise, it is perhaps too much to expect that everyone is an expert in identifying threats and vulnerabilities. Therefore, it is common to appoint one person to have the role of *security champion*. Working in a team means that each participant has different roles, which is more rational and efficient than when everyone has the same responsibility.

In addition to being a member of the development team with development tasks, the security champion is the person in the team who has a special responsibility to see the software from the perspective of those who work in IT operations and CyberOps where attacks and exploitation of vulnerabilities must be handled. This can be a challenging role, given the traditional distinction between IT operations and software development.

People who work with security in IT operations typically have a limited understanding of how software developers work, and their challenges in creating software that is free of vulnerabilities. On the other hand, software developers often underestimate how bugs in the lines of code they produce can create security challenges for IT operations. The security champion can bridge this gap. A software development team can increase their ability to produce software that is secure by design if they have a security champion on their team.

The benefit of having a dedicated security champion is faster and more comprehensive identification of potential security vulnerabilities, which can then be removed early and result in higher quality code. It is also an advantage that a security champion has insight into or participates in security work in IT operations to see the security challenges from both sides. Experience from there can thus be brought back to software development. The purpose is to prevent costly vulnerabilities that need to be fixed later in the software lifecycle. This is secure software by design in practice.

11.5 Identification of Threats During Software Development

Many threats and vulnerabilities are only possible to identify during the implementation and coding of systems. In secure agile software development, threat modelling is done as part of each sprint cycle, as illustrated by the activity "identify and prevent threats and vulnerabilities" in Fig. 11.4 in Sect. 11.4.3. Aspects of threat modelling during software development are described in the following sections. Threat modelling as part of general risk assessment is described in Sec. 19.4.2

11.5.1 Threat Modelling

> Threat modelling consists of mapping out possible ways in which a system or process can be abused and attacked.

A system can be attacked in an infinite number of ways, but not all of them are equally relevant and plausible. The challenge is precisely to identify the most relevant threats. Attacks can occur through all interfaces to the system in the environment where the system is in operation. Figure 11.5 shows an example of an online environment with a front-end web server connected to back-end logic and a database, which has a variety of interfaces through which an attack could potentially happen.

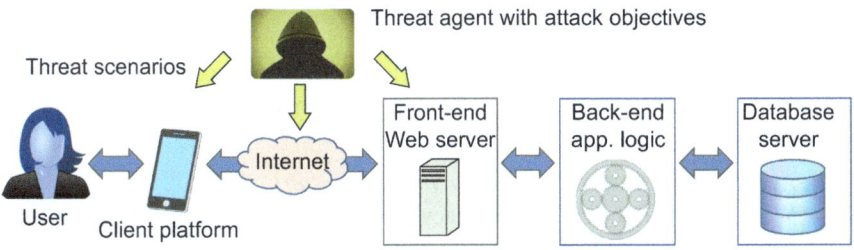

Fig. 11.5 Possible attacks through interfaces to a system and its environment

The development team must make every effort to identify relevant threat scenarios (possible ways of attacking), which can be challenging. A simple approach is to think like an attacker and describe the attacker's objectives as *attacker stories*, in the same way that the user's objectives can be described as user stories. The next two sections describe similarities between user stories and use cases on the one hand, and between attack stories and threat scenarios on the other.

11.5.2 User Stories and Use Cases

Requirement specification in agile software development is largely based on describing user stories. For each sprint cycle, the development team selects one or more user stories to implement.

- A user story looks at the system from the user's perspective and describes what the user wants to do. This can be expressed with the general sentence below:

    As [user], I want [function] to achieve [result].
    Let us assume that the agile development team is working on implementing an online pharmacy called *WebNetPharma*. User stories can represent process goals for customers or for employees, who all want to do different things with the system. For customers, a specific user story can be expressed like this:
    As a registered customer of WebNetPharma, I want to be able to order prescription medication.

- A use case looks at the system from the perspective of the designer and developer. A use case is a description of a set of interactions between elements in a system, between systems, or between systems and actors, as shown in Fig. 11.5. The result of playing a use case is that a goal is achieved for an actor or the system.

11.5.3 Attacker Stories and Threat Scenarios

Leaving aside the sinister aspects of a threat actor, and looking only at the threat actor as an entity with goals, these goals can be described as *attacker stories*, which in structure resemble user stories.

- An attacker story looks at the system from the attacker's perspective and describes what the attacker wants to do. This can be expressed with the general sentence below:

    As [attacker], I want [function] to achieve [result].
    We assume that the online pharmacy WebNetPharma has been put into operation and that the attacker wants to attack this. Attacker stories can represent goals for threat actors who want to do different things with the system. A specific attacker story can be expressed as follows:
    As an attacker, I want to steal the identities of customers of WebNetPharma to see what drugs they buy.

- A threat scenario (misuse case) considers the system from the perspective of a threat actor who designs ways for a system to be attacked. A threat scenario is a description of a set of interactions between modules in a system, between systems, or between systems and actors, as shown in Fig. 11.5, where parts of the interactions are controlled by a threat actor. The result of a threat scenario is that

11.5 Identification of Threats During Software Development

the threat actor achieves an attack goal, which from the victim's perspective is a security incident with negative impacts.

By viewing a threat scenario in the same way as a use case, it can be easier for the development team to identify different threat scenarios. Another method of identifying threats as part of software development is STRIDE, which is described in the next section.

11.5.4 STRIDE Threat Modelling for Software Development

STRIDE is a threat modelling approach that is performed as part of software development, typically in conjunction with design or coding. STRIDE is designed by Microsoft and is relatively prevalent in application development environments. The word STRIDE is the initial letter for six categories of security threats, as shown in Table 11.1.

Table 11.1 STRIDE threat modelling

Initial	Threat category	Description	Desired property
S	**Spoofing** Identity theft	*Can an attacker pretend to be somebody else by stealing another's identity?*	Authenticity
T	**Tampering** Tampering	*Can an attacker modify data processed by the system?*	Integrity
R	**Repudiation** Denial	*Can an attacker deny an attack without us being able to prove who did the attack?*	Non-repudiation, traceability
I	**Information disclosure** Data theft and leakage	*Can an attacker gain access to confidential and personal data?*	Confidentiality
D	**Denial of service** Denial of service	*Could an attacker block or diminish system availability?*	Availability
E	**Elevation of privilege** Extended accesses	*Can a logged-in attacker get more access than authorized for the account?*	Access control

The right-hand column of Table 11.1 specifies the desired security property that each threat potentially could breach. STRIDE is useful for brainstorming and identifying threats to a system. Put another way, STRIDE is intended to help answer the question of how somebody could attack the system you are developing.

STRIDE can typically be used as part of the activity *"identify and prevent threats and vulnerabilities"* in Fig. 11.4 in Sect. 11.4.3 that describes the process for secure, agile software development. Of course, STRIDE can also be used in conjunction with the process for secure by design, typically as part of phase 3 (design) and 4 (implementation) illustrated in Fig. 11.1 in Sect. 11.3.

11.6 Application Security

Application security focuses on security in client and server apps that communicate over the Internet. Most often, applications communicate over the HTTPS protocol with port number 443 (or the HTTP protocol with port 80) according to the Internet architecture described in Sect. 6.1. Continuous innovation produces increasingly advanced online services that communicate using the HTTPS protocol, and maintaining adequate security is a major challenge as complexity increases.

11.6.1 Web Applications' Exposure to Threats

In computer networks that allow HTTPS (and HTTP) traffic, the outer firewall must necessarily have open port number 443 (and port 80), which means that applications on front-end web servers are directly exposed to the entire Internet through these ports. At the same time, front-end web servers communicate with back-end servers and databases. Although that traffic is typically filtered through internal firewalls, an internal firewall will typically allow certain types of traffic from front-end web servers to back-end servers and databases because it is a necessary part of the application's function.

If vulnerabilities exist in applications in the front-end web server, it is easy to imagine how an attacker could exploit such vulnerabilities and then attack back-end servers, as shown in Fig. 11.6.

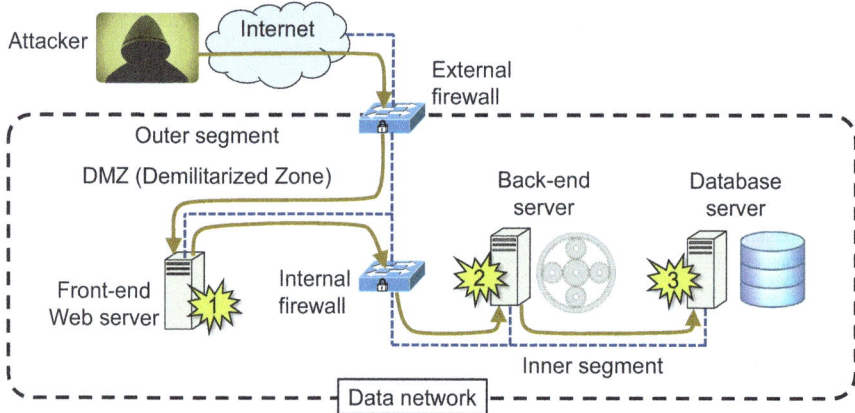

Fig. 11.6 Attacks through vulnerabilities in web applications

Clients communicating with application servers are also exposed to these types of threats because they can be attacked when they contact vulnerable applications.

Example of attacks based on the pattern of Fig. 11.6 are for example SQL-injection and XSS which are described in Sect. 11.7.

11.6 Application Security

Several initiatives have addressed the challenge of maintaining strong security in Internet applications, of which OWASP is probably the most prominent.

11.6.2 OWASP: The Open Web Application Security Project

OWASP[4] is a non-profit organization that aims to improve the security of applications and services on the Internet. With open-access resources, organizations and end users can get advice, guidance, and tools to strengthen application security and make informed decisions about real security risks in applications.

Companies, educational institutions, and individuals from all over the world are involved in the OWASP community, which produces freely available information materials, methods, tools and technologies.

OWASP has several parallel projects, some of which have been going on for a long time and are very mature, while others are more immature. It is up to volunteers to start and participate in projects. Two of the most mature projects are OWASP Top 10 and OWASP ASVS.

11.6.3 OWASP Top 10

The OWASP Top 10 is a guide that ranks the ten most critical security risks for web applications. The guide also provides advice on how vulnerabilities and risks can be prevented. The guide has been prepared by security experts from all over the world. The risk is ranked and based on the frequency of security flaws detected, the severity of the vulnerabilities and potential impacts. The purpose of the guide is to explain the most widespread security risks and how they can be prevented, so that developers of applications can avoid the vulnerabilities and minimize the risks. At regular intervals of a few years, OWASP revises the list of what they consider to be the ten most serious security risks. The most recent revision was conducted in 2021, and ranks the following application security risks:

1. **Broken Access Control:** Restrictions on what authenticated users are allowed to do are often not enforced properly. Attackers can exploit these flaws to gain unauthorized access, such as access to other users' accounts, to read sensitive information, or to modify data.
2. **Cryptographic Failures:** Such errors may arise from weaknesses in the implementation or configuration of cryptographic functions, cryptographic keys, or user error. Since cryptographic systems are supposed to protect sensitive data, such errors inevitably lead to exposure of sensitive data such as credit card numbers and passwords.

---
[4] https://owasp.org/

3. **Injection:** Injection failure occurs when manipulated input data is sent to an application as part of a command or request. The attacker's manipulated data can trick the application into performing unintentional and unauthorized actions. Different variants of injection are SQL, NoSQL, OS and LDAP injection.
4. **Insecure Design:** This category represents missing, inadequate, or weak security design. Insecure design is not the same as insecure implementation, they have different causes and must therefore be prevented in different ways. A secure design can still have implementation flaws that lead to exploitable vulnerabilities. Nor can an insecure design be fixed with a perfect implementation. Secure design is a culture and methodology that continuously evaluates threats and ensures that your system or application is designed to withstand known attack methods.
5. **Security Misconfiguration:** Misconfigured security is very common and difficult to avoid completely. It could be an insecure default configuration, unintentional open cloud storage, misconfigured HTTP(S) headers, or error messages revealing sensitive information.
6. **Vulnerable and Outdated Components:** These may be outdated modules, application libraries, and APIs that lack security updates, or where the manufacturer has stopped making updates. Inadequate updating and long update cycles can often lead to vulnerable components. In the case of known vulnerabilities that are not patched expediently, the organization will remain vulnerable for long periods.
7. **Identification and Authentication Failures:** Authentication and management of sessions are often implemented incorrectly, allowing attackers, for example, to compromise passwords, keys, or cookies to steal identities or take over control of a session.
8. **Software and Data Integrity Failures:** Such security flaws can occur when the application or infrastructure does not protect against integrity violations. An example of this is when an application fetches plugins, libraries, or modules from unreliable sources and content delivery networks (CDNs). Such violations can also occur when automatically updating software without authenticating the new version. This allows attackers to upload their own malicious updates. The solution to this risk is typically to use digital signature of data and software.
9. **Security Logging and Monitoring Failures:** Inadequate logging and monitoring, combined with lack of or ineffective integration with incident response, allows attackers to commit mischief without detection or traceability. Studies show that the average time to detect a security incident is over 200 days from the time the attack first began.
10. **Server-Side Request Forgery:** SSRF is a type of attack that abuses the server so that the client (which is the attacker) can retrieve, or manipulate, information that the server technically has access to, but to which the client (attacker) should not have direct access. Various controls can prevent such errors, e.g. by the server validating that all requests from clients comply with the security policy.

Staying on top of the most critical security risks and vulnerabilities is very important for anyone who develops web applications and manages a website. The OWASP Top 10 is a good starting point for creating awareness on this topic, and should be mandatory curriculum learning for all application developers. To illustrate typical risks described in OWASP Top 10, we describe SQL injection and XSS in Sect. 11.7.

11.6.4 OWASP ASVS

OWASP ASVS (Application Security Verification Standard) is a guide for technical security testing, which also defines a list of requirements for secure development. The primary goal of OWASP ASVS is to define best practices for safe development and security testing. The standard focuses on controls that protect against vulnerabilities described in the OWASP Top 10. The guide provides a basis for establishing a level of confidence in the security of web applications. ASVS can be used for:

- *Security Benchmark*: Provides application developers and application owners with a benchmark to assess the degree of security assurance of their web applications.
- *Secure Development Guideline*: Provides guidance to developers on how to build applications to meet security requirements.
- *Procurement requirements*: Provides a basis for specifying security requirements when ordering and purchasing systems and application software.

11.7 Examples of Attacks Against Applications

There are many different attack methods, and threat actors are constantly working to develop new ones. In this section, we describe two classic attack methods: SQL injection and a variant of XSS (Cross Site Scripting).

11.7.1 SQL Injection

SQL (Structured Query Language) is a machine-readable language with commands for interacting with a DBMS (Database Management System). A DBMS is the application used to manage *databases* containing tables and collections of structured data. SQL injection is an attack against a DBMS that consists of manipulating SQL commands to perform operations on a database in an unintended way. The attack is possible if SQL commands take parameters directly from user input. If an attacker knows or can guess how to build the SQL command, it is possible to send manipulated input data that causes the original SQL command to become an attack command. This could allow the attacker to extract sensitive data from a database, typically usernames, passwords and email addresses, modify the database, or gain unauthorized access to systems.

Imagine a login process where the user specifies their user ID and password. This typically results in applications building up an SQL command that is sent to the DBMS to check if the username and password are correct. The SQL command below is an example:

```
SELECT * FROM users WHERE userid = '$userid' AND password = '$password'
```

The command is built with user-input `$userid` and `$password`, and makes a lookup in the database named `users` that returns data for all matches with the correct combination of `userid` and `password`. We also assume that the application the user gets logged in if the DBMS returns a hit from the database. DBMS returns no results if the combination of `userid` and `password` is not present in the database.

Let us assume that the attacker enters the following text as the user ID, which is then added to the parameter `$userid`:

```
' OR 1=1--
```

The first character ` is an apostrophe that terminates the encapsulation with apostrophes upon the construction of the SQL command. The subsequent logical expression `OR 1=1` is always `TRUE`, which is thus interpreted as a match at the lookup. The last two characters `--` (two hyphens) are metacharacters which means that subsequent data should be interpreted as comments. Thus, the rest of the command is commented out, no matter what is written there.

The manipulated SQL command would thus look like this:

```
SELECT * FROM users WHERE userid='' OR 1=1-- ' AND password='$password'
```

The request will yield `TRUE` for every single user ID in the database, and DBMS will return data for everyone. The first user ID in the database is often the admin user, so the attacker is typically logged in as an administrator.

SQL injection is a well-known attack, and at the same time a solved problem that can be easily prevented by standard controls. However, there are still many web applications that are vulnerable to SQL injection, and many attacks still take place this way.

The most common method of preventing SQL injection consists of employing parameterized commands. A typical process for building parameterized SQL commands goes as follows:

1. *Prepare*: First, the application builds a template for the SQL command that will be sent to the DBMS. Certain parameters are still unspecified (marked "?" below):

```
SELECT * FROM users WHERE userid = '?' AND password = '?'
```

2. *Compile*: Then, the DBMS compiles the command template without executing the command. The structure of the command is thus fixed and cannot be manipulated by injection.
3. *Execute*: Finally, the application binds values for the parameters in the template and allows DBMS to execute the SQL command.

Parameterized SQL commands prevent SQL injection because the values of the parameters are added after the template is compiled with a fixed structure. Manipulated input values, as in the example above, would result in an SQL command with the expected structure, but with parameters that would not match anything in the database.

11.7.2 XSS: Cross-Site Scripting

Websites often send JavaScript to the client to provide good functionality on the web page. From a security perspective this seems risky because JavaScript is active code that can potentially cause mischief to the client, but that is usually unproblematic as long as the JavaScript comes from trustworthy websites. However, there are ways for attackers to trick websites into sending malicious JavaScript to clients, and that is precisely what XSS does. If a website (and browser) is vulnerable to XSS attacks, it is possible for attackers to inject JavaScript malware into clients simply when a user is visiting a (vulnerable) website. This is a form of drive-by attack, meaning that the victim user is only visiting a website, i.e. driving by.

There are various forms of XSS, where three common types are persistent (stored) XSS, reflected XSS, and DOM-based XSS. As an example, persistent XSS is described below.

Imagine a website where users can provide input that is stored and made visible to other users. This could be, for example, a website where users can ask questions and give answers to each other. An attacker could attempt to send a JavaScript in response to a question. If the website is vulnerable to XSS, the JavaScript will be stored and made visible to anyone visiting the site. All clients/browsers that visit the website will thus receive the JavaScript. If the browser is unable to filter out the JavaScript, the script will be executed. This is illustrated by the simple scenario in Fig. 11.7.

Fig. 11.7 Persistent XSS

There is no single control that prevents all types of XSS attacks. In general, both server and client filter input to protect themselves from manipulated data and malicious scripts. The OWASP Top 10 describes in detail methods for avoiding XSS vulnerabilities.

11.7.3 Security in the cloud

Using the "cloud" as a computing platform is gradually becoming the most common way to provide and use digital services. Therefore, it is important to understand what consequences it has for cybersecurity to put services in the cloud. Before diving into the security aspects, it is helpful to explain key concepts around cloud infrastructures.

11.7.4 Cloud services

Sourcing is the choice of who will perform a work task. The sourcing strategy is thus the strategy that underlies how the business makes such choices. Outsourcing means that companies separate tasks they have previously performed themselves to third parties, such as manufacturing parts for a product or assembling finished products. In principle, most IT services can be outsourced, such as system development, administration, operation and maintenance of hardware, software platforms, and applications. Traditionally, such services were provided by consulting firms based on a delivery model with customized contracts, dedicated IT resources and a close business relationship between customer and supplier.

Cloud services have many similarities with traditional outsourcing of IT services, but cloud services follow a more agile delivery model. The peculiarity of cloud services is that the customer has a less close business relationship with the cloud provider, and resources at the cloud provider (such as hardware, software,

11.7 Examples of Attacks Against Applications

data storage and physical buildings) are shared among several customers at the same time. A definition of cloud services is given by NIST:[5]

> *Cloud computing is a model for enabling ubiquitous, convenient, on-demand network access to a shared pool of configurable computing resources (e.g., networks, servers, storage, applications, and services) that can be rapidly provisioned and released with minimal management effort or service provider interaction.*
> NIST SP 800-145

The three service models of cloud computing are: *Infrastructure-as-a-Service (IaaS), PaaS (Platform-as-a-Service)* and *SaaS (Software-as-a-Service)*, which are explained below. IT environments without cloud solutions are called *On-Prem (On Premises)*. The set of hierarchical layers in an IT infrastructure is called a *IT stack*. The same stack is called a cloud stack when reflected in the IaaS, PaaS, and SaaS cloud models, as shown in Fig. 11.8.

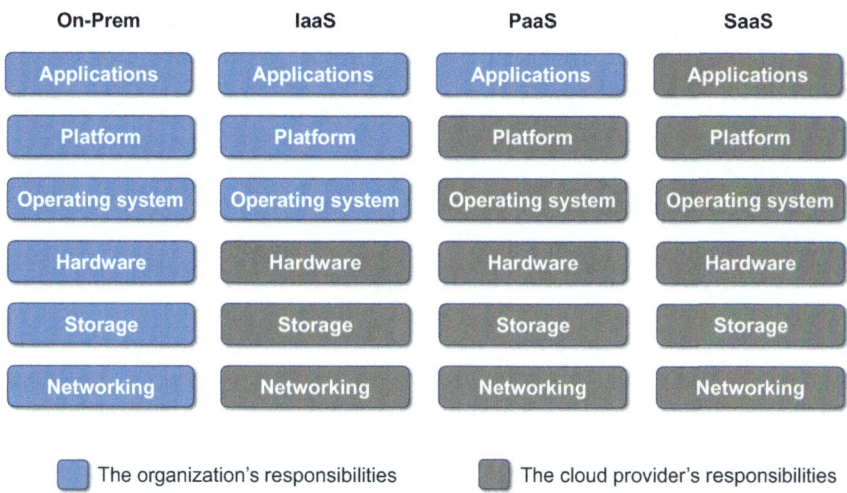

Fig. 11.8 Shared responsibility for layers in the IT stack in different cloud models

Who is responsible for each layer in the stack depends on the cloud model in general, but details and variations are of course specified in the SLA.

On-Prem means that the organization owns and operates hardware and software itself "on-premises" i.e. located in the office buildings of the organization or some other location they control.

---
[5] https://csrc.nist.gov/publications/detail/sp/800-145/final

The IaaS model offers (virtual) machine and storage resources. IaaS is used by IT architects to design and set up your company's IT infrastructure. The enterprise must install and operate operating systems and develop/install applications to operate services.

The PaaS model offers a platform for developing and operating applications. This platform consists of tools for developing, installing, and running applications. PaaS is used by companies that develop/install and operate their applications, but at the same time do not want to operate the systems where the applications run.

The SaaS model offers ready-made applications to the customer. SaaS products are services we as end users know best, such as Google Apps, Gmail (for email), Dropbox (for storage), and Office 365 for productivity and collaboration. SaaS offers standardized services that your business can immediately put in production.

SaaS services are managed exclusively by the cloud provider, limiting how a customer can tailor the product to their needs. If the customer needs more flexibility, the alternative is to purchase a PaaS or IaaS solution to be able to build and install custom applications on top of a development platform or on top of a virtual machine.

A cloud provider that offers services on the open market is called the public cloud, in the same way that a store or café is open to everyone. A cloud provider that only provides services to a closed group of customers is called a private cloud, in the same way that a private club only provides access to members. A hybrid cloud is seen from the customer's perspective and means that the customer uses services from both public and private clouds.

11.7.5 Cloud Security

Cloud security is the protection of data, applications, and infrastructures from attacks in the cloud. Many aspects of cloud security are the same as any on-premises IT architecture. At the same time, there are differences that are important to be aware of. An important difference is that security functions can also be virtualized, and that the cost of acquiring e.g. a virtualized firewall is far lower than acquiring a physical firewall.

Intuitively, it seems scary to store and process the company's data in giant data centres that are spread around the world, and that are operated by third parties. Several, if not all, cloud providers have segmented their infrastructure into regions, allowing customers to more easily control where data is stored and processed. It is also important to note the difference between "user data" and "user credentials". Although the user data is stored in a segment that is limited to a specific region, the federated IAM often mean that user credentials are stored globally.

At the same time, it is reassuring that cloud providers are typically large professional players with abundant expertise and resources for secure operations, who in most cases can maintain higher security assurance than most organizations are capable of. However, it is important to also be aware that cloud providers are

11.7 Examples of Attacks Against Applications

continuously exposed to attacks and that they too experience security incidents. Such events do not necessarily apply to the entire cloud service, but for example one or more SaaS services. Cloud providers regularly publish operational statistics, and it is the customer's responsibility to keep up with the status.

Security in the cloud is a shared responsibility between cloud provider and customer. Figure 11.9 shows a simplified overview of shared responsibility for security functions in different models. The overview is inspired by Microsoft.[6]

To better understand what it means for the cloud provider to be responsible for system security in Fig. 11.9, it is useful to be aware that system security is based, among other things, on virtual machines, as illustrated in Fig. 3.9 in Sect. 3.6. The cloud provider owns and operates both hypervisor hardware and operating systems, and this operating function often takes place from a country other than where the hardware is physically located. At the same time, personnel with physical access to the hardware will technically also have access to virtual hardware, systems and data.

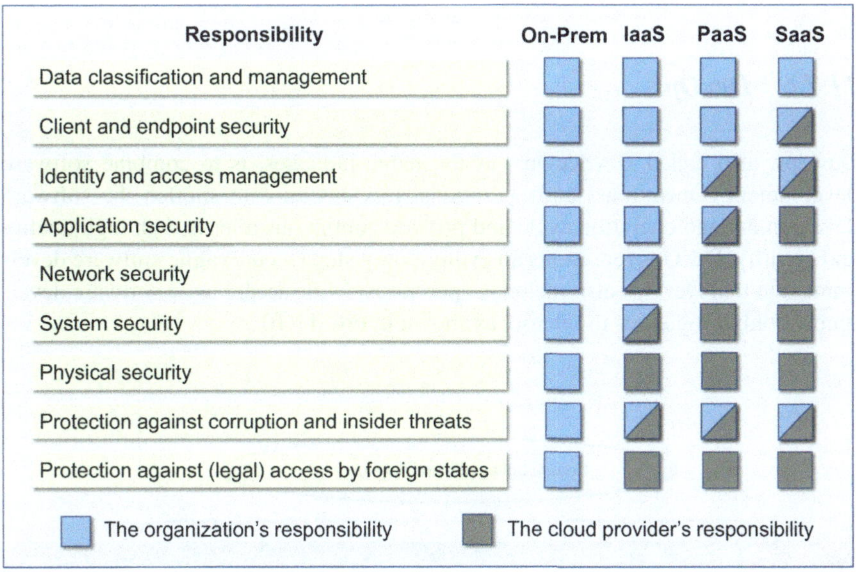

Responsibility	On-Prem	IaaS	PaaS	SaaS
Data classification and management	■	■	■	■
Client and endpoint security	■	■	■	◪
Identity and access management	■	■	◪	◪
Application security	■	■	◪	■
Network security	■	◪	■	■
System security	■	◪	■	■
Physical security	■	■	■	■
Protection against corruption and insider threats	■	◪	◪	◪
Protection against (legal) access by foreign states	■	■	■	■

□ The organization's responsibility ■ The cloud provider's responsibility

Fig. 11.9 Shared responsibility for security in different cloud models

The bottom two responsibilities for information security in Fig. 11.9 are not described in Microsoft's model for shared cloud security responsibilities, but is included here because these security challenges are specifically relevant for cloud services.

---

[6] https://docs.microsoft.com/en-us/azure/security/fundamentals/shared-responsibility

The potential for corruption and insider threats is increased in the cloud because privileged employees of the cloud provider can technically access customers' data and applications. The likelihood of such threats is often correlated with the degree of instability in countries and regions as well as the (security) culture within the cloud provider's organization. Specifically, the challenges relate to the question of who can technically access the data, and how a cloud service can help prevent technical access from being misused, by having controls and guidelines in place that can prevent and detect irregularities and deviations.

Another security challenge is the extent to which a cloud provider is legally obliged or can be forced to let government authorities get access to customers' data and applications when required by law. This applies not only in cases of government surveillance and spying, but may also apply in cases of criminal investigation. It can be difficult to know to what extent governments can demand access in general, and perhaps even harder to know whether it has actually happened, because that type of access takes place in secret.

11.7.6 DevOps

DevOps, also called DevSecOps, as the name indicates, is to combine software development (Operations) and operations. DevOps aims to shorten the software development and operation cycle and provide continuous release with high quality and security. DevOps represents an evolutionary step (secure) agile software development in that DevOps also includes operations, while secure agile software development only progresses to release, as shown in Fig. 11.10.

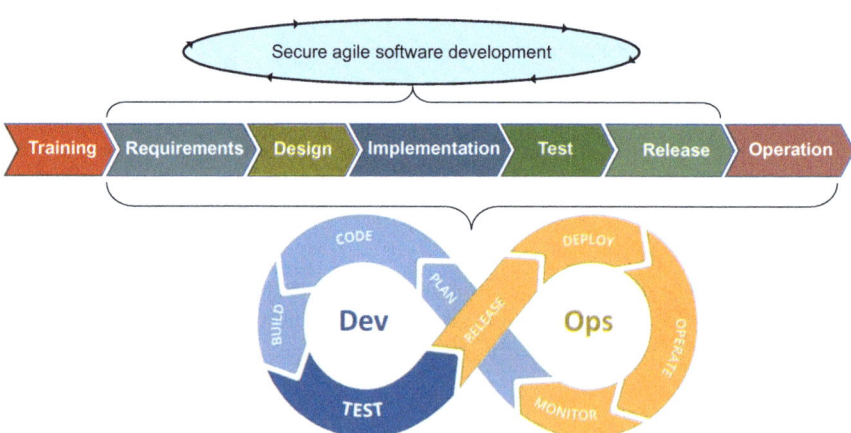

Fig. 11.10 The correspondence between secure agile software development, security by design and DevSecOps

DevOps is a totally cloud-based approach. All tools needed throughout the DevOps cycle reside in the cloud, from development tools and testing tools to tools for deploying, operations and monitoring. Cloud models for DevOps must be either IaaS or PaaS, with each model having its advantages and disadvantages.

At its core, DevOps is an automation of secure agile development methodology. The idea is to give developers the opportunity to implement new requirements and put the new version of the software into production in a very short time. At the same time, DevOps considers security in a more automated way and at an early stage of the cycle. Term *Shift-left* in the context of DevOps means that there is more focus on security on the left side of the figure-of-eight in Fig. 11.10 (during development and testing), with the result that there will be fewer security vulnerabilities and incidents to deal with on the right side of the figure-of-eight (during operation).

11.7.7 Cloud Security Alliance

The Cloud Security Alliance (CSA)[7] is an organization with over 80,000 members worldwide dedicated to defining and raising awareness of best practices for securing cloud services. CSA applies the expertise of its corporate and individual members in the private and public sectors to offer research, education, certification, seminars, and products related to security for cloud services. CSA is independent of private individual companies, creating a neutral network of expertise where different parties can work together to create and maintain a sustainable ecosystem for cloud services.

Cybersecurity is often mentioned as the biggest source of uncertainty for companies that want to adopt cloud services. CSA publishes guides and checklists that are useful for increasing their client competence regarding security among users and customers of cloud services, and for cloud providers to better adapt their products to expected security requirements. A useful guide is eg. *Software Defined Perimeter (SDP) and Zero Trust* from the Cloud Security Alliance (CSA 2020).

Since cloud services form part of an organization's supply chain, it needs to consider supply chain risks as described in Sect. 2.1.6.

11.8 Tasks

1. Secure by design
    (a) What does it mean that something is secure by design?
    (b) What are the phases in Microsoft's SDL (Secure Development Lifecycle)?
    (c) Mention what should be done in general, and what a business should do to ensure good cybersecurity training (1st phase).

---

[7] https://cloudsecurityalliance.org/

2. Secure agile system development

    (a) In what way is agile development different from system development with the waterfall method?
    (b) How can security be integrated into agile system development?

3. Security in the cloud

    (a) Mention factors that can give cloud solutions stronger security than on-prem solutions.
    (b) Mention factors that can give cloud solutions weaker security than on-prem solutions.
    (c) Explain the meaning of "Shift-Left" in DevSecOps.

Chapter 12
Physical Information Security

The time to repair the roof is when the sun is shining.

John F. Kennedy

Hacking remotely from somewhere on the Internet is not the only way information assets can be compromised. There are also physical threats that must be considered, where logical controls used for system, application, and network security would be ineffective. Controls for physical security include safety measure for staff, physical access control and intrusion detection, physical site design and shielding of equipment, and the continuity of power supply and other support systems. Physical security mechanisms protect people, data, equipment, systems, facilities, and information assets in general. Building cybersecurity without physical security would be like building a house on sand. Interestingly, the trust anchor for many cybersecurity controls is typically some form of physical security, like protecting an authentication token or storing a cryptographic key in physical hardware which must be locked away.

The learning objectives of this chapter are to understand physical security as a necessary link in the whole information security chain, and to become familiar with important security controls used to achieve specific goals for physical security.

12.1 Physical Security Goals and Threats

Threat modelling as part of risk management is described in Sect. 19.4.2, and consists of identifying ways in which the organization's information assets could be attacked and harmed. For physical security, threat modelling follows the same general principle which is to ask: *"how can our physical infrastructure and assets be harmed?"* Before trying to find good answers, we might first have to ask: *"what are the physical assets we want to protect, and what are their security goals?"* To answer that question, we can list obvious things and security goals, such as:

(a) Safety of staff and their working environment both onsite and offsite,
(b) Controlled authorized access of people to areas within the organization's sites,

(c) Physical and technological shielding of the organization's sites and equipment,
(d) Continuity of technical support systems such as electricity, gas, water, and ventilation.

The general threats to these physical security goals are the same as the general threats to assets as illustrated in Fig. 1.1 in Sect 1.2. These threats are briefly described below.

1. Natural environmental events can be floods, earthquakes, tsunamis, storms, landslides, extreme temperatures (hot or cold), volcanos, and fires. Given sufficient intensity, these threats can certainly breach the security goals mentioned above.
2. Technical faults can originate from within the organization or from outside, and there is of course a wide range of potential technical fault. A technical fault from inside representing a physical threat could be when an electrical short circuit causes fire. A technical fault from outside representing a physical threat could be when there is an electricity outage at the electricity supplier. These threats primarily can to a lesser or greater degree breach all the security goals above.
3. Adversarial attacks are e.g. unauthorized physical access, forced entry by firearms and explosives, physical tampering and destruction by insiders, vandalism, brake-ins and theft.
4. Human errors occur because making mistakes is human. Employee errors are often the cause of physical damage.
5. Politically driven action are e.g. strikes, riots, civil unrest, terrorist attacks and warfare.

The relationship between the above-described threat sources and physical security goals is illustrated in Fig. 12.1.

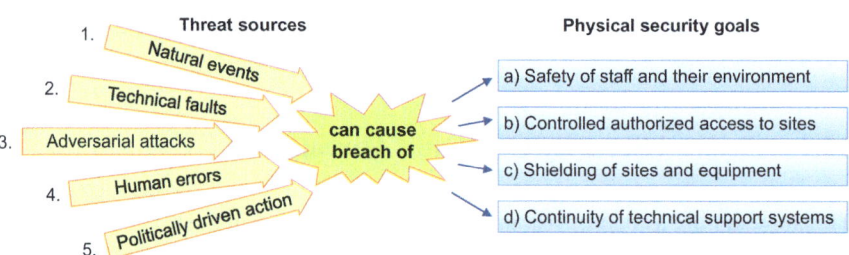

Fig. 12.1 Relationships between threats and physical security goals

The various controls for physical security can be preventive, detective and corrective controls, in the same way as for cybersecurity controls illustrated in Fig. 1.6 in Sect 1.5. In the sections below we describe physical security measures according to the four physical security goals mentioned above.

12.2 Safety of Staff and Their Environment

Life safety goals should always take precedence over all other types of security goals. Safety deals with the protection of life and assets against fire, natural disasters, and devastating accidents.

12.2.1 HSE: Health, Safety and Environment

HSE is an interdisciplinary field covering the study and implementation of environmental protection and safeguard of people's health and safety, especially in organizations, but also in public spaces. It is what organizations must do to make sure that their activities do not cause harm.

From a safety perspective, HSE involves the identification of potential workplace hazards and the implementation of controls for reducing the likelihood of occurrence, the ability to detect, and the capacity to handle accidents and hazardous situations. It also includes training of personnel in accident prevention, accident response, use of equipment and protective clothing, and emergency readiness in general.

The HSE discipline is well regulated in most countries. HSE managers must identify and understand relevant EHS regulations, which then must be communicated through the organization in order for suitable measures to be implemented.

12.2.2 Weighing Priorities for Safety Against Security

While the safety of staff must always have the highest priority, the prioritization between physical security goals brings dilemmas. Safety measures to protect people sometimes result in weaker physical security for controlled authorized access to assets. When a door lock is electronically controlled, the organization must decide how the lock should behave in the case of a power failure. There are two options:

- **Fail safe:** The door unlocks when electric power is removed. This gives the greatest freedom for people to escape whenever needed such as in the case of fire or other safety hazards. However, this creates a vulnerability whereby an attacker could deliberately cause a power failure to unlock doors and thereby get unauthorized physical access.
- **Fail secure:** The door locks when electric power is removed. This ensures that unauthorized will not get access in the case of power failure. However, this can make people trapped and can restrict their freedom to escape in the case fire or other safety hazards.

Obviously, fail-secure locks must be used with care. They can be used for sensitive high-security areas like vaults, IT equipment rooms, or areas with hazardous materials, where one might expect that attackers could trigger or exploit power failure to gain unauthorized access. It must still be possible to open with a mechanical override, like a regular key controlled by a few authorized individuals.

12.3 Controlled Authorized Access

Physical access control ensures that only authorized staff can enter sites and restricts their movements within the site and movements from one security zone to the next.

12.3.1 Secure Entry Points

There should be a limited and well-defined set of entry points to sites. For guests there should be a single entry point with a front desk and procedures for sign-in before entering the site. Depending on the sensitivity of the zone, the policy can require guests to be escorted at all times inside the site.

Access mechanisms include the use of access cards, biometrics or two-factor authentication such as an access card combined with a secret PIN, as described in Chap. 8. Mantraps (double security doors) should be considered for access to sensitive areas.

Access points such as delivery and loading zones where external people can enter parts of the premises should be controlled and, if possible, isolated from information assets to avoid potential unauthorized access to such assets.

12.3.2 Physical Security Monitoring

Electronic security monitoring can be used to detect and deter unauthorized physical access. While security guards traditionally have been monitoring security cameras, automated monitoring based on AI is increasingly taking over that job, which only alerts security guards when something suspicious has been detected by the AI system. We humans are unable to focus and concentrate on video surveillance monitors over a long time, and will typically fail to notice suspicious activity on the monitors even if we sit right in from of them. AI monitoring systems are also cheaper than human security guards, of course. Recordings from video surveillance are valuable for investigating incidents and crime. Video surveillance is also called CCTV (Closed-Circuit TV).

Security monitoring can also include detection of sound and vibration, as well as contact detectors that trigger when a contact is made or broken for entry points such as doors and windows. Sound detectors can recognize steps, voices, the sound of breaking glass or other relevant sounds which can trigger an alarm to alert security guards.

Detection of people can be based on infra-red cameras which trigger an alarm when an object with temperature above a certain threshold passes through their field of view. Detection of motions can be based on ultrasound or laser beams which get disturbed by objects moving through the controlled zone.

Monitoring and video recording of people must be done in compliance with local laws and regulations including data protection legislation such as GDPR described in Sect. 10.5.

12.4 Shielding of Sites and Equipment

Physical sites and equipment need shielding against unauthorized access, spying, physical interference, theft, damage and destruction. In theory, attackers can use any physical vector such as physical force, lockpicking, electromagnetic signals and radiation, visual observation by people or cameras, sound perception by people of microphones, laser beams, magnetic fields, electric fields, and quantities of liquids and gases. Threat agents have a large repertoire of methods to choose from, and hence a large variety of controls would be needed to protect against all these attack vectors. As part of threat modelling and risk assessment, each organization must assess relevant types of attack for which controls might be needed.

12.4.1 Physical Perimeter Defences

Perimeter defences must be implemented based on layers, which means that they should work together in a tiered architecture. The principle is that if one layer fails, other layers are still able to protect the assets. The layered design can best be understood by starting from the outer perimeter toward the asset. For example, around the site there can be a fence, then the walls of the building, then entry points with access control, then a guard, then internal zones with card access, and finally locked server rooms and safes. Camera surveillance and alarms can be used around each layer. So, if an attacker were able to climb over the fence, climb up the wall and force open a window, they would still have to circumvent several layers of controls such as alarms and locked internal doors before getting their hands on valuable information assets.

The strength of each of the perimeter layers should be specified in accordance with the security requirements related to the assets to be protected. For each layer, the principle of the weakest link applies, because a smart thief takes the path of least resistance. For example, there is little purpose of having an extremely fortified and secure door if the surrounding walls are made from lightweight plaster boards than can easily be penetrated. Between a low-security zone and a high-security zone, additional barriers and perimeters might be needed. Many office spaces have dropped ceilings, meaning that the interior walls only extend to the dropped ceiling and not to the true ceiling. In this case, interior walls can be bypassed if the intruder is able to lift a ceiling panel and climb across the dropped ceiling to other rooms.

The organization can consider preparing for additional physical security measures during situations of heightened threats.

12.4.2 Working in Secure Areas

Organizations should define policies for working within secure areas. Below are some aspects to consider.

(a) Clean desk: Sensitive information, whether it is stored on paper or on electronic storage media, should be locked away when not needed and in the absence of the owner or custodian. In that way the sensitive information is not at risk when unauthorized staff pass by during office hours, or when cleaning or maintenance staff enter the office space outside of office hours. The storage facility can be a safe, cabinet, desk with locked drawers, or a separate locked room.

(b) Clear printer: Printers in common office areas should be configured with an authentication function, so only the originator can get their printouts and only when standing next to the printer. Alternatively, the policy can specify that the originator must collect outputs from printers immediately.

(c) Locked device: Endpoint devices should be logged off or locked when not in use or when unattended to prevent the device from being misused or the screen from being observed by unauthorized persons in the working environment. The timeout for automatic logout should be configured according to need.

(d) No filming or recording: There should be a policy for carrying and using equipment for photographic, video, and audio recording, such as cameras or mobile phones with recording capabilities. If such recording is prohibited by policy, additional strict enforcement could require all such equipment to be left outside of secure site areas.

(e) Spotting intruders: In large organizations it is impossible for staff to know everyone, and to always tell whether somebody is a genuine employee simply by recognizing their face. Staff should be trained in spotting potential intruders, and how to act when suspicion arises. This is described in detail in Sect. 13.6 on physical social engineering.

12.4 Shielding of Sites and Equipment 277

12.4.3 Security of Assets Off-Site

In today's modern working environment, organizations typically allow their employees to carry equipment outside of secure sites, or to use their personal devices for work (BYOD). Below are some guidelines to consider for the protection of devices and information assets when used off-premises.

(a) Consider the security of the off-site working environment, which e.g. could be your home, a hotel room, the premises of another organization, a café, a train, an airport or an airplane. The security check at airports is an area where you must pay particular attention to your IT devices as they sit on the X-ray conveyer belt while you might be held up for a body scan.
(b) Never leave equipment unattended in public and unsecured places. Do not put a laptop in check-in luggage when flying.
(c) Follow the device manufacturers' instructions for protection, and for implementing location tracking and the ability for remote wiping of devices whenever relevant.

12.4.4 Protecting Against Physical and Environmental Threats

Physical and environmental threats are e.g. fire, floods, landslides, earthquakes, explosion, civil unrest, toxic waste, as well as extreme heat and extreme cold. Below are elements to consider during the planning of the site to mitigate such threats.

(a) Consider the location, such as appropriate floor in an office, and potential for flooding or exposure to other external environmental threats.
(b) Consider urban threats regarding the potential for criminal activity, political unrest, or terrorist attacks. The site and its immediate environment should be designed in a way that reduces urban threats. For this purpose, CPTED (Crime Prevention Through Environmental Design)[1] is the framework to follow. Secure environmental design can also take aesthetics into consideration. For example, instead of installing concrete or steel bollards, it is possible to install statues or water fountains that serve as both as physical barriers and as aesthetic features.

12.4.5 Emission Security

Attackers do not necessarily need to enter a site or steal equipment to get to sensitive information. A fundamental aspect of sites and devices is that they emit signals in the form of sound, electric currents, light and electromagnetic waves that can be

---

[1] https://www.cpted.net/

picked up by sensitive receptors placed at the exterior. TEMPEST is a specification from the NSA and NATO regarding the interception of information systems through leaked physical signals, including unintended electromagnetic or electrical signals, sounds and vibrations. TEMPEST covers methods of spying on others as well as methods for shielding equipment from such espionage. For example, it is relatively easy to intercept text and images displayed on a computer screen several hundred metres away just by using a directionally sensitive antenna that captures electromagnetic radiation from the screen. Methods for shielding are also known as emission security (EMSEC). Emission security is related to side channels described in Sect. 3.10. Possible attacks and mitigation strategies are briefly described below.

(a) Electromagnetic emission from computer screens can be picked up from several hundred meters away with sensitive antennas and receptors, and if there is no obstruction in the way. It is then possible for attackers on the outside to regenerate what is shown on the computer screen. Solid walls of concrete will dampen the signals, but only a room built as a Faraday cage will totally block emissions. Carrying cases can also be Faraday cages, where e.g. smartphones can be stored and completely hidden from mobile base stations and other radio-detection equipment. Remember that your smartphone keeps communicating with nearby base stations even if you switch it off. Faraday cages are described in Sect. 7.1.1.

(b) Oral conversation generates soundwaves which cause minute vibrations in windows and objects in the room. Attackers on the outside can point laser beams on windows or other objects, where the perturbations in the reflected light caused by the vibrations can be detected. It is then possible for the attackers to regenerate the sound of the conversations. A mitigation strategy against such attacks is e.g. to ensure that sensitive conversations are held in rooms without exterior windows. Unfortunately for some executives, this means that they cannot discuss sensitive business matters in their top floor corner office.

12.5 Continuity of Technical Support Services

Ensuring safety for staff, controlled authorized access and robust perimeter security represents the foundation for physical security. However, it is also essential to ensure reliable power supply, air conditioning, and water. Addressing these support services can be considered part of physical information security, because their malfunction or disruption could negatively affect information security in the organization.

12.5.1 Uninterruptible Power Supply

An uninterruptible power supply (UPS) provides continual electric power when the mains power fails in some way. A UPS is typically used to protect equipment such as computers, network routers, whole data centres or other electrical equipment where an unexpected power disruption could cause serious business disruption or hazards. UPS capacity varies in size from protecting a single computer to powering entire data centres or buildings. The battery power backup of most UPSs is typically relatively short, i.e. from several minutes up to a few hours, but provides sufficient time for starting a generator or for properly shutting down sensitive equipment. The mains electric power is provided as AC (Alternating Current) which means that the voltage varies as a sinusoid with frequency of 50 or 60 Hz depending on the country. Although voltage alternates with the AC frequency, AC power is characterised by its average voltage which typically is 110 or 220 V, depending on the country. AC power can be converted to DC (Direct Current) power with a fixed voltage used to charge batteries, typically 12 or 24 V, and back again to AC power. Quality problems with mains AC power supply that a UPS can handle include the following:

- Fault: momentary interruption in power
- Blackout: prolonged interruption in power
- Sag: momentary drop in the power voltage level
- Brownout: prolonged drop in the power voltage level
- Spike: momentary increase in the power voltage level
- Surge: prolonged increase in the power voltage level
- Frequency instability and noise: Deviation from normal mains AC frequency

The main types of UPS systems are illustrated in Fig. 12.2 and briefly described below.

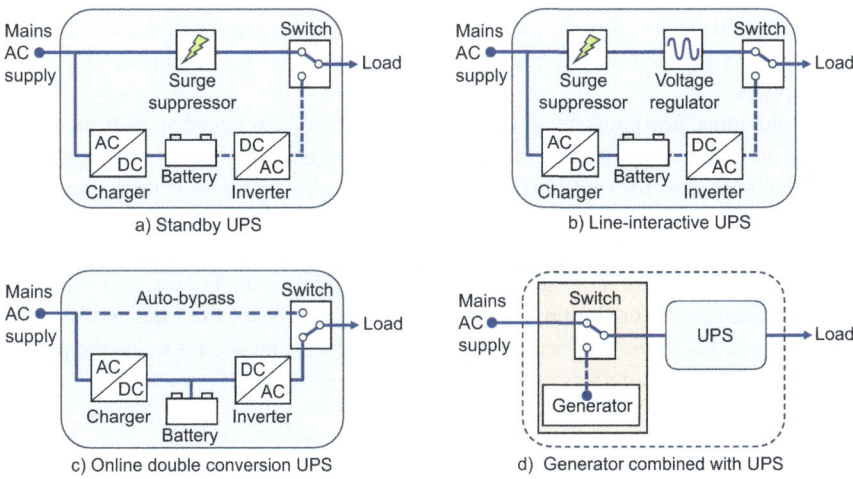

Fig. 12.2 The main types of UPS systems

(a) Standby UPS: AC power passes directly through under normal conditions, and the UPS switches to battery mode when a power failure is detected. A standby UPS protects against spikes caused e.g. by lightning but does not eliminate voltage fluctuations during minor sags and surges, nor can it correct frequency distortions.
(b) Line-Interactive UPS: Similarly to standby UPS, AC power passes directly through under normal conditions, and the line-interactive UPS switches to battery mode when a power failure is detected. In addition, line-interactive UPS provides conditioned power, eliminating voltage fluctuations.
(c) Online double Conversion UPS: The mains AC power is converted to DC and back again to high-quality AC power, free from voltage fluctuations and other distortions common in mains power supply. A disadvantage is that a small fraction of power is consumed by the double conversion.
(d) The combination of UPS and a power generator running on diesel or petrol provides reliable continuous power in case of long-term outage of the mains power supply. The UPS provides intermediate backup power until the generator can be started.

Because digital business processes are so essential in today's world, power failures can have much greater consequences than 25 years ago. Organizations should identify critical systems that absolutely need continuous power supply, and then plan how uninterruptible power can be provided for those systems.

12.5.2 *Water, Gas, Air Temperature and Humidity*

Running water is standard in most sites, and some sites also have gas supply for heating and cooking, and steam/hot water supply for heating. There will of course be problems if these utility services are interrupted. The organization should assess the reliability of each of these services, and be prepared for situations of service interruption.

In cold areas, heating is necessary at all times to prevent water pipes from freezing and bursting. The location of the shutoff valve must be known, and someone must be responsible for shutting of water if a water pipe bursts, or if large quantities of water are leaking for other reasons. Rooms with water must have positive drains, which means that overflowing water will flow out and not in.

Similarly, the location of the shutoff-valve for gas must be known, and someone must be responsible for shutting of the gas line e.g. in case of fire. Site and security personnel should know the locations of shutoff valves, and should know the procedures to follow in emergency situations.

In case of exterior flooding, water can enter sites and building. The placement of IT equipment should be done in a way that reduces exposure to flooding. For example, it might not be a good idea to place IT equipment in the basement.

Computing equipment requires appropriate temperature and humidity for reliable operation. Too high humidity can cause corrosion, and too low humidity can cause static electricity. Discharge of static electricity onto electronics can turn devices into junk in an instant. Too low temperatures can cause moving mechanisms to slow or stop, and too high temperatures can damage electronics and will make cooling fans run constantly which consumes power.

For optimal operation of electronic devices, the relative humidity should be kept in the range between 40% and 60%, and the temperature in the range between 65 °F and 75 °F. (18–24 °C).

12.6 Tasks

1. Video surveillance

    A company considers to implement a video surveillance system that will monitor a large area outside the facility.

    (a) What types of legislations does the company need to consider?
    (b) How can the video streams best be analysed in real time to detect suspicious activity?
    (c) What is the value of keeping video recordings?

2. Emission security

    (a) Why should sensitive conversation not be held in a room with external windows?
    (b) How can an attacker detect what is displayed on a computer screen even if the screen is not visible to the attacker?
    (c) How can you make sure than your smartphone cannot be detected by radio signals?

3. Uninterruptible power supply

    The main types of UPS systems are Standby UPS (S-UPS), Line-interactive UPS (LI-UPS), Online Double Conversion UPS (ODC-UPS), and Generator with UPS (G-UPS). In the table below, tick the boxes for the type of power supply quality problems that a UPS system can handle. For G-UPS, assume that the UPS is an Online Double Conversion UPS (Table 12.1). A suggested population of the table is available on the website of the Digital Security Group at UiO: https://www.mn.uio.no/ifi/english/research/groups/sec/instructional

Table 12.1 Power supply quality problems and the various types of UPS systems that potentially can solve them

	S-UPS	LI-UPS	ODC-UPS	G-ODC-UPS
Fault				
Blackout				
Sag				
Brownout				
Spike				
Surge				
Frequency instability and noise				

Chapter 13
Security Culture

> *It's easier to manipulate people rather than technology.*
>
> Kevin Mitnick

We can have a highly secure system with server applications that have been developed under the principles of "secure by design" running in a well-protected network segmented with "zero-trust architecture". However, all this would be futile if system administrators click on links and open attachments in every phishing email they receive. Culture and human factors are of crucial importance for cybersecurity. Employees in an organization often represent vulnerabilities that can be exploited by external threat actors to attack the organization. Employees can themselves become threat actors that harm their own organization. There are many things we can do to prevent such things from happening.

In this chapter, you learn about security culture in general, and specifically about behaviors that support cybersecurity within the organization, about the insider threat, and about how we can "harden" employees can harden to become resistant against social engineering. Finally, you learn about security usability and stages of security learning.

13.1 Definition of Security Culture

Organizational culture is the shared values, norms and perceptions of reality that the members of an organization have. Values are what is perceived as important and worth striving for, while norms are what is considered acceptable and unacceptable in terms of attitudes, actions and behaviors. *Cybersecurity culture* is part of the general organizational culture.

Since this book is about cybersecurity, we will henceforth use the truncated term "security culture", implicitly meaning "cybersecurity culture". Often the term "cybersecurity culture" is used in the same sense. There are many different definitions of security culture, where ENISA's definition is given below [41].

> *Cybersecurity Culture (CSC) of organizations refers to the knowledge, beliefs, perceptions, attitudes, assumptions, norms and values of people regarding cybersecurity and how they manifest in people's behaviour with information technologies.* (ENISA)

As the above definition is quite general, it is useful to elaborate on specific aspects on security culture. The *Security Culture How-To Guide* describes the following seven core dimensions of security culture in organizations [42]:

- **Attitudes:** Employees' feelings and beliefs about the various activities affecting cybersecurity in the organization. This includes feelings of fear or lack of fear, and attitude towards risk to oneself, to colleagues, to the organization and to external parties.
- **Behaviors:** Employees' actual or intended activities and risk-taking that may have a direct or indirect impact on cybersecurity. This includes personal integrity, which is each employee's compass for ethical and lawful behavior according to norms and rules.
- **Cognition:** Employees' awareness, verifiable knowledge, and understanding of practices, activities, and self-efficacy related to cybersecurity. This could be, for example, employees' understanding of cyber threats and sensitivity of different categories of information.
- **Communication:** The quality of communication channels used by employees to communicate security related matters, and how the communication promotes their perception of belonging, and their active support for reporting incidents.
- **Compliance:** Employees' knowledge of the existence and content of security policies, and the degree adherence to such policies and loyalty to the organization.
- **Norms:** Employees' perceptions of and adherence to informal and unwritten codes of conduct related to cybersecurity in the organization. This may, for example, be conscious or unconscious expectations of security behaviour regarding the use of ICT. This also applies to attitudes towards colleagues' respect of security policies.
- **Responsibility:** Employees' understanding of their roles and duties as a critical factor for sustaining cybersecurity in the organization, and how they can jeopardize cybersecurity if they do not fulfill this responsibility. This includes security in the handling of passwords and other authenticators, as well as secure handling of the organization's IT equipment.

According to Verizon's 2023 Data Breach Investigations Report,[1] 74% of data breaches stem from manipulative social engineering techniques and human error or oversight. Hence, security culture matters. We can have the most security-hardened platform running in a strictly segmented network with encrypted data transmission

---

[1] https://www.verizon.com/business/resources/reports/dbir/

and storage. But if system administrators with privileged access authorizations click on every phishing email they receive, then it does not matter how secure everything else is. The aim of improving the organization's security culture is to strengthen the human link in the security chain by making make secure behaviour is an integral part of every employee's habit and conduct in their day-to-day actions.

13.2 Building the Security Culture

Good organizational culture, i.e. when employees are positively motivated by common values and objectives, is a prerequisite for a good security culture. In other words, if the organizational culture is broken by diverging goals and lack of respect for policies and norms, it will be difficult to build a strong security culture. Frameworks for building security culture can be found in e.g. [41] and [42]. Below, we simply list a set of relevant actions for building security culture.

(a) Explain in a positive way to employees why cybersecurity important.
(b) Explain employees' responsibility for cybersecurity.
(c) Highlight and reward positive security behavior.
(d) Implement simple routines for reporting and follow-up.
(e) Educate employees about the importance of reporting.
(f) Foster an open reporting culture that does not blame people for making mistakes.
(g) Motivate employees to increase their own competence in cybersecurity.
(h) Give regular briefings and hold mandatory follow-up campaigns and tests on various security challenges, such as social engineering, the threat landscape and the threat from insiders.
(i) Conduct exercises in the field of personnel security.

Be aware that the controls above have different benefits depending on individual employee attitudes. Below is a simplified categorization of employee attitudes into four categories.

1. Motivated
2. Indifferent
3. Disgruntled
4. Secret insider agent for a foreign state

It is impossible to adopt a uniform strategy suitable for enhancing knowledge, motivation, attitudes and behaviour with regard to cybersecurity for all four categories of employees. Employees in category 4) are probably incurable, and at the same time difficult to detect.

It is important to recognize that even the best can be deceived, for example with phishing attacks. Therefore, the organization should never attribute blame for security incidents when an employee has caused a security incident due to lack of awareness, and no one should be described as stupid or naïve. The organization should have a plan in place to shield employees who may have inadvertently caused serious security incidents. This will contribute to points (e) and (f) above. Psychological safety is what you want to achieve, and a culture of fear is what you want to avoid.

Security culture is an integral part of the organizational culture. Working with the security culture should therefore also be integrated into the work of building a good organizational culture. For example, an attempt to establish an authoritarian security culture would most likely be a failure if the organizational culture is anything but authoritarian.

13.3 The Insider Threat

An insider threat is an employee or other person with a legitimate user account and access authorizations who abuses those privileges to attack the organization, either out of their own motives or on behalf of external actors, such as foreign states. Insider threats can be very potent because the attacker is familiar with the organization's information resources and is authorized to access to some of these resources. Knowledge of security processes and mechanisms can also make it easier to carry out an attack and prevent the attack from being detected. The insider threat can involve fraud, theft of confidential or commercially valuable information, theft of intellectual property, and sabotage against computer systems. A concise description of the insider threat can be found in chapter 6 of Paul Martin's book: *The Rules of Security: Staying Safe in a Risky World* [43].

Insider threat statistics show that this is a significant security issue. Large organizations with many employees are often exposed to insider threats, while this occurs less frequently in small organizations with few employees. In any case, the insider threat is an issue that every organization should be aware of and that should be part of the organization's risk management.

Potential groups of people who could become insider threats are all employees, including senior executives and middle managers, as well as customers, visitors, business associates, subcontractors and consultants. All of these groups have certain access privileges that could potentially be abused. The fight against the insider threat aims to ensure that (authorized) access privileges are only used according to policy, i.e. that privileges are not abused.

Technological controls to prevent the insider threat may include mechanisms and procedures that check and monitor the compliance of operations carried out with policy. The next section addresses the prevention of insider threats by strengthening the personal integrity of employees and other individuals who have any form of access.

13.3.1 Personal Integrity

During a hiring process, the organization can assess personal integrity as an aspect of a job applicants. Even if the assessment concludes that an applicant has high personal integrity there and then, there is always uncertainty associated with such assessments, and employee attitudes will often change over time. An employee who initially has high personal integrity may, under certain circumstances, lose their integrity and power of judgement to become an insider threat. It will be of great benefit to the organization to counteract such negative changes in the personal integrity of employees.

Discussing insider threats can be a very sensitive topic that the organization must consider and treat with caution and prudence. The wrong approach can quickly do more mischief than good. For example, if the organization treats all employees as potential insider threats, the suspicion can paradoxically create conditions that amplify such threats.

The goal must be to create a culture that stimulates the personal integrity of all employees and others who come into contact with the organization. Many elements can be used to prevent insider threats and maintain personal integrity, some of which are mentioned in the points below.

- All employees should undergo awareness training that also addresses personal integrity and attitude towards policy compliance. The training should also address the consequences of (deliberate) breaches of policy.
- Insider threats have a dynamic with underlying causes and typical patterns, often with recognizable indicators. Warning of an insider threat may be changed behaviour characterised by unhappiness and irritability, and that an employee starts working outside normal working hours. Ensuring that employees understand these dynamics and indicators is crucial to recognising it in themselves and in others, and for strengthening the individual's moral judgment and attitudes regarding their own and others' actions.
- A culture must be created where employees do not have to fear being scrutinized. Assigning blame can be counterproductive and can discourage incident reporting. However, creating an environment that does not assign blame does not mean that individuals do not have to be accountable for their actions. Individuals must be given the opportunity to acknowledge and account for an incident, at the same time as they are involved in identifying the underlying causes and in coping with the consequences. Such an approach will contribute to good handling of the incident, and to reduce the potential for recurrence among the person concerned and other employees with insight into the incident. A more detailed description of these aspects is given in Sidney Dekker's book: *The Field Guide to Understanding 'Human Error'* [44].
- Policy reminders are important. If it has been years since an employee participated in security awareness training, the learning that remains is probably quite faded. If staff is not called in for regular reviews, then probably neither has the organization's management. If, for example, the employees see that the compa-

ny's security policy is still signed by the previous director, who left several years ago, this can have a demotivating signal effect.
- Employees with broad access privileges should receive special training and follow-up (and pay).
- Provide support for, and follow-up of, employees in special situations, such as:
  - For employees in highly trusted positions,
  - In case of conflicts and situations where an employee may feel unfairly treated,
  - When an employee has personal problems or problems in their private life,
  - In the event of a change in job situation and
  - When an employee leaves the organization (see the next section on the responsibilities of management).
- There are very limited ways in which an employer can or should monitor or monitor employees' private lives and personal relationships. At the same time, these are factors that can significantly affect the employee's personal integrity, and thus the security of the organization. Arrangements should be made so that employees can seek advice if they have problems of any kind that are not directly related to the work situation and which they feel it is useful to discuss with the employer. The framework around such an arrangement must be secure with clear policies on confidentiality and privacy. Guidelines on monitoring and privacy in the workplace aiming to comply with the GDPR has been published by the UK Information Commissioner's Office[2]

There is a category of insider threats where the above controls probably have little effect, namely secret agents acting on behalf of an adversary such as a foreign state. Already established attitudes and influence from the adversary make it very difficult to influence the threat actor's personal integrity. It is important to be aware that this type of insider threat can be influenced by extortion from a foreign state. For example, a foreign state may threaten actions against the person (the threat actor), or against the person's family members living in the foreign state. If this type of insider threat is suspected, it may be useful to seek advice from the national security authority in your country.

It can be uncomfortable to talk about personal integrity in the workplace because, in a way, it suggests that some of the audience may lack this. However, it should not be a taboo topic. With careful introduction and sensible explanation, the topic of insider threats should be open for discussion, because it can have great significance for cybersecurity.

---

[2] https://ico.org.uk/for-organisations/uk-gdpr-guidance-and-resources/employment/monitoring-workers/

13.3.2 Management's Responsibility for Handling the Insider Threat

The management has an important role in dealing with the insider threat. Management should give clear signals, show clear attitudes and set a good example for behaviour that demonstrates personal integrity.

Dissatisfied employees are significantly more likely to become insiders than satisfied employees. While it is completely unacceptable that dissatisfaction should lead to insider threats, it is still important to ensure that employees do not end up in situations or roles that generate acute dissatisfaction or general dissatisfaction over a long period of time. This is important not only for preventing insider threats, but also for the well-being and productivity of the entire organization. Management has a particular responsibility to ensure that employees thrive and are satisfied in their everyday work. Statistics and research show that the likelihood of insider threats increases significantly when employees leave. The most common reasons for leaving an organization are:

- Voluntary resignation
- Retirement
- Redundancy due to downsizing or reorganization
- Dismissal due to misconduct

Being made redundant during downsizing or reorganization, and the sense of injustice that it often entails, are significant factors when employees become insider threats. While it is completely unacceptable that the sense of injustice should to lead to insider threats, management should place great emphasis on ensuring "fairness" when redundancy is necessary.

It is essential to have a good dialogue with employees who leave as a result of a redundancy process. Managers should be aware that those who have to leave and those that remain often expect different managerial behaviour during the downsizing process.

Access rights policies should be designed depending on the ways in which an employment relationship ends and on the type of organization. Approaches to managing access rights can be the following:

1. Former employees retain limited privileges.
2. Orderly and agreed deletion of all privileges.
3. Immediate deletion of all privileges, security guard escort to the exit.

Upon termination of an employment relationship, agreements may be reviewed with regard to what information the former employee may bring with them and use with other organizations, quarantine period, restriction on working for competing companies, etc.

13.4 Social Engineering

Kevin Mitnick is a well-known expert on *social engineering*. He started as a criminal hacker around 1990, precisely by engaging in social engineering, was arrested and sentenced to several years in prison, and is now a prominent consultant who conducts consulting and gives talks on cybersecurity and social engineering. Based on his experience, he stated the following[3]:

> *The biggest threat to the security of a company is not a computer virus, an unpatched hole in a key program or a badly installed firewall. In fact, the biggest threat could be you.*
> ...
> *What I found personally to be true was that it's easier to manipulate people rather than technology.* (Kevin Mitnick, BBC online, 2002)

Social engineering takes place through many channels, where we distinguish between *techno-social engineering* and *physical social engineering,* where the latter takes place physically face-to-face, or through real-time communication channels such as video conferencing or telephone.

13.5 Techno-social Engineering

Techno-social engineering is typically based on technical contact with victims, such as email, SMS, social networks, chat and via websites. A combination of different channels is often used. A very common attack vector that employs social engineering is sending phishing emails.

13.5.1 Phishing Attacks

Phishing is a techno-social attack where threat actors use fake emails to trick people into sharing sensitive information or executing malware. Typical phases of phishing attacks are:

1. Attacker sends phishing email, which might get through the spam filter and land in the victim's inbox.

   - It is increasingly difficult for attackers to get through spam filtering and email security mechanisms such as SPF, DKIM, DMARC. Hence, they are constantly forced to adapt and improve their methods.
   - The content of the email must be sufficiently credible to trick the victim into believing that it is a genuine email. Generative AI based on LLMs (Large Language Models) gives a strategic advantage to attackers for generating

---
[3] http://news.bbc.co.uk/2/hi/technology/2320121.stm

credible email content. Using AI for deception is described in Sect. 2.1.9 and in Chap. 15.

2. The victim is deceived and performs the proposed action in the message. Examples of actions are the following:

   - Visit a fake website and provide your user ID and authenticator (password and one-time PIN).
   - Respond to phishing emails with sensitive information.
   - Install and/or execute malware.

3. The attacker exploits of the victim's action to proceed with the next step in the attack chain.

   There are different categories of phishing attacks:

   - **Mass phishing: Mass** attacks with large volumes of emails meant to reach as many people as possible.
   - **Spear-phishing:** Targeted attacks targeting specific individuals or companies. By using OSINT and or AI, attackers can customize the message and make the deception harder to detect.
   - **CEO fraud:** A type of spear phishing that targets "big fish", including high-profile individuals or those who have authorizations or access to transfer large amounts of money.
   - **Clone phishing:** Copy of a legitimate and previously delivered email, with original attachments or links replaced with malicious versions. The email is sent from a spoofed email address, making it appear to come from the original sender or another legitimate source.

13.5.2 Detection of Phishing Attacks

In most cases, it is relatively easy to detect cases of phishing and just delete it right away. However, sophisticated attackers are capable of articulating phishing email that even experts are unable to detect.

Typical signs of phishing are:

- Spelling errors (e.g. passward), lack of punctuation or poor grammar
- An Internet link that differs from the one that appears or is hidden
- threatening language that requires immediate action
- Unwarranted requests for personal information
- Announcing that you have won a prize or lottery
- Donation requests

However, spearphishing typically is well designed and contain none of the elements above. In general, always be skeptical and use common sense when opening email. STOP–THINK–ASK is a useful rule of thumb. To STOP means that you should never click on links or open attachments without thinking. To THINK means to consider whether this might be phishing and look for suspicious details. To ASK means that you should ask others if you have any doubts, or simply delete the email if it seems unimportant anyway. Some organizations have policies for reporting phishing, so report it if you think it is phish.

Be extremely careful about clicking on links in an email. Use your computer mouse to hover over each link to verify the actual address, even if the message appears to be from a trustworthy source.

Pay attention to the URL and check for incorrect spelling or if the Internet domain is different from what you expect (e.g. G00GLE.com instead of GOOGLE. com). Consider navigating to websites "manually", i.e. by typing instead of clicking and following links in messages.

Examine websites to assess if they are genuine. Criminal websites can appear identical legitimate websites. Check that the website uses HTTPS before filling in sensitive information on a website. Keep in mind that HTTPS does not say anything about whether the website is criminal or not – it just says that the connection is encrypted, as explained in Sect. 5.3.1.

13.5.3 When Realising that You Have Been Phished

If you suspect that you have taken any action in a phishing email or on a fraudulent website, contact the helpdesk or other relevant entity immediately.

If you may have provided sensitive, personal, or financial information to a fraudulent website, change the password of the affected account on the real website immediately. If you use the same password for multiple accounts and websites, change it for each account – and never reuse the same password in the future.

Look for signs of identity theft by reviewing bank and credit card statements for possible unauthorized payments and activities. If you notice anything unusual, contact the bank immediately. Consider reporting the attack to the police.

For social networks, we are often permanently logged in through multiple devices, such as laptops and mobile phones. If an account is suspected to have been hacked, i.e. that an attacker has obtained the password, you must immediately change the password and log out of the account on all devices. If you are already locked out of your account, you will need to try contacting the website for assistance, which unfortunately can be frustrating but sometimes allows you to regain control of your account.

13.6 Physical Social Engineering

Physical social engineering is, for example, when an attacker enters the building of a business and attempts to manipulate employees. The attacker's goals could be to obtain verbal information, get people to take actions, steal documents and materials, install spyware such as cameras and microphones, or install components in the computer network that could tap information or provide unauthorized access. Physical face-to-face communication with the victim provides the most opportunities for the attacker, but video conferencing and phone calls also have characteristics similar to face-to-face communication. Not least, telephone and video calls allow for the use of advanced voice or image manipulation with deepfake as described in Sect. 2.1.9. On the other hand, phone and video communications have other limitations. In an environment with direct physical communication, the attacker can play on many different strings to trick the victim and execute attack steps. The next two sections describe direct physical social engineering strategies and how an organization can defend itself against such attacks.

13.6.1 Physical Social Engineering Attack Strategies

Physical social engineering requires the attacker to have a certain talent as an actor to play a role that appears credible to the victim. This is also why such people are called *con artists*, which expresses that it is an "art" for which the attacker requires talent. Below is a set of techniques used for this type of social engineering.

Neuro-linguistic programming: This is a pseudoscientific approach to human communication according to which language and neurological processes can influence behavioural patterns. In this way, an attacker can influence a victim by mirroring the body language, tone of voice, tone of voice and word usage of the victim. In a business context, industry and corporate jargon is an important part of the social glue. The effect of neuro-linguistic programming is to induce an emotional connection with the victim on a subconscious level, making it easier to influence the victim. Neuro-linguistic programming is also a technique used by salespeople to influence customers.

Build trust: Depending on the country, region and culture, people are generally naturally helpful and trusting. This basic trust can be reinforced and exploited for social engineering. An attacker can build trust by asking seemingly innocuous questions during normal conversations, and then slowly and imperceptibly exploit trust by asking for increasingly sensitive information. In this way, the attacker can learn corporate jargon, names of key personnel, names of projects and departments, names and locations of servers and applications. If you are the attacker, one trick to get the victim to trust you is that you first cause a problem without revealing that it was you, and then offer help to fix it. For example, you can disconnect the LAN cable from a workstation when no one is looking and be nearby when the victim complains that the network is down. Talking negatively about competitors/enemies and positively about partners/friends makes the victim think you are one of them.

Induce emotions: Emotions are closely related to behavioral patterns, which means that changing emotions can interrupt current behavior and produce a different type of behavior. If a person usually tries to consider arguments rationally, sudden strong emotions can make the person less vigilant and less able to identify misleading arguments. For example, the attacker may induce excitement/joy ("you've won an award"), fear ("you're going to lose your job") or confusion ("the project manager says X but the head of department says the opposite").

Information overload: A person's ability to interpret information and evaluate arguments can be diminished by the attacker providing too much information. The attacker's goal is to create a cognitive and mental blindness, for example by describing many relevant and irrelevant aspects as well as providing arguments from unexpected angles, which creates a high mental load. When the load exceeds a certain limit, the victim will to some extent stop interpreting the information and no longer evaluate the arguments rationally.

Reciprocation: All human beings have an inherent instinct to return favors, which can be exploited for social engineering. This instinct kicks in even if the first favor was not requested, and may apply even if the return favor is more valuable than the first favor. If the victim has received a favor from the attacker, the victim will be more easily manipulated by the attacker. *Double disagreement* is an attack technique in which the attacker first asks for two favors that the victim is initially unwilling to give. If the attacker then withdraws the request for one favor, the victim will unconsciously feel some kind of obligation to give the other favor. *Anticipation* is a technique whereby the request for a favor is associated with the promise that the attacker will become a future ally and will reciprocate the favor in the future.

Diffusion of accountability and moral duty: This attack strategy is about making the victim believe that they do not need to feel accountable for the type of actions the attacker requests, and instead making the victim feel that it is a moral duty to satisfy the attacker's request. For example, the attacker could demonstrate that the security policy is illogical, has adverse aspects, is outdated, and that it is typically ignored, which is an argument for not following the security policy.

Commitment creep: People tend to follow through on commitments they have taken on, even when they feel that it may be unwise. Withdrawing commitments can be interpreted as having a weak personal character, something people would rather avoid. One attack technique that takes advantage of this tendency is to create a situation in which the attacker first makes a relatively harmless request that the victim carries out, and then makes a serious request that logically resembles the first one so that the victim feels obligated to carry it out as well. For example, the attacker may first ask to use the victim's workstation for a simple task. The next day, the attacker asks for the victim's password to log on to another workstation. If the victim is reluctant to give up the password, the attacker can say that in practice it is the same as what they did the day before, and that there is therefore no reason to refuse.

Authority: We have a strong tendency to obey authority, as documented by the famous Milgram experiment.[4] It is generally considered rude to challenge the authenticity of authority. There are many ways to assume a false authority, such as by falsely issuing or falsely claiming to be a chief executive, manager, or to represent an external organization. It takes acting talent to be able to do this convincingly, and con artists have such talent.

13.6.2 Defense Against Physical Social Engineering

In order to defend against physical social engineering, as described in the previous section, organizations should understand the typical attack patterns and establish defenses. A guide to establishing an effective defense against physical social engineering is published by SANS.[5] The guide is called A *Multi-Level Defense Against Social Engineering*, and is illustrated in Fig. 13.1. Although the guide is primarily aimed at physical social engineering, it is also relevant to techno-social engineering.

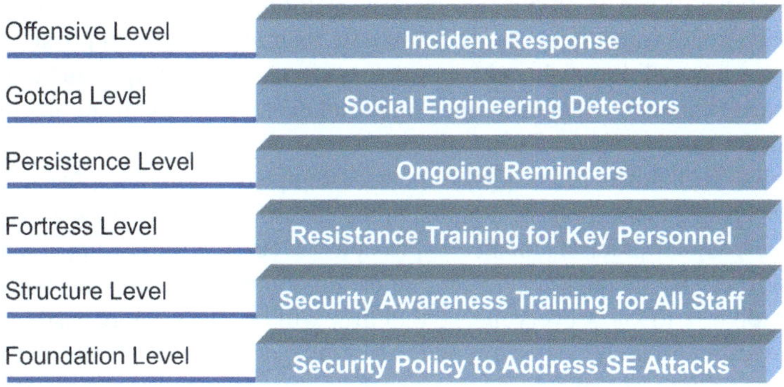

Offensive Level	Incident Response
Gotcha Level	Social Engineering Detectors
Persistence Level	Ongoing Reminders
Fortress Level	Resistance Training for Key Personnel
Structure Level	Security Awareness Training for All Staff
Foundation Level	Security Policy to Address SE Attacks

Fig. 13.1 Multi-level defense against physical social engineering

Each level of the model in Fig. 13.1 is described below.

1. **Foundation.** A policy for dealing with physical social engineering must be formulated. Without it, employees will have no guidelines to follow. For example, the policy may provide guidelines for verifying identity, affiliation and authority, whether physically, over the phone or through video meetings. The policy should prohibit practices similar to social engineering attack patterns, such as requesting passwords over the phone. The policy should, among other things, define situations where it is mandatory to call back to verify identity or affiliation.

---

[4] https://en.wikipedia.org/wiki/Milgram_experiment
[5] https://www.sans.org/reading-room/whitepapers/engineering/multi-level-defense-social-engineering-920

2. **Structure.** All employees must undergo awareness training to detect and manage physical social engineering. Elements of the training include understanding typical tactics and recognizing attacks, knowing when to say "no", knowing what type of information is sensitive, understanding the dangers of casual conversations, learning healthy scepticism, and understanding that "uniforms are cheap". The term "uniforms are cheap" means that it is easy to acquire fake props such as reflective vests and workwear to play the role of a repairman, or an elegant suit to play the role of a manager. Fake access cards are easy to make, and even if they don't work, other employees will usually help open the door. Awareness of policies should make the employee feel that the only choice is not to be manipulated.
3. **Fortress.** Key personnel who are particularly exposed to physical social engineering are, for example, those who sit at the front desk and those who take care of the help desk, IT administration and customer service. These need reinforced training to form a stronghold against social engineering. Approaches can be updates on new types of attacks they can expect to be subjected to, hardening through being subjected to simulated attacks, and self-awareness of what typical mental weaknesses we humans have that allow us to be misled.
4. **Persistence.** Even if employees have been through awareness training once, their awareness must be sustained, otherwise their ability to resist physical social engineering will diminish and become less effective. Fresh-ups can be done by repeated reminders on posters, messages and exercises in the form of simulated attacks.
5. **Gotcha.** Techniques are needed to detect attempted or ongoing physical social engineering attacks. In large organizations, it is impossible to know all your colleagues, which precisely makes it possible for an attacker to impersonate an employee. One technique is to let one or more employees who have as part of their job task to know everyone, so that they will be able to more easily identify an attacker. Anomaly reporting and a centralized log of suspicious events can make it easier to identify typical patterns of physical social engineering. When receiving phone calls, a policy may require a call back, a requirement to ask key questions, and a "please wait" requirement to give the recipient of the call time to log and assess the call. Cunning detection methods include asking false questions to test how a suspect responds, or having an "alarm account" with user IDs and passwords placed on yellow stickers around the office landscape.
6. **Offensive.** In the event of a (suspected) attack, it is important that employees know how to act and behave. When it comes to physical social engineering, safety is the number one priority. No ordinary employee should have to expose themselves to physical hazards, so security personnel should be notified immediately. Nevertheless, the reaction of employees should be vigilant and determined. Employees who are in direct contact with the suspect may attempt to stay with the person until a security guard arrives. If possible, other nearby employees should be notified. It is of course possible that it is a false alarm, and the policy should describe how such cases are handled. An employee who has been suspected should be well taken care of when the case is solved. In any case, this is something all employees should be informed about during awareness training, so that they are not completely unprepared to be suspected.

13.7 Security Usability and Security Learning

Usability is the property of a system which enables a user to perform a task correctly, efficiently and safely. *Security usability* is something similar, only that in this case the system is a security function. Thus, we can define security usability like this:

> *Security usability is the property of a security function which enables a user to perform a security task correctly, efficiently, and safely.*

We could have wished that security usability was nothing more than ordinary usability, i.e. the user did not need any expertise in cybersecurity to perform security functions. In other words—we would like cybersecurity to be completely transparent (invisible) to the user. However, in many cases we do need a certain level of security expertise to perform a security function correctly,

The difference between usability and security usability can be illustrated with an example. If a system has poor usability, the user will typically not be able to perform the task at all, or at least not easily or correctly, which can probably be noticed immediately. If the system has good usability, but at the same time has a security function with poor usability, the user will be able to perform the task, but in an insecure way. The user will typically see that the task has been performed correctly, but may not notice that it has been done in an insecure manner.

User interfaces for common applications have many security features that we must understand and relate to. Browsers have padlock icons that we must be able to interpret, we get pop-up warnings to accept or reject certificates, and we must learn to manage passwords securely.

Metaphors are very common in the IT industry, and especially in the IT security industry. It talks about firewall, virus, Trojan horse, key, honeypot, etc. Sometimes the metaphors provide correct associations, allowing users to intuitively understand a security concept correctly. Other times, the metaphors can give misleading associations that cause users to misunderstand a security concept. As an example, the term "firewall" might lead the user to believe that a firewall is completely impenetrable and blocking everything, when in reality it is a traffic filter governed by rules (Sect. 6.4). "Trust" in website certificates can lead the user to believe that a website that has a certificate therefore is trustworthy, when in reality it only means that the domain name can be authenticated (Sect. 5.3.1). The correct interpretation of these terms must be learned, otherwise the user may be misled.

A good example of metaphor confusion is an experiment on the use of asymmetric cryptography with public and private keys, described in the article *Why Johnny Can't Encrypt* [45]. In the experiment, participants were asked to select the correct keys to send a confidential message and an authenticated message. Many chose to use their private key to send a confidential message because they thought that "confidential" and "private" belong together. The reason for this error lay in misguided metaphors.

Good metaphors are important for learning, but it is not always possible to create good metaphors for advanced security concepts. It may be better to avoid metaphors than to use bad ones. Instead, one should define new security concepts and give them a descriptive semantic content that must be learned. PKI (Sect. 5.2) and forward secrecy (Sect. 4.7.4) are concepts that are not based on metaphors and that must be learned in order to understand what they mean.

Security learning is important in order to use security technology correctly. Figure 13.2 illustrates the three stages of security learning that people typically go through.

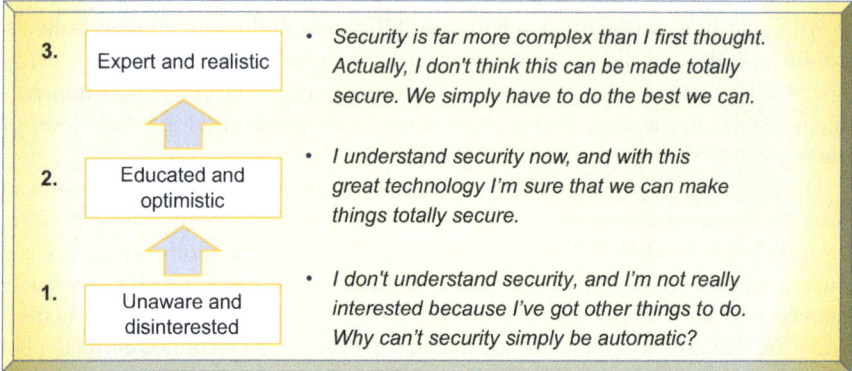

Fig. 13.2 Stages of security learning

This book is intended to support security learning in a way that makes you become an expert and be realistic about what is possible to achieve. The importance of expertise in cybersecurity is discussed in Christian Espinosa's book: *The Smartest Person in the Room: The Root Cause and New Solution for Cybersecurity* [46].

13.8 Tasks

1. Building a security culture

    (a) Consider the aspects of security culture described in Sect. 13.1. Cite an example of poor safety culture for each aspect.
    (b) Based on the examples from a): Propose controls to strengthen security culture.
    (c) Propose possible strategies suitable for enhancing knowledge, motivation, attitudes and behaviour with regard to cybersecurity for the following categories of employees in an organization:

    - Responsible
    - Indifferent
    - Disgruntled
    - Secret agent acting on behalf of a foreign state

How can different strategies be applied when it is uncertain which category each employee fits into, or should one try to design a strategy that suits all categories of employees?

2. The insider threat

   (a) In the period from being hired to leaving a company, when is an employee most likely to become an insider threat?
   (b) Controls to strengthen attitudes and personal integrity can help prevent employees from becoming an insider threat. What is one type of insider threat that is unresponsive to this type of control?

3. Social engineering

   (a) Describe ways to use social engineering for
       - Installing malware on the PC of the CEO of a company
       - To gain unauthorized access to the building of a company.
   (b) Imagine that employees constitute human "IDS" (Intrusion Detection System) against social engineering attacks. What would be a "false positive" and a "false negative" detection of social engineering? Discuss the potential consequences of false positive and false negative detection of social engineering.
   (c) What should you do if you think you have been deceived by social engineering?

Chapter 14
Cybersecurity Readiness, Security Testing and Audit

By failing to prepare, you are preparing to fail.

Benjamin Franklin

In the early years of computer security, all the attention was focuses on preventing incidents from happening. However, no matter how much effort an organization puts into implementing preventive security controls, cyber incidents will happen. A prominent initiative to establish cyber readiness was the establishment of the first CERT (Computer Emergency Response Team) in 1988 in response to the Morris worm incident which affected network servers in research institutions mainly in the US. In those days, business processes were not online and were therefore not affected by the Morris worm. In today's world where everything is online, business processes are easily disrupted by cyber incidents. And when organizations experience a cyber incident, they cannot expect that a CERT will come to the rescue. They are on their own, to a large extent. To be prepared, every organization must have a plan and a certain capacity for handling cyber incidents.

In this chapter, you will learn about cybersecurity contingency planning which is how organizations plan for handling cyber incidents while keeping the business running as best they can. This includes to establish a cyber incident response plan which deals with the incident itself, as well as business continuity and recovery plans which deal with how to keep the business running during an incident and how to recover from it. You will learn about digital forensics which is often done as part of investigating cyber incidents, and which is required for evidence to be presented in court. Finally, you learn how pentesting and security audit are done to assess the strength of cyber defences and the level of regulatory compliance.

14.1 Background for Cyber Contingency Planning

Contingency planning covers e.g. incident response, business continuity, disaster recovery. A contingency is defined as a future event or circumstance which is possible but cannot be predicted with certainty. For our purpose, this possible event is a cyber incident. The purpose of cyber contingency planning is to be as prepared as possible when a cyber incident occurs, so that the extent of damage can be limited as much as possible. Without contingency planning, a cyber incident that is initially insignificant can grow to produce disproportionately large and negative impacts. Contingency planning is specified as a requirement by cybersecurity management frameworks such as NIST CSF (Cybersecurity Framework) and ISO/IEC ISMS (Information Security Management System) described in Chap. 18.

So far, this book has mostly described methods and principles for safeguarding assets against attacks, which is about preventing or reducing the likelihood of security incidents occurring, or for detecting when attacks occur. However, it is impossible to stop all cyber attacks and impossible to prevent all incidents. When a cyber incident is detected, it is important that the business has already established a capability for readiness and resilience. The activities for the response and further handling of security incidents are called *incident management*.

Risk management described in detail in Chap. 19 involves estimating the likelihood and impact of potential security incidents to compute the levels of risk for each type of incident. Another way to analyse and visualise risk is with the help of bow-tie diagrams which show the correlation and sequence of steps between causes and impacts of an incident. A separate bow-tie diagram is used to visualize each type of incident.

A bow-tie diagram visualizes: (1) relevant threats sources, (2) likelihood-reducing preventive controls (which may already be, or which could be implemented), (3) the specific type of incident, (4) impact-reducing mitigation controls (which may already be, or which could be implemented) and finally (5) the possible negative impacts that may result from the incident. Possible impacts of cyber incidents are various types of losses as described in Sect. 19.4.3. The bow-tie diagram in Fig. 14.1 shows a non-specific incident as an example and illustrates the logical sequence of events from left to right.

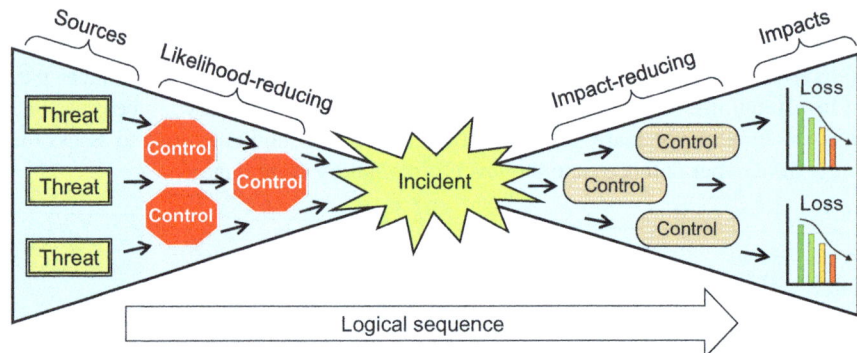

Fig. 14.1 Bow-Tie diagram to visualize the risk of an incident

The left side of the diagram represents both causes/threats and possible preventive actions that can reduce the likelihood of the incident occurring. The right side represents possible controls that can reduce the impact and limit the damage given that the incident occurs, despite preventive controls.

Bow-tie diagrams are often used in safety analysis to analyse unwanted physical events such as fires, floods and accidents. The diagram can also be used to visualize the relationship between prevention and management of cyber incidents.

In the early years of information security practice until roughly around year 2000, security practitioners mostly focused on the left side of the bow-tie diagram, that is, on controls to prevent security incidents. The thinking was that with sufficiently strong security controls, all incidents could be avoided. Then sometime after 2000 the saying was *"there are two types of companies: those that have been hacked, and those that will be hacked"*. As a result, the security community started to also focus on the right side of the bow-tie diagram, which is covered by cyber contingency planning and incident management. Every organization has cyber vulnerabilities, and given that cyberthreats are ubiquitous the saying nowadays is *"there are two types of companies: those that know they have been hacked, and those that have been hacked but not detected it yet"*.

14.2 Contingency Planning Principles

In contingency planning the primary focus is to capture and structure a set of activities that are effective for handling incidents. It is also important to consider how those activities can be structure in a rational and efficient way. Below are four guidance principles for contingency planning [47, p. 61] aimed at establishing preparedness for civil disasters and emergencies.

1. **The principle of responsibility.** *The entity responsible for an area in normal conditions is also responsible for the prevention, readiness, and implementation of the necessary actions in case of emergencies or disasters.*

    The term "area" is interpreted as the organization's own activities and business processes, which thus means that each organization must have its own contingency plans.

2. **The principle of similarity.** *The organization of people and resources used during emergencies should be as similar as possible to that which operates in normal conditions.*

    This principle states that if an incident occurs within a specific department or business process, then the incident should be handled by the very people who normally work in that department or who manage that business process.

3. **The principle of proximity.** *Incidents should be handled at the lowest possible level in the organization.*

    Crisis management can be very resource-intensive, and if small incidents that happen frequently were to be escalated to crisis management every time, it would take a toll on the organization's resources. This principle shall make incident and crisis management as effective as possible, and shall make rational and efficient use of the organization's resources. Of course, the management of a major crisis will necessarily involve external parties.

4. **The principle of collaboration.** *Authorities, private enterprises and/or government agencies have an independent responsibility to establish collaboration with relevant entities and agencies in the field of prevention, readiness, and crisis management.*

   This principle states that each organization must identify relevant external entities that can assist with warning and detection of incidents, and for response during incidents and crises. Contacts and agreements must be established with these actors so that they can quickly be called upon when needed.

Although the above principles were defined for national civil security, they are also applicable for cyber contingency planning. However, a typical difference between a civil disaster and a cyber incident is how it is detected. Civil disasters are obvious to see, but cyber incident are often difficult to detect, precisely because threat actors try to be stealthy. Hence, organizations must establish a cyber detection capability in connection with contingency planning. Monitoring and detection can be bought as a service, which is often a good option for small organizations. Anyway, the morale is to have a clear assignment of responsibilities in contingency planning, which means knowing who does what when disaster hits.

14.3 Technical Concepts of Cyber Contingency Planning

Various technical terms are used to quantify temporal objectives for cyber contingency planning. Figure 14.2 shows the connection between these terms, which are further explained below.

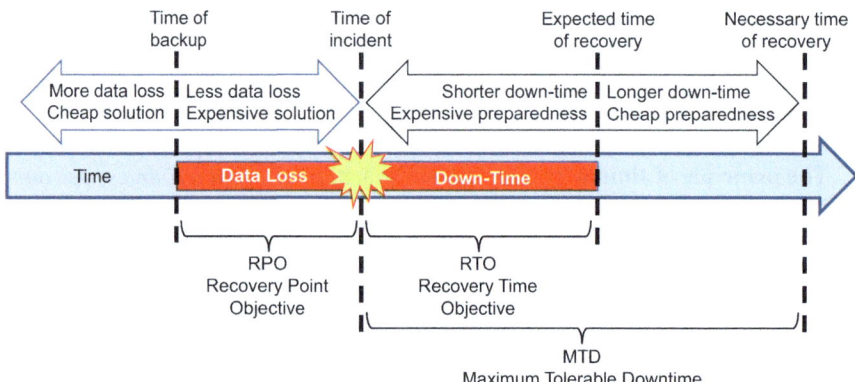

Fig. 14.2 Time aspects for cybersecurity response and recovery

- **RPO (Recovery Point Objective)** is the acceptable length of time since the last backup. Backup routines must be dimensioned to satisfy the RPO. If a storage device fails, files are accidentally deleted, or as the result of an intentional malicious action, it should be possible to restore files or databases that are at least as fresh as the RPO from backup. The organization must be prepared for the possibility that more recent data with a lower age than RPO may be lost in the event of an incident. The shorter the RPO, the more costly the backup solution becomes.
- **RTO (Recovery Time Objective)** is the time the organization expects to need after a cyber incident to restore or resume a specific function, process, or service. Hence, resources for incident response and recovery must be dimensioned to satisfy the RTO. If it is just a matter of retrieving backup, that will be the time needed to retrieve and install the backup. In the case of incidents that disrupts operations and services, that will be the time needed to recover. The shorter the RTO, the more expensive it will be to establish adequate incident response and recovery capacity with the necessary people and expertise.
- **MTD (Maximum Tolerable Downtime)** is the maximum amount of time that processes or services may be disrupted before it is considered to have disastrous consequences. Determining MTD is based on its impact on business functions, and analysis of expected lost revenue and other costs incurred for each hour, day, or week a given that processes or services are disrupted. MTD is specified for critical processes and services as part of the BIA (Business Impact Analysis), see below. MTD is used to specify a continuity strategy, where two options are DRP (Disaster Recovery Planning), and BCP (Business Continuity Planning) which are explained below.
- **DRP (Disaster Recovery Planning)** is the process of planning for the recovery of systems and processes that have been damaged after an incident, so that production and services return to normal operation. For example, if EFTPOS (Electronic Funds Transfer at the Point Of Sale) systems in a retail chain are disrupted as a result of a cyberattack, a decision to invoke DRP means that the cyber incident must be addressed and the systems restored so that the EFTPOS systems work again. In some cases, it may take longer than MTD, so it may be necessary to also plan for alternative procedures.
- **BCP (Business Continuity Planning)** is the act of planning for alternative procedures to maintain the operation of production and services while recovery of damaged systems and processes is ongoing. With alternative procedures, capacity and quality will usually be reduced, but this may be a better alternative than disrupting processes and services completely. For example, if EFTPOS systems in a retail chain do not work as a result of a cyberattack, BCP may involve customers paying with cash or with online payment systems such as PayPal or ApplePay. A decision to implement BCP is typically made if it is assumed that recovery of normal operations with DRP will take longer than MTD.
- **BIA (Business Impact Analysis)** is done to identify mission-critical functions in the sense that a disruption would give a very negative impact. This should cover (a) prioritization of the most vital business functions and define MTD and security requirements for these, (b) identify dependencies between ICT functions and (c) define RTO and RPO, as well as the ability to operate with reduced service

levels by using the BCP. It is useful to define categories of MTD for all mission-critical functions.

Category 1: Critical processes,	MTD = 0–12 h
Category 2: Very important processes,	MTD = 13–24 h
Category 3: Important processes,	MTD = 1–3 days
Category 4: Less important processes,	MTD = more than 3 days

14.4 Cyber Contingency Planning

Cyber contingency planning shall prepare the organization to handle small and large cyber incidents, so that handling can be carried out in the best and fastest possible manner. This step is at the same time part of the secure-by-design process described in Sect. 11.3. In particular, before a critical application is released, the organization must specify continuity and recovery plans. The 10-step guidance to cybersecurity published by the British NCSC mention the following elements of readiness planning[1]:

- The team for establishing the CIRP (Cyber Incident Response Plan) should include the IT security team, but will also include legal, HR and Public Relations staff, and possibly suppliers and vendors. Consult senior management when specifying plans for critical decisions and media handling during major incidents.
- Cyber incident response plans should be linked to and integrated with the organization's other disaster recovery, business continuity and crisis management plans. Establish contact and agreements with relevant third parties so that they are ready to provide support if required. These may be sector CERTs, national CERT, suppliers, and MSSP (Managed Security Service Providers), i.e. providers of SOC and CSIRT services.
- Roles and responsibilities must be clearly defined and understood, with appropriate training. For this, a RACI chart can be used [1, p. 215]. The acronym RACI is derived from the four key roles: Responsible, Accountable, Consulted, and Informed. Below is a simple example RACI chart with four activities and four individuals.

Table 14.1 Example RACI chart

	Alice (Analyst)	Bob (Owner)	Claire (CISO)	David (Coord.)
Analyse and assess severity of incident	R	C	A	I
Activation of the CIRP	C	C	R, A	I
Shutting down process	C	R	A	I
Contact with external parties	I	R	C, A	R

Legend: R Responsible, A Accountable, C Consulted, I Informed

---

[1] https://www.ncsc.gov.uk/collection/10-steps

- Specific individuals should be empowered with clear terms of reference to make decisions and manage any incident that may occur. Contact details of key personnel must be readily available, also on paper. Alternative communication channels must be provided.
- Establish structured detection methods and alert reporting channels. All alerts should be sent without delay to those responsible for assessment and triage.
- Define criteria for invoking the CIRP and for escalation to senior management.
- When disaster hits and chaos erupts, it is not the time to improvise. Having prepared playbooks is therefore essential for ensuring that activities are coherent from the start. Of course, no incident is the same, so there must be room for deviating from the playbook.
- Describe categories of situations where the technical team can handle incidents autonomously, according to the principle of proximity described in Sect. 14.2.
- The CIRP should include basic guidance on legal or regulatory reporting requirements, such as reporting breach of data protection to the national DPA according to GDPR (Table 14.1).

Testing and practice can be done in different ways, from abstract to practical. For example, some variants of testing/practice are the following:

- *Checklist test*: The CIRP is reviewed by affected units for comments on improvement.
- *Structured review*: Representatives from affected entities will come together to review the plan. Such a test needs little preparation, and can be done in a workshop lasting a few hours. A structured review may also include training for management regarding decision making, coordination and communication.
- *Simulation test*. Representatives from affected entities come together to practice responding to cyberattacks on systems and business functions in a simulated environment. Such exercises requires extensive preparation, and typically take the whole day.

14.5 Cyber Incident Management

Incident management is an organized response to a security incident. The purpose is to manage the situation in a way that limits damage and minimizes recovery time and losses.

The technical aspects of incident management are called incident *response,* which is typically performed by a dedicated team with support from IT operations and other support functions in the organization. Organization of incident response varies widely from business to business. In a business that has a SOC (Security Operations Centre), incident response will start in the SOC. An operations centre for a large organization handles hundreds of alerts and several incidents every single day. If the incident is serious, or escalates above a given level, a separate team is established, and even then, there is often another threshold for deciding on activating the CIRP and notification to management.

For some sectors, a threshold has also been defined for when the organization must notify the authorities of an ongoing incident. Then it can be a matter of hours, and not, for example, 3 days after the event.

In major, serious incidents, there will be multiple teams. The organization typically establishes a crisis management team, and if necessary, a crisis team is also appointed to relieve crisis management of obtaining relevant information, prepare regular situation reports and dispatch all routine communication. In this way, crisis management can concentrate on core problems, planning, coordination and decisions. This applies to all types of serious incidents, and not only cyber incidents. For businesses that have this in place, it is a matter of course to include cybersecurity incident handling.

Incident response mainly covers IT aspects in terms of incident response, but also other aspects. Incident management is a broader process in which technical IT functions collaborate with decision-makers, PR and communications representatives, and legal representatives.

In case of inadequate handling of any of the aspects, an initially small security incident can have disproportionately large impacts. With inadequate technical handling, small incidents can grow out of proportion. In weak decision-making, an otherwise competent team may waste time and resources on unnecessary activities or may work inconsistently. In the case of weak or poorly thought-out communication, external parties may distort press coverage of the incident, which may damage the organization's reputation. In the event of inadequate handling of evidence, it may be difficult to establish a legal case against the attackers, even though it may be known who was behind the attack.

It is primarily important to respond quickly when an incident is detected, but it is also important to ensure good handling of the other aspects of the whole management process. The incident management process consists of phases that each contains a set of steps, as shown in Fig. 14.3.

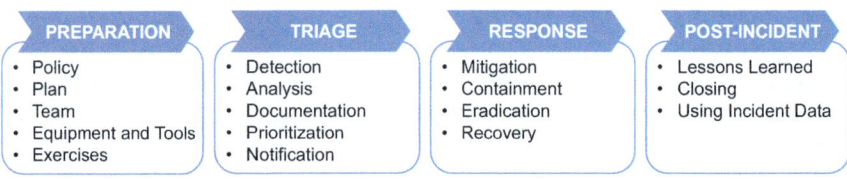

Fig. 14.3 Incident management process

The process of Fig. 14.3 is approximately the same as that described in NIST's *Computer Security Incident Handling Guide* [48] and in SANS' *Incident Handler's Handbook* [49]. The difference between these two frameworks is mainly that SANS splits the RESPONSE phase so that the steps Containment, Eradication and Recovery become separate phases.

The figure may give the impression that it is a complete sequential process. In practice, several of these steps are typically worked on in parallel and in iterations. The phases and steps of the process in Fig. 14.3 are described below.

14.5.1 Preparation

Preparation is to establish a cyber response capability so that the organization is ready to handle incidents when they occur. The points below describe the main elements of the preparation phase.

- **Policy**
  The IR policy must of course contain a statement of management commitment to having an IR capability. Furthermore, the policy should define roles, responsibilities, and levels of authority. The point about authority is particularly important e.g. when it comes to deciding whether do disconnect servers and applications, as this can have serious implications for business operations.

- **Plan**
  The IR plan (IRP) should describe specific procedures of the IR process, such as the steps in phases 2, 3 and 4 shown in Fig. 14.3. Following standard operating procedures (SOPs) is important for minimizing errors, particularly those that might be caused by stressful incident handling situations. The plan must contain contact lists of people in various roles, and contact points for authorities. Reporting to external parties can be required by law or can be considered necessary for other reasons. Reporting to other organizations under the TLP (Traffic Light Protocol) can be done as part of collaboration and CTI (Cyber Threat Intelligence) sharing.

- **Team**
  The IRT (IR Team) can be organized in different ways. The optimal model depends on the size and nature of the organization. Large organizations working in IT typically opt to have a dedicated IR team of own staff. For medium-size organizations it can be more economical to outsource parts of the IR work to an MSSP (Managed Security Service Provider). For small organizations, a fully outsourced IR team might be the most economical. Handling cyber incidents can be very stressful. Hence, it is important to let team members get enough rest to prevent burnout. An attack tactic could precisely be to launch detractor attacks as an attempt to cause burnout in the IR team before the real attack takes place.

- **Equipment and Tools**
  As the quote says: *"A good tool improves the way you work. A great tool improves the way you think."* (Jeff Dunteman), it is important that the IRT is well equipped. An IR jump kit is a portable case containing equipment that may be needed when investigating incidents. For example, a jump kit typically includes a laptop loaded with software tools (e.g., packet sniffers, digital forensics), blank

media for backups and storing evidence, hard-bound paper notebook, camera, audio recorder, chain-of-custody forms, evidence tags and evidence storage bags, to preserve evidence for possible legal actions. The laptop used to perform packet sniffing, malware analysis, and other actions could potentially contaminate the laptop. Hence, this laptop should be scrubbed, and all software reinstalled before it is used for another incident. This laptop should of course not be used for reading email, writing reports and performing other general computing activities.

- **Exercises**
  The IRT does not suddenly become good at responding, it can only happen through training and experience. Exercises are essential to check that the IRP works as intended, and to improve the IRT's ability to adequately handle real cyberattacks. Types of exercises can range from table-top exercises to simulated cyberattacks. Red-teaming, and TIBER (Threat-Intelligence Based Ethical Red-Teaming) are examples of training with simulated cyberattacks.

14.5.2 Triage

Triage is the process of investigating received alerts before deciding whether it is a real incident that needs to be handled, or if the alert is just a false alarm. Incidents are never the same, and some require more investigation than others. The IRT is often divided into tiers, where Fig. 14.4 shows an example.

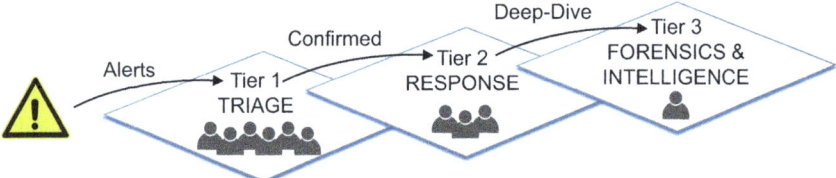

Fig. 14.4 Tiers of incident handling

As the figure indicates, there are typically more people involved with triage than people responding to confirmed events. Tier 3 comes into action when an incident requires deeper investigation. The points below describe the main elements of the triage phase.

- **Detection**
  Incidents come in many forms, which also influences how they are detected. Incident detection can e.g. based on (i) Intrusion Detection System (IDS) or other monitoring that classifies network traffic or system events as harmful or normal, (ii) internal anomaly reporting, or (iii) reporting by internal or external

parties.. Some incidents have overt signs that can be easily detected, whereas others are almost impossible to detect. Different types of incidents can be recognised by typical indicators. Hence, the IRT should be familiar with typical indicators of the most common attack vectors, as e.g. described in Sect. 2.1. Detection can also rely on CTI (digital threat intelligence), which is described in Sect. 16.2.

- **Analysis**
To obtain the greatest possible understanding of what has happened and what should be done next it is necessary to analysis available signs. The volume of signs of potential incidents is typically very high, where most alerts are false alarms. A sign of an incident can be a *precursor* or an *indicator*. A precursor is a sign that an incident may occur soon. An indicator is a sign that an incident may already have occurred or is currently occurring. Digital forensics are included in this step, see Sect. 14.6. The analyst needs deep, specialized technical knowledge and extensive experience for proper and efficient analysis of available data to make quality assessments. Unfortunately, having human analysts scale poorly with high numbers of incidents to be analysed. There is thus a great market for security vendors to innovate and offer increasingly more advanced tools to better classify incidents.

- **Documentation**
Continuous documentation of relevant facts leads to a more efficient, more systematic, and less error-prone IR process. Every step taken from the time the incident was first detected to the closing of the case should be documented with a timestamp and signature from the incident handler. Not only is this important to keep oversight and control over the response process and learn from it later; every piece of evidence regarding the incident can be used as evidence in a court of law if legal prosecution is pursued. Whenever possible, incident handlers should work in teams of at least two, where one person performs the technical tasks while the other records and logs events. A hard-bound notebook is an effective and simple medium for this, but laptop can also be used.

- **Prioritization**
Once an incident has been classified as a real incident, its prioritization for handling is critical. Prioritization should be based on the relevant factors, such (i) business impacts, and (ii) expected recoverability from the incident based on estimated time and resources needed. An incident with a high business impact and low recovery effort is naturally prioritized for immediate handling. However, some incidents may not be simple to recover from in a technical sense. Some incidents may need a more strategic-level response—for example, there is no easy recovery from an incident where gigabytes of sensitive data have been exfiltrated and publicly posted on the dark web, since the data cannot be un-exposed. In such cases the handling of the incident may be to develop a strategy for preventing future breaches and to create an outreach plan for alerting those individuals or organizations that are affected by the exposed data.

- **Notification**
  As soon as an incident has been confirmed, notifications must be sent to the appropriate individuals so that all involved will be able to fill their roles. The incident response policy should include provisions for incident reporting—at a minimum, what must be reported to whom and at what times. Different types of information is provided to different parties, depending on need. In the EU, the national Data Protection Authority shall be informed within 72 h of an incident with a data protection impact. It may be necessary to provide status updates to certain parties, even in some cases to the public and the entire organization. The team should plan and prepare different communication methods, and select the most appropriate method depending on recipient and the type of information. Reporting to the police may be relevant when the incident is characterised as serious crime. A challenge when reporting to the police may be seizure of equipment and that the IRT loses control of the investigation. On the other hand, police have access to various resources that can be very useful for the analysis phase of an incident.

14.5.3 Response

Response is the active intervention in the organization's infrastructure to stop the incident and remediate the compromised system. The points below describe the main elements of the response phase.

- **Containment and mitigation**
  The purpose of containment and mitigation is to prevent the incident from worsening and to ensure that the attack does not spread further. This is easier said than done. Simply unplugging infected systems or stopping compromised processes might stop the attack, but it could by itself have negative impact of business processes are interrupted. Use discretion to determine whether servers, applications, and systems should be stopped or isolated (set offline) or continue to run and be online. It may be necessary to allow infected systems to continue running to maintain critical and important business applications. Here, virtual platforms can have certain advantages. Ideally, virtual machines can be stopped and stored for analysis, and new VMs can be quickly spun up to maintain critical services.

- **Eradication**
  After the incident has been contained, eradication is necessary to completely remove all incident components within the organization's domain. This activity may consist of identifying and removing vulnerabilities, upgrading/patching the system, removing malware, and correcting system configurations. Often, this cause will be a threat actor that has unauthorized access and has installed malware or changed configurations. The goal of the eradication is to remove all of that. The knowledge gained in the analysis is very important input to this phase.

In order to evict the threat actor, one must be absolutely certain to (i) have full control of what the actor has done to ensure knowledge of any backdoors, and (ii) to have a robust infrastructure with good detection capabilities that enable repeated intrusion attempts to be detected. It takes time and a lot of effort to get certainty about these two points. For that reason it sometimes takes months and even over a year before the organization is ready to execute and complete the eradication. If the infected infrastructure is damaged beyond repair, the organization may start to build a new infrastructure and gradually phase over operational functions into this new and controlled environment before eradication on the old infrastructure is complete. Recovery is then actually carried out ahead of eradication.

- **Recovery**
  The work to bring systems and business processes back to normal operation normally starts after eradication, but sometimes starts before or during eradication, as mentioned above. It is important that recovered systems have no traces of the incident or infection, and that exploited vulnerabilities have been removed to prevent similar attacks in the future. Recovery may include restoring systems and data from backups, rebuilding systems from scratch, changing passwords, and hardening systems and tightening network security policies (e.g., firewall rulesets and access authorizations). Hardening systems consists of reducing their vulnerability surface by removing unnecessary programs and functions. More comprehensive system logging or network monitoring are often introduced to improve detection capabilities. This is sensible, as attackers typically will try to attack similar resources resource again.

14.5.4 Post-Incident

- **Lessons-learned**
  There is great potential for continual improvement of the IR process by reflecting on how an incident was handled. Holding a "lessons-learned" meeting after a major incident, and maybe periodically for lesser incidents, is a valuable opportunity for improving the organizations security posture and the IR process itself. The meeting should be held within a few days of the end of the incident.
- **Closing**
  The step of closing the incident is normally decided by the system owner. Incident response teams and others involved hand over the final report, optionally with additional information and formats. An example of additional information is that the threat and incident are described as CTI in a machine-readable format, such as MISP or STIX, described in Sect. 16.2.4. In that way, CTI about the incident can be shared with other organizations to make them better prepared to stop the same kind of attack.

- **Using Incident Data**
  Data collected from handling incidents can be valuable in many ways. For example, risk assessment can be made more objective when estimates of incident likelihood and impact can be bases on statistics from real incidents. Data science and ML (Machine Learning) can be used to extract new knowledge from large amounts of incident data. For example, ML tools can learn to classify threat actor behavior in order to better identify and prioritize incidents. There is also a great potential for developing automated threat response tools by learning from how the IRT has worked during previous incidents. LLMs (Large Language Models) can be developed based on the collection of own incident reports and reports received from external organizations. Such LLMs can then be used by incident handlers when investigating future incidents.

In the event of an incident, the enterprise is generally well prepared if it has established and exercised an IRP. Contingency plans must be available on physical documents, which must also be practiced in use. Building and maintaining a dedicated incident response team can be costly. Alternatively, the business can purchase commercial services like MSS (Managed Security Services), MDR (Managed Detection & Response), and XDR (Extended Detection and Response). There are different types of security teams, the most common of which is SOC (Security Operations Centre) and CSIRT (Computer Security Incident Response Team). A SOC provides infrastructure monitoring and incident notification. A CSIRT handles reported incidents. These concepts are mixed to some extent, allowing a SOC to also drive incident response. There is a growing market for such services with an increasing number of providers operating at national and international levels.

14.6 Digital Forensics

Digital forensics includes analysis of material contained in digital devices and network logs, usually in the context of security incidents and cybercrime. The investigation may be to discover what has happened, or it may be aimed at proving or disproving a hypothesis. Digital forensics can be included as part of incident management.

Digital forensics can be divided into several sub-branches, depending on the type of technology and digital devices affected. Below are examples of digital forensics that practitioners may encounter in their investigations:

- Network forensics focuses on monitoring and detecting evidence in network traffic.
- Wireless forensics is a part of network forensics aiming at collecting and analysing wireless network traffic.
- Disk forensics is to collect and analyse evidence from digital storage devices such as hard drives, phones, USB memory sticks and DVDs.
- Database forensics is to collect evidence from and investigates databases.

14.6 Digital Forensics

- Email and message forensics focuses on collecting and recovering messages before analysis.
- Malware forensics is to identify malware and reverse-engineer their functionality.
- Memory forensics is to collect and investigate the data in a computer's cache memory or RAM.

The digital forensics process consists of a set of steps as e.g. illustrated in Fig. 14.5 and described below.

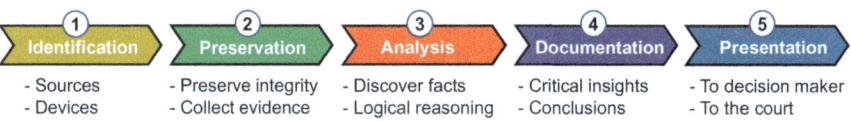

Fig. 14.5 Steps in the digital forensics process

1. **Identification**

    This step consists of detecting an event of interest, e.g.:
    - "My computer is acting weird."
    - "Someone has posted our entire database on the Dark Web."

    From here, the investigator needs to identify relevant sources of information that hopefully will make it possible to find out what happened and derive conclusions.

2. **Preservation**

    Make sure that the evidence is not destroyed by negligent handling. You may need to say, "Don't turn off your computer!" The collection of all potential evidence must be done in *the order of volatility*—first volatile data such as RAM, then hard drives, and lastly inert media such as DVDs. Use write-blocking and make multiple copies of disks where one copy is defined as "master copy". It's important to make sure you have an uninterrupted *chain of custody,* which means that the collection and handling of evidence is done in a legally justifiable manner that prevents evidence from being compromised. The chain of custody must be documented, which is necessary to be accepted as evidence in any court case.

3. **Analysis**

    Collected data is examined to uncover relevant elements, which may be to find hidden files and partitions on a disk. Through analysis of collected material and logical reasoning, the investigator tries to arrive at facts that can prove or disprove hypotheses. It may be required to use two different tools to perform the same analysis to confirm the results of the analysis. The steps of the analysis must be repeatable. Make a chronological record of assumed events.

4. **Documentation**

    The process and method of data collection and analysis must be described in a way that is repeatable. The reasoning leading to the main findings and conclusions must be described.

5. **Presentation**
    This step consists of presenting evidence and conclusions of the analysis to the system owner, to the police, or to the court.

In the case of live collection, the evidence is retrieved from a system where the microprocessor is running. In the case of post-mortem collection of evidence, it is obtained from a storage medium where the system is switched off. Post mortem improves integrity preservation and does not affect data. However, volatile data can be lost in the process of shutting down a system. Live collection makes the collection of volatile data possible, but also affects the data. In the event that the hard disk is encrypted, it is better to retrieve the information from the disk while the system is running, because the encryption/decryption key is located in RAM.

A chain of custody in digital forensics can be achieved by creating a chronological record showing the seizure, storage, handling, transportation, analysis and disposal of the evidence, whether physical or electronic. Because evidence can be used in court to convict persons of committing crimes, it must be handled in a particularly reassuring manner to avoid possible erroneous allegations of falsification or misconduct. If such allegations can be supported by arguments of lack of notoriety, it could prejudice the prosecution's case and lead to an acquittal, even if the prosecution is otherwise convinced that the accused is guilty.

14.7 Security Testing of Systems, Networks, and Enterprises

The best way to determine if a business has sufficiently robust protection against attacks is to subject the business to simulated attacks. Such attacks can be cyberattacks or attacks on physical buildings or other security barriers. There are several different methods, some of which are described in the following sections.

14.7.1 Pentesting

A penetration test, usually just called pentest or ethical hacking, is a simulated attack against an organization, authorized by management and carried out to evaluate the security of the organization's networks applications and systems. Usually, the system owner and IT operations are informed that the pentest is being carried out. As the first step in pentesting, a mapping of networks and systems is usually done through exposed interfaces. After this, a vulnerability scan will often be performed. Vulnerability scanning can be performed manually, or using partially automated programs that can identify if a server is vulnerable and could potentially be exploited in an attack. The vulnerability may, for example, be that the server lacks security updates, or uses outdated operating system or other software.

The next step in pentesting is precisely to exploit identified vulnerabilities to explore the potential for unauthorized access to system functions and data. In this

14.7 Security Testing of Systems, Networks, and Enterprises 317

step, a plan is made for which vulnerabilities to attack, and a goal can be defined for the attack. The goal may, for example, be to gain unauthorized access to the information that may be on the system, which would represent a breach of confidentiality.

It is common to define pentest as white-box testing or black-box testing, depending on how much information the pentester is given. White-box testing means that the pentesters receive in advance all relevant information about the systems to be tested, which can be complete network drawings. Black-box testing means that the pentesters have no description of networks and systems available, so they have to explore this as far as possible. In any case, both white and black-box testing will normally involve a process of examining the system, as network drawings and other documentation often turn out to be outdated. Vulnerabilities that the pentest reveals must be reported to the system owner. The report can also assess the risk as a function of vulnerabilities' potential attack surface, measured against the severity of impacts and operational problems that the attack may cause in the relevant servers and applications. The report can also suggest countermeasures to mitigate risk.

14.7.2 Red-Teaming and Blue-Teaming

A person, or group of people, performing pentesting toward a target is called a *red team*. On the other hand, a *blue team* is the group of people whose role is to defend against attacks and stop the red team. In practice, blue teams consist of those who usually handle security incidents in an organization. As with pentesting, the company must authorize simulated attacks from red teams before the test starts. The difference between pentesting and red-teaming is that in red-teaming, the blue team is not informed that there will be simulated attacks. In this way, red-teaming is a realistic test of the company's cyber readiness, including both technical, procedural, human and management aspects. As the blue team is not informed about the test in advance, there is a *white team* who are informed that the organization will be exposed to simulated cyberattacks. Red-teaming is more realistic than just pentesting, precisely because IT operations and the blue team are not informed in advance. One disadvantage of red-teaming is the red team's need to remain undetected, which means that red-teaming typically reveals fewer vulnerabilities than pentesting. Assuming that a system contains 100 vulnerabilities that can be exploited in a cyberattack, a red team will normally only seek to find a subset of vulnerabilities to get in, while a pentester wants to find all 100 vulnerabilities.

As with pentesting, red-teaming will use vulnerability mapping to assess possible attack methods. In addition, both red-teaming and pentesters can conduct initial reconnaissance, which means gathering all possible relevant information about the attack target. Reconnaissance is the 1st stage of the CKC (Cyber Kill Chain) as described in Sect. 16.1.2.

OSINT (open-source intelligence) is typically part of reconnaissance, and is used by threat actors to identify targets, to determine how targets can be attacked, and to be used as part of an attack. To prevent or impede stage 1 of the CKC, businesses and individuals should consider limiting or reducing the amount of information that can be produced with OSINT, for example by limiting information posted on websites and social networks. There are automated tools for OSINT, including SHODAN[2] to find IoT units, Maltego[3] for mapping and visualizing data from open sources, and The Harvester (part of Kali Linux), which is a tool used to collect email addresses and domain names related to an (attack) target. The world's largest search engine Google is also an excellent OSINT tool, where the term *Google hacking* means to use Google for OSINT. A physical form of OSINT is *dumpster diving*, which consists of looking for paper documents from a company or private individual in garbage bins and dumpsters. Controls against dumpster diving include secure shredding.

A variant of red-teaming and blue-teaming is practicing cyberattacks and defenses in a *Cyber Range*, which is a synthetic IT environment with networks, servers and applications. Red teams attempt to penetrate the network, and take down servers and applications, while blue teams attempt to detect red teams' activities, eradicate them, and keep servers and applications up and running. The advantage of cyber ranges is that it does not touch real business applications. However, an exercise in a cyber range is not a security test, it is just a training exercise for cyber operations.

14.7.3 TIBER

TIBER (Threat Intelligence-Based Ethical Red-Teaming) is the most advanced form of pentesting. TIBER-EU is a framework developed by the European Central Bank (ECB) to test financial institutions' ability to detect and defend against advanced cyberattacks. Gradually, TIBER is also being used in other sectors.

In addition to being a form of red-teaming as described in the previous section, the special feature of TIBER testing is that it also employs targeted cyber threat intelligence as described in Sect. 16.2, so that the red team can mimic real APTs in the most realistic way possible. As TIBER is a very comprehensive method for security testing, it is also very expensive to implement. Businesses must consider whether the benefits of TIBER outweigh the costs. TIBER is a relatively new framework, where businesses currently do not have as much experience as with pentesting and red-teaming.

---

[2] https://www.shodan.io/

[3] https://www.maltego.com/

14.8 Security Audit

An audit evaluates to what degree and organization is adhering to a set of criteria, where "criterion" is audit jargon for something that the organization is supposed to comply with. In general, an audit is a systematic process executed by a competent professional to evaluate controls and processes against the criteria, and then to write a report on the effectiveness of these controls and processes regarding how well they satisfy the criteria. The findings shall be objective in the sense that a new audit done by a different auditor should reach the same conclusion. The two parties in an audit are the auditor and the auditee, where the auditor should be independent of the auditee. The focus of audits can be to evaluate the degree of compliance with a set of internal or external criteria. Three types of security audits are (1) legal security compliance audit, (2) security standards audit, and (3) technical security audit. These are described next.

1. Legal security compliance audit is based on external criteria in the form of security laws, regulations, and contracts such as SLA (Service Level Agreements). Depending on where the organization is located, there are national laws and regulation such as the US federal regulations HIPAA (Health Insurance Portability and Accountability Ac) and SOX (Sarbanes-Oxley Act. In the EU, GDPR is a prominent example of assessment criteria for data protection which apply to all EU/EEA countries.
2. Security standards audit is based on external criteria such as standards and industry norms that the organization is obliged to follow or has decided to follow. This could be security norms such as the CSA (Cloud Security Alliance) CCM (Cloud Control Matrix). This could also be for the purpose of formal certification of an organization, e.g. for having an ISMS (Information Security Management System) according to the standard ISO/IEC 27001 (see Sect. 18.5.2) or for receiving a SOC 2 security attestation for cloud service providers.
3. Technical security audits can review and test firewall configurations, malware and antivirus protection, password policies, data loss prevention, authentication, access control and other components in an organization's IT security program. The assessment criteria can be internal policies or external technical standards and industry norms. A technical audit can include vulnerability scanning and pentesting as described in Sect. 14.7. Human factors such as security awareness can also the subject of pentesting as described in Sect. 13.6. Finally, the physical infrastructure of the organization can be the subject of pentesting and audit.

The auditor produces a report with observations, recommended changes, and assessments of other aspects of the organization's information security practice. The audit report may describe specific gaps in compliance with regulations, in compliance with security controls according to standards and best practice, may point out security vulnerabilities and may even reveal previously undiscovered security breaches.

These findings can then be used to inform the organization regarding how to reduce security risk and to identify points of improvement to reach compliance. Auditors will typically rank their findings in order of priority to help the organization with prioritizing its resources for following up the findings of the report.

14.9 Tasks

1. Continuity
   (a) Based on the bow-tie diagram—which element(s) specifically is/are related to continuity?
   (b) What is the most common security control for continuity?
   (c) What is a BIA (Business Impact Analysis)?

2. RPO, RTO and MTD
   (a) What is RPO and what assessments does the agency need to make to choose shorter or longer RPOs?
   (b) What is RTO, and what assessments does the company need to make to choose a shorter or longer RTO?
   (c) What is MTD and how is MTD determined for specific functions and processes?

3. Response teams
   (a) What are possible models of organizing an incident response team?
   (b) How can a RACI chart be used as part of contingency planning described by the UK NCSC (National Cyber Security Center)?
   (c) Why is it important for an organization to have pre-established collaboration with external agencies for handling potential cyber incidents?

Chapter 15
AI and Cybersecurity

All models are wrong, but some are useful.

George Box, British statistician

AI will have radical consequences for cybersecurity. Threat actors will weaponize AI to launch increasingly potent and destructive attacks. AI-based tools can create advanced deceptions and deepfakes that are already being used in real attacks, and the volume will only increase. AI-based threat orchestration will scale up the attack volume through automated bots, and will target more divers types of organizations independently of language, country, and region. This is scary. To counter the threat of offensive AI, companies and research organizations are currently investing Big in the development of AI-based cyber defence technologies. Defenders need expertise in offensive uses of AI to better understand how defensive AI should work. In addition, the AI systems themselves can be the target of attacks, and hence we need to develop techniques for the defence of AI systems.

The first learning objective of this chapter is to understand the basic principles of AI (Artificial Intelligence) and ML (Machine Learning) . Then you learn how AI is being used for both offensive and defensive purposes. You also learn how AI systems themselves have vulnerabilities that can be exploited by threat actors, with possible ways to protect AI systems against such attacks.

15.1 Introduction to AI

The birth of artificial intelligence as a scientific discipline can be traced to Alan Turing's work on computation during the 1940s, and to the Dartmouth workshop in 1956 which was attended by a group of mathematicians, including Claude Shannon who invented information theory, see Sect. 4.10.2. John McCarthy, the organiser of the workshop, coined the term "artificial intelligence" when writing the funding proposal for the workshop. Since then, the AI field has gone through multiple cycles of optimism followed by periods of disappointment known as "AI winters". Early AI paradigms were based on the idea that all available knowledge in the world could be codified syntactically as data, where logic programs could perform deductions to generate new knowledge. When a seemingly promising paradigm like this turning out to be flawed, the AI winter set in until a new promising paradigm was proposed a few years later. Interest in AI increased after 2012 when *machine learning* outperformed previous AI techniques, and the early 2020s saw the beginning of the AI boom with impressive machine-learning based AI applications in all sectors.

15.1.1 Artificial Neural Networks (ANN)

Machine learning (ML) is the general term to denote the study of automated methods for designing or configuring a computer program (the machine) to perform a given task. More specifically, ML is usually based on *artificial neural networks (ANN)*, and in the following we will use the term ML in that sense. The term ANN comes from the fact that these techniques are biologically inspired by the way neurons are interconnected in animal brains. An artificial neuron is simply a variable called "node" in a computer program. *Deep learning* means that the ANN consists of multiple (at least two) intermediate layers of nodes, as shown in Fig. 15.1.

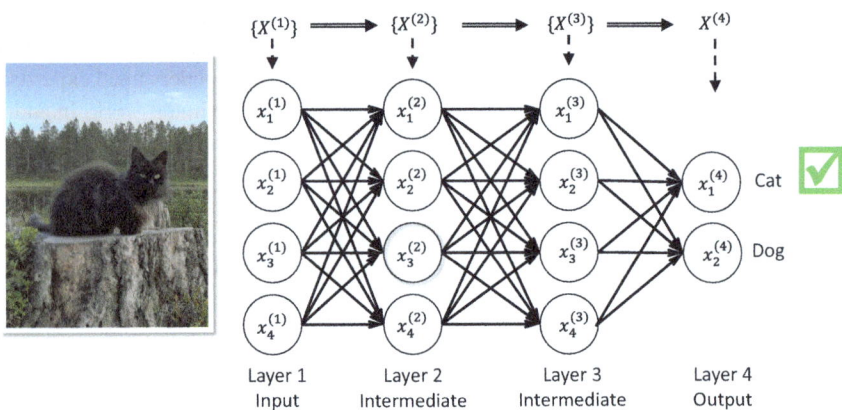

Fig. 15.1 Toy example of deep ANN for classifying images of cats and dogs. (Cat photo by the author, CC BY-SA)

The edges between nodes are weights that can be thought of as cells in a spreadsheet with fixed values. The nodes can also be thought of as spreadsheet cells with values computed as the sum of incoming weights. When the sum of incoming weights to a node reaches a specific threshold value, then that node *fires*, which is analogous to how a neuron in the brain sends out signals to neighbouring neurons. To fire means that the weights of outgoing edges from the node are added to the respective nodes in the next layer. The cascaded firing of cells from layer to layer is called forward propagation. If a node does not fire, then no weights are forwarded to the connected nodes in the next layer. In practice, storing and summation of

15.1 Introduction to AI

weights are not done with an "office spreadsheet", but with specialised software or hardware for fast propagation of node firing and weight summation through the ANN. With reference to Fig. 15.1, the *training* phase consists of automatically testing the ANN with many images of cats and dogs, and gradually tuning the weights of the edges so that the ANN gets better and better at telling cats from dogs. Once the training is completed, the ANN is a *model* which can be put into operation for image classification, i.e. to tell cats from dogs in our example. Note that in most other technical disciplines, a "model" represents a blueprint for building something that works, whereas in the ML jargon a "model" is the thing that works. For image classification, the input layer typically represents the matrix of pixels of the image, which of course can be in the millions (i.e. megapixels), not just four nodes like in our toy example. Said simply, an input node fires if the pixel is bright, and does not fire if the pixel is dark. In real applications, the relative brightness and colors are also taken into account, of course.

If node $x_1^{(4)}$ fires in the ANN of Fig. 15.1, it means that the model has classified the image as representing a cat. It may seem mysterious that an ANN can tell which animal it is without even being disturbed by the background. One way to explain it is that somewhere in the intermediate layers there are nodes that fire when there is something pointy on the image, which could indicate pointy ears of a cat. When many intermediate nodes fire as a result of subtle indicators typical of cats, the ANN will in the end fire the output node $x_1^{(4)}$ for cat. However, for some of the complex ANN models e.g. used for the LLMs (Large Language Models, see below), even their designers are unable to completely explain how the models can produce meaningful outputs. Researchers typically experiment with different training methods, and if the resulting model produces good results they cannot always explain why.

15.1.2 Machine Learning Paradigms

The approaches for learning are traditionally divided into supervised, unsupervised, and reinforcement learning, as described below.

- **Supervised learning:** Training data for supervised learning consists of *labelled data* where a label indicates the correct class or prediction for given data example. Obtaining labelled data can be expensive if the process of labelling data cannot be automated. With reference to the example of Fig. 15.1, who is going to label all the training images of cats and dogs?
- **Unsupervised learning:** Training data for unsupervised learning does not contain labels, i.e. there is no "correct" class or prediction to learn. Instead, the training consists of learning patterns in the training data, which typically is a fully automated process. LLMs are trained with unsupervised learning, where the goal is to learn patterns in the text corpus, such as how sentences typically are structured.

- **Reinforcement learning:** The training for reinforcement learning is based on giving a "reward" to the model for each output it produces as a response to stimulus in a dynamic environment. More precisely, the reward is a feedback value automatically generated as a function of the quality or correctness of the output. The model will gradually improve the quality of output to obtain better rewards. The model can be considered complete when the output quality has reached an acceptable level. Alternatively, the learning can continue while the model is in production. No dataset is needed for training, as the training data are dynamically and automatically generated during learning. ML models for playing chess are trained in this way.

The training of large ML models does not necessarily require training data to be stored in a centralised database. Federated learning is a way of training a model in a decentralized way with datasets stored in different data centres that can be owned by different stakeholders. Motivations and concerns of federated learning can be data privacy, data minimization, data access rights, and supply-chain risks. Transfer learning is a training technique where knowledge (i.e. model parameters) learned from one task is re-used to train a model on a related task. For example, for image classification, knowledge for recognizing cars could be applied when training a model to recognize trucks. This is possible because cars and trucks have similarities, e.g. they have tires and wind screens, and they drive on roads.

15.1.3 Training Methods and Model Tasks

Under the learning paradigms supervised, unsupervised and reinforcement learning, there are different training methods that can produce ML models for a variety of tasks, where Fig. 15.2 gives a very simplified overview.

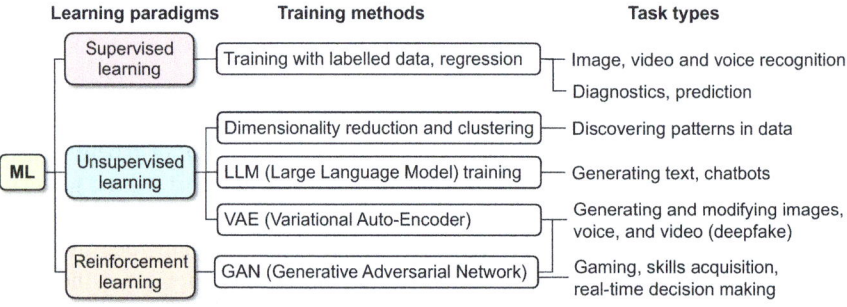

Fig. 15.2 Machine learning technologies and task types

The training methods and model tasks illustrated in Fig. 15.2 are briefly described below.

15.1 Introduction to AI   325

- **Training with labelled data**: These techniques produce ML models that can recognise and classify things. Training with labelled data means that during training the model is told what the correct classes are. As an example, a medical cancer classification tool can be trained by showing many X-ray images, and each time tell the model if there is cancer or not. The cat-and-dog classifier in Fig. 15.1 is also an example of a classifier which has been trained with labelled data. These models can recognize the classes in the data it has been trained on, such as images, speech, music, data traffic patterns in computer networks, or objects seen in the video stream of an autonomous-driving car. For prediction tasks, the training is done with historical data, where the labels can be known real events that result from things described by the historical data. The internal weights of the ANN are gradually adjusted to minimize the difference between its output classes and the actual labels in the training data. Once the training is complete, the model can be used to recognise things and predict future events. Limitations of training with labelled data is that the models can only recognise and predict similar things to what exists in the training data, and that classification and prediction will be unreliable if the model is presented with something that has never been seen before.
- **Dimensionality reduction and clustering**: The data dimension refers to the number of variables, columns, or inputs in the training dataset. Too high dimensionality makes it difficult to learn anything from the data sets. Hence, dimensionality reduction is an essential aspect of machine learning by making the data more manageable. This helps avoid the "curse of dimensionality", removes redundancy in the data, and reduced data storage and computation time. After dimensionality reduction comes clustering which refers to grouping unlabelled data based on similarities or differences. The aim of clustering is to discover patterns in raw data and group them accordingly.
- **LLM:** As the name "Large Language Model" says, LLMs must be trained with huge amounts of text to produce meaningful output. In the Big-Tech business of LLM-training, two powerhouses have emerged as leaders: BERT (Bidirectional Encoder Representations from Transformers) from Google, and GPT (Generative Pre-trained Transformer) from Open AI. These two are in a race for world dominance. The term "transformer" means that the training transforms the text corpus into statistics about semantics in the text corpus. When generating new text, these LLMs simply selects words sequentially where each next word is the most likely from the statistics of the training corpus. LLMs are an essential component of Natural Language Processing (NLP), i.e. technology for understanding and generating human text and voice.
- **VAE:** The term "Variational Auto-Encoder" can best be explained with the typical application of modifying images. It consists of two parts, the encoder and the decoder. The encoder compresses the image to a lower-dimensional latent space, which means that it represents the image as a set of features, such as face and hair of a person. The decoder decompresses the features to recreate the full image again. Assume that we have trained the encoders and decoders for images of person A and person B. It is then possible to "face swap" by recreating the image

of person A using the decoder for the image of person B, i.e. so that the image of person A now has the face and hair of person B.
- **GAN:** The term "Generative Adversarial Network" can best be explained with the typical application of image generation. GAN models are trained with two adversarial ANNs: a generator and a discriminator. The generator first produces a synthetic image. Then the discriminator does a comparison between the synthetic image and a real image to judge if the synthetic image looks as real as the real image. If the synthetic image fails the test, then the generator tries again with a new and more realistic synthetic image. This goes on until the synthetic image passes the test, whereby the GAN model training is complete. The term "network" refers to the fact that the two adversarial ANNs send data to each other.

The training of LLMs makes it possible to create conversational bots which we can talk to and get assistance from. The interface to such bots can be through text, voice, video, and even physical robots. NLP is fundamental element of such assistants.

In the first years of machine learning, ML applications were unimodal in the sense that they could only handle a single category of data such as text, images, sound, video, or tabular data, but not several at the same time. Since 2023 we have seen the emergence of multimodal AI that can handle several data categories at the same time. GPT-4 is a multimodal LLM model launched in 2023, and is one of the optional engines of ChatGPT. The main new feature of GPT-4 is its ability to take images as input in addition to text prompts.

15.2 Offensive Use of AI

Offensive uses of AI refer to the utilization of AI technologies to enhance and automate offensive cyber operations. These techniques leverage AI algorithms and capabilities to develop malware, launch attacks, exploit vulnerabilities, infiltrate systems, and execute harmful actions. The sections below briefly describe the main threats with possible mitigations.[1]

15.2.1 Deepfakes

We recognize people by their faces and voices in the physical world. In the virtual world we are also accustomed to assume that the voice we hear over the phone, or the face we see in an online meeting, actually belongs to the person we already know. However, we can no longer take this for granted. Deepfakes make it easy to imitate speech and faces in video meetings, forcing us to be more sceptical about who we think we are seeing and talking to online. Threats posed by deepfakes are

---
[1] https://www.computer.org/publications/tech-news/trends/ai-fighting-ai

described in Sect. 2.1.9. The term "deepfake" refers to the use of deep learning to synthetically create images, video and sound. Deepfakes are typically produced with GAN and VAE technology described in Sect. 15.1.3.

With GAN it is possible to create synthetic images and sound with such high quality that it is impossible to tell that its "fake". VAE can be used for face and sound swapping with high quality, which makes it possible to impersonate somebody in a video or phone conversation. Combined with LLM transformers the attacker can talk in any language, which dramatically expands the boundaries of deepfake-based social engineering crimes.

To mitigate these types of threats, organizations must put in place safeguards to verify the identity of participants and the authenticity of their video, voice and chat messages. In theory, this could be done either informally by prompting participants during the call, or through technical means using MFA techniques described in Sect. 8.6. However, this is still uncommon, and no best-practice methods currently exist. In 2022 the US Department of Homeland Security has published a report on measures to combat deepfake fraud [13]. One could hope that a deepfake can be detected through technical means, but unfortunately that only works with poor-quality deepfakes. When training with GAN makes deepfakes as realistic as the real thing, then it is theoretically impossible to detect that it is a deepfake just by observing it. Another approach could be for states to legally ban deepfakes, and some states do, to some degree. The European Union is evaluating possible regulatory alternatives with the proposed AI Act. The UK has banned non-consensual explicit sexual deepfake images. In the US, Virginia has enacted a law against pornographic deep fakes. California and Texas have enacted laws against deepfakes used for disinformation to influence elections. China was the first country in the world to enact a law against deepfakes, with its 2022 law which prohibits the use of deepfakes that are considered harmful to the society in China.[2]

15.2.2 Malware Engineering

Coding copilots are widely used for efficient software development, and the same type of tools can of course also be used for efficient malware development. This amplifies the capacity of threat actors to rapidly develop new types of malware. It also makes it possible to automatically develop malware that can exploit new (zero-day) vulnerabilities as soon as they emerge. For malware attacks there is often a race to develop and leverage exploits during the time window before vulnerabilities are patched and removed. GPTs can also learn to make malware that evades traditional protections, that mutates and adapts to make it difficult for defenders to detect and mitigate. While legitimate coding copilots might refuse to generate code for

---

[2] https://www.afcea.org/signal-media/cyber-edge/chinas-deep-fake-law-fake

malware, resourceful threat actors can train their own coding copilots without such restrictions.

To mitigate this type of threat, defenders need to apply similar methods to speed up their defences. Vendors must shorten the time to develop patches for vulnerabilities, organizations must shorten the time to install patches, and there must be automated sharing of CTI and courses-of-action as described in Sect. 16.2.4. However, this is largely still work in progress.

15.2.3 Attack Automation

Automated AI-enhanced hacking tools are on the wish list of threat actors. AI algorithms can scan networks, identify weaknesses, and launch attacks with cyber speed and global coverage. If AI-based attack tools reach a certain level of maturity and sophistication it can lead to devastating data breaches, financial fraud and system infiltrations.

Threat actors have started using AI to enhance the ability of their botnets to evade detection and coordinate attacks. AI also enhances the CC infrastructure of botnets, making it able to analyze target network behavior and reconfigure its attack tactics to adapt to cyber defenses.

LLM transformers can provide attack recipes, just like it can for cake recipes. While legitimate GPTs might refuse to provide attack recipes in the same way as it refuses to tell you how to make bombs, resourceful threat actors can train their own GPTs without such restrictions. While an attack recipe in itself is not an attack tool, it can feed into automated tools for attack orchestration.

Most cyber incidents today start with a phishing attack (see Sect. 2.1.1). While the success rate of general phishing spam is very low, the success rate of spear phishing is considerably higher. The cost of creating credible spear-phishing copies has traditionally been a limiting factor for attackers. With LLM transformers combined with AI-based information gathering, this barrier is removed. For each victim who takes the bait, AI-driven bots can automatically take over to complete the attack. This can include using bot deepfakes for exchanging messages or for phone or video conversations with the victims. A well-orchestrated AI-driven attack platform can in this way scale up the attack capacity with several orders of magnitude compared to traditional manual attack processes.

Mitigating this type of attacks will require the combination of multiple, if not all, techniques in the cybersecurity toolbox. Email security with SPF, DKIM and DMARK increases the difficulty of sending phishing email. Measures to verify people's identities and confront possible deepfake entities reduces the attackers' success rate with such techniques. CTI sharing and AI-driven automated courses of cyber-defence actions are general techniques to prevent large scale cyber attacks.

15.3 Defensive Use of AI

Although the previous section about offensive uses of AI also describes mitigation strategies, this section gives a structured overview of relevant applications of AI for cyber defence. Defensive uses of AI refer to the utilization of AI technologies to enhance and automate components of defensive cyber operations. Most providers of cybersecurity solution providers are already using AI. These techniques leverage AI algorithms and capabilities to automatically identify threats and vulnerabilities, analyse malware, detect and investigate attacks, situational awareness and decision making to automatically execute responses to attacks. The sections below briefly describe the main cyber-defensive uses of AI.

15.3.1 Detection of Malicious Activities

AI tools can monitor and analyse network activities to establish benchmarks of safe or regular activities, and thereby detect activities that may be deemed anomalous or potentially harmful. AI can also monitor and analyse user behavior along the same principles, with the potential of detecting possible cases when a user has become a victim of social engineering, or has become an insider threat as described in Sect. 13.3. Discriminative and predictive ML models are suitable for detecting malicious activities, i.e. for classifying types of activity and for predicting future events. Model training can be done with examples of both normal and malicious network activities and user behavior. Additional analysis in more depth can be triggered if during usage-time the model detect patterns that neither resemble normal nor malicious activity.

15.3.2 Malware Detection and Analysis

The traditional way of malware detection is by *signatures*, which is either a hash value of a piece of malware, or a defined sequence of software functions indicative of malware. Signature-based malware detection is very efficient and gives few false positives, but it typically suffers from too many false negatives, i.e. it is not good at spotting new types of malware that have not been seen before. On the other hand, AI-based tools can do better at detecting known and unknown types of malware, such as file characteristics, code patterns, and behavior to determine if a file or script introduced to the system is safe or malicious.

15.3.3 AI Support for Cyber Threat Intelligence

AI has the potential to radically improve the capability, capacity and coverage of cyber threat intelligence CTI which is described in Sect. 16.2. By automatically gathering information from relevant sources, including the dark web, AI can identify emerging threats, correlate indicators of compromise, and present actionable CTI.

AI makes it more efficient to crawl online sources and parse through vast amounts of textual data to effectively structure text using NLP. Language is not a barrier, as text from sources in different languages can be analysed and categorized based on machine translation.

Human analysts can use AI as a capacity amplifier in threat hunting. AI-based tools can assist in sifting through enormous datasets to automatically uncover and analyse indicators at the technical and tactical layers of CTI, enabling analysts to focus more on strategic CTI and decision making. The deluge of security alerts and event information, usually with a large percentage of false positives, can be efficiently and effectively filtered to prioritize the most urgent alerts.

AI-based automated production, sharing and consumption of CTI is considered as a condition for enabling different organizations in a sector, country or region to configure defences against a new threat in cyber-relevant time, i.e. in minutes and hours, instead of days and weeks. An attack against one organization can then quickly be turned into protection for all the others. This will rob the threat actors for their ability to run attack campaigns against many organizations.

15.3.4 Automated Response and Mitigation

Once an incident has been detected and confirmed, AI tools can feed into tools for automated orchestration of response and mitigation actions. Vendors are developing AI-based SIEM (Security Information and Event Management) tools for automated response and mitigation as described in Sect. 16.2.3. The AI engines in these tools generate *playbooks* that can be automatically executed, thereby stopping threat actors as early in their attack scenarios.

Moreover, AI-based security chatbots based on NLP can assist human security analysts in evaluating situations and for making the right decisions. Organizations with limited cybersecurity expertise can use cybersecurity chatbots for guidance to stay in control of the situation during an incident.

15.4 Vulnerabilities and Exploitation of AI Systems

ML models have a training phase and a usage phase, where different attacks are relevant for each phase. Training-time attacks are executed during the training of ML models. This is typically done by the attackers providing false training data which results in inadequate or vulnerable ML models. Another attack method is to directly corrupt the model parameters after the training. The training can also be corrupted unintentionally if the training data contain bad examples from a security perspective. Usage-time attacks are executed while the ML model is in production, e.g. by manipulating an ML based application to misbehave through malicious prompts and stimulation. Another category of attacks is of course traditional cyber-attacks against the hardware and software platforms of AI systems, but we do not consider such attacks in this chapter. The NIST report "Adversarial Machine Learning" gives an overview of threats and corresponding mitigation for AI systems [50]. The sections below briefly describe the main threats and vulnerabilities with possible mitigations.

15.4.1 Poisoning Attacks

A poisoning attack is when an attacker injects corrupted data and parameters used for training. The attacker might be able to control part of the training data, their labels, the model parameters, or the program code of the ML algorithm used for training, resulting in a corrupt ML model being produced. This is thus a training-time attack. The aim of the attack is to cause the model to misbehave during usage time. For discriminative and predictive ML models, poisoning attacks will cause incorrect classifications and predictions, possibly leading to wrong decisions made by automated systems or by humans, which could be very serious. For generative AI, poisoning attacks will cause the generation of wrong, misleading, low quality or inappropriate output.

Poisoning attacks can be considered as a Trojan attack which gives the attacker a backdoor to exploit the poisoned ML model during usage time. For example, it could open for evasion attacks (see below) to trigger integrity violations, trigger wrong classifications and predictions, as well as trigger generative models to reveal sensitive information about the training data.

Detecting poisoning attacks can be difficult, and analogous to finding a needle in a haystack. An AI model is of course just an computer application where the principle of secure by design should be followed during the whole life cycle, as described in Chap. 11. In addition, the AI model should be thoroughly tested to see if it is possible to detect or trigger unexpected or unacceptable behaviour.

15.4.2 Polluted Learning

A fundamental property of ML based on training data (both supervised and unsupervised learning), is that the ML models will be no better than the training data. Thus, ML based on training data is unsuitable for applications where it is expected or required that the model will be better than the training data. For example, if a coding pilot has been trained with software that contains security vulnerabilities, then it cannot be expected that the pilot will produce secure code. On the contrary, it will necessarily produce vulnerable code.[3]

Pollution of training data is not the result of intentional malicious intent, in contrast to poisoning described above. Instead, the presence of bad examples in the training data can be the result of failure to obtain quality training data and poor-quality management of the training process.

Mitigation strategies include quality control of the training data. However, obtaining quality data for supervised and unsupervised learning is typically expensive. It would for example require cleansing the software corpus for security vulnerabilities before being used to train coding pilots. This is a daunting task because it is impossible to find all vulnerabilities. All software contains security vulnerabilities, so in theory there would be nothing left for training. Assuming that the creators of a coding pilot have made a reasonable effort to remove security vulnerabilities from the training data, but new vulnerabilities are discovered after the completing the training, then the only way to remove the bad influence of the vulnerabilities is to scrap the model, fix the vulnerabilities in the training data, and train a new model. This cycle never ends, which necessarily will make it very expensive to ensure that coding pilots produce secure-by-design software.

15.4.3 AI Supply Chain Risks

As ML models get increasingly large, there is a growing market for pre-trained models that can be used directly or that can be fine-tuned with new datasets for specific applications. This creates supply-chain risks if ML-model providers can make malicious modifications of pre-trained models, if the creators of ML models are attacked, if the creators use polluted or poisoned training data, or if the creators use inadequate training methods, resulting in the distribution of vulnerable and/or compromised ML models. When modifications are intentionally malicious, the resulting ML models can be considered as Trojans which later can be triggered to disturb their performance, force incorrect processing, generate incorrect or low-quality results, generate vulnerable software, or leak confidential data.

---

[3] https://snyk.io/blog/copilot-amplifies-insecure-codebases-by-replicating-vulnerabilities/

15.4 Vulnerabilities and Exploitation of AI Systems

Mitigation strategies include testing and verification of third-party ML models, as well as other security measures of C-SCRM (Cyber Supply Chain Risk Management) described in Sec. 2.1.6.

15.4.4 Evasion Attacks

Evasion attacks against ML models were discovered around 2013 when researchers were able to trigger wrong classification of images by modifying the images in a way is not even visible to humans. For example, it can be trivial to trick an ML model to classify a car as an ostrich by making invisible perturbations to the image of a car [51]. An evasion attack thus consists of making small intentional perturbations to the input (often imperceptible by humans) which causes a wrong classification or prediction by a discriminative or predictive ML model.

The ability to mount evasion attacks must be considered as a fundamental flaw which should not exist in ML models. However, such weaknesses are common in current ML models, and they can be exploited with potentially serious consequences. To develop mitigation methods is an ongoing research field in the ML community, where the goal is to ensure that models are robust against evasion attacks. Approaches are e.g. to improve the learning during the training phase or improve detection of weaknesses during testing.

15.4.5 Privacy Attacks

ML models are typically built with training data that is either commercially or privacy sensitive or confidential. In addition, an ML model itself has business value, is considered as IP (Intellectual property) and is typically not available for free. Two types of privacy attacks are membership inference and model extraction.

- **Membership inference attacks** consist of providing specially crafted input to the ML model, for which the results can be used indirectly to determine whether a particular record or data sample was part of the training data. This could e.g. be to determine if medical data of a given person was part of the training data, which then would reveal personal medical information about that person.
- **Model exfiltration attacks** consist of repeatedly querying the ML model using different valid inputs to progressively determine the model structure, and possibly to create an equivalent ML model. This attack represents a form of reverse engineering of the ML model. For example, it is typically expensive to train an ML model with supervised learning. A model exfiltration attack against an image recognition classifier could be to automatically prompt it to classify many images, and use the results as labels for supervised learning of the attacker's own classifier.

It is challenging, and even impossible to totally prevent privacy attacks. Possible mitigation strategies include throttling responses to queries, or limit the number of queries from particular endpoints, to slow down attempts of membership inference and model exfiltration. For data privacy of people, *differential privacy* (DP) is an approach to train ML models with data about groups of individuals while blurring data about specific individuals. This is done by making arbitrary small changes to individual data that do not change the general patterns of the group.

15.4.6 Prompt Injection Attacks

Prompt injection attacks consist of providing specially crafted prompts to a GPT model to alter its intended behavior. Motivations for such attacks could be to bypass safeguards against producing results containing mis/disinformation, propaganda, unethical and hurtful content, sexual content, malware (code), or phishing content. Because the list of prohibited results is often explicitly crafted by the model creator, attacks to bypass such safeguards are also called jailbreak attacks.

When using GPT models, many have found that it is easy to give prompts that trigger inadequate responses, and the creators of such models are doing the best they can to prevent such things from happening. However, it is challenging to mitigate prompt injection attacks because the parameters for inadequate responses are hidden deep inside the models, and cannot easily be found and removed. This fundamental weakness of GPTs means that they can be a "loose cannon" which of course will limit their legal and ethical applicability. If a country enforces censorship policies, prompt injection to GPT models can provide a hidden channel to obtain censored information. A government can therefore consider that certain GPTs represent an unacceptable risk, and may want to ban or restrict their use.

15.5 Tasks

1. Offensive use of AI
   (a) What is the challenge of detecting deepfakes?
   (b) Will legislation against deepfakes prevent treat actors from using deepfakes in cyberattacks, e.g. for social engineering? Explain your answer.
   (c) Is it possible to prevent the use of coding copilots for generating malware? Explain your answer.

2. Defensive use of AI
   (a) How can AI-tools mitigate insider threats? Discuss legal and ethical considerations.

15.5 Tasks

    (b) How can AI strengthen the principle of "attack against one organization is protection for all the others"?
    (c) Discuss the potential for using GAN (Generative Adversarial Network) for training offensive and defensive AI bots.
    (d) What could a "cyberwar of bots" look like?

3. <u>Vulnerabilities of AI</u>
    (a) Describe how poisoning attacks can affect AI-based security tools.
    (b) Give an example of how a poisoning attack against an AI-based security tool can be done.
    (c) Describe how evasion attacks can affect AI-based security tools.

Chapter 16
Cyber Operations

> *We know they are lying, they know they are lying, they know we know they are lying, we know they know we know they are lying, but they are still lying.*
>
> Aleksandr Solzhenitsyn, Russian writer

Cyber operations, or CyberOps, is a rather vague term used with varying meanings in the context of attacks and defenses within digital infrastructures. CyberOps could mean activities to coordinate and strengthen the ability to withstand cyberattacks, which include *cyber threat intelligence* (CTI). The term cyber operations can also have the meaning of offensive tactical and strategic cyber activities carried out by states or by groups sponsored by states. CyberOps are also related to cyber warfare which refers to the use of cyber attacks, hacking, and digital espionage as a means of conducting warfare or achieving strategic objectives in cyberspace. Cyber warfare is a growing concern for governments, businesses, and individuals worldwide, as the reliance on digital technologies and connectivity continues to increase. Efforts to regulate cyber warfare are ongoing at regional and international levels, but have so far achieved very little.

In this chapter, you learn about advanced cyber threats and the typical steps that threat actors execute as part of advanced cyberattacks. You also learn about cyber threat intelligence and how defenders can share and use this information to strengthen their cyber defences. Finally you learn about cyber warfare, the relative lawlessness of the Internet, and the important role that Big Tech companies have for cyber operations.

16.1 Advanced Cyber Threats

Advanced cyberattacks are carried out by resourceful threat actors. The next two sections describe what is meant by the terms APT (Advanced Persistent Threat), and *cyber kill chain*, which is a model for targeted cyberattacks, and a model for how such can be stopped.

16.1.1 APT: Advanced Persistent Threat

An APT is a threat actor or grouping which often is a nation-state entity or which is sponsored by a nation-state. However, there are also APTs that are only criminals and have no connection to a state other than that the persons reside in a country. Such APTs are driven by profit, can be as advanced and persistent as state APTs, and can even be protected by the state in the sense that they are allowed to operate freely as long as they do not attack targets in their own country. In addition, some politically motivated actors may be referred to as APTs. They are typically not tied to a specific state, but are driven by an ideological conviction with significant resources from organizations that support the same ideology.

An APT is not described in terms of an organization name and address, but as a grouping with an observed *activity profile*. The term APT was first used by the US Air Force in 2006, and has gradually become a general term for the most resourceful threat actors. An APT is *advanced* because it has at its disposal ample resources for intelligence, expertise, and tools to develop exploits, to control infrastructures, and to carry out attacks.

An APT is *persistent* because such an actor has long-term goals, as opposed to opportunistically seeking only short-term gain for itself. As a rule, objectives are set by overarching political or strategic units. Its long-term approach means that an APT does not give up if it encounters resistance, but that it has the stamina to execute attacks at a slow pace, which means that such attacks often "go under the radar" and remain undetected. One objective is typically to achieve sustained access over several years to infiltrated systems and networks, where access may lie unused in anticipation of the need to exploit it.

An APT is a *serious threat* precisely because such an actor has motivation both through long-term objectives and capacity through state resources or generating profit through cybercrime, and has professional employees who have a full-time job conducting offensive cyber operations.

Attacks by an APT typically target specific countries and sectors, such as ministries, defense, finance, industry, telecommunications, healthcare, and energy.

The average duration of an APT's unauthorized accesses, i.e. the time it takes for an APT attack to be detected and stopped, is over 100 days. With such long access, the attacker usually has time to complete the entire attack cycle and achieve all or large parts of their objective.

Several organizations maintain lists of the most active APT groupings, where a relatively comprehensive overview is maintained by *MITRE ATT&CK* described in Sect. 16.2.2.

16.1.2 Cyber Kill Chain: A Model for APT Attacks

The *Cyber Kill Chain (CKC)* describes the steps of a typical targeted cyberattack executed by an APT, which can cover a period of several years. The term "kill" refers to the goal of the attack which metaphorically is to kill. Since a complete chain of steps is needed for an attacker to reach its goal, it also means that the attack can be interrupted and prevented at any step in the chain—the earlier, the better, of course. If in step 1 of the CKC the attacker is unable to perform sufficient reconnaissance to identify an organization as a target of attack or determine how to attack the organization, that organization will not be attacked in the first place. Figure 16.1 shows the seven steps of the CKC, with the rough estimates for the time required for the completion of each step.

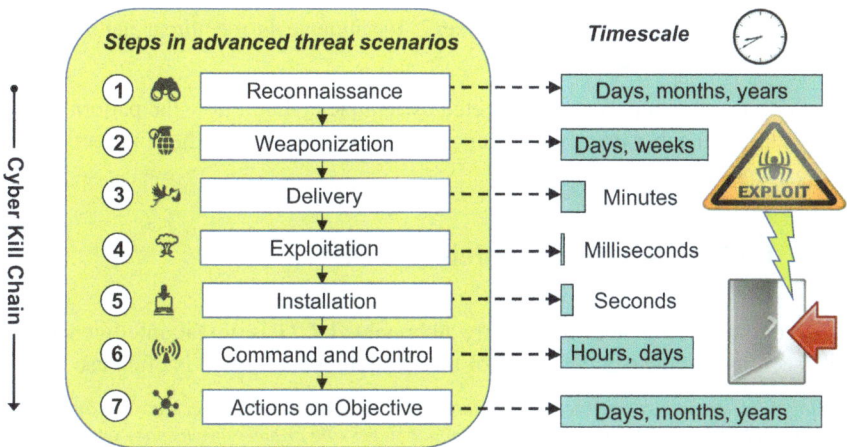

Fig. 16.1 The steps in the cyber kill chain, with their corresponding timescales

The CKC model is useful for analysing an attack by an APT. All the steps are theoretically detectable, but the first two are typically very difficult to detect. An APT attack is targeted, which means that it starts with reconnaissance against specific designated organizations, which are later attacked with, for example, spear phishing and malware. Being able to form a complete picture of the attacker's actions in the different phases of a targeted attack is very important to make a complete eradication and take measures to protect against new attacks. The steps in the CKC are the following:

1. **Reconnaissance:** The threat actor gathers information about the potential victim to find vulnerabilities and opportunities for an attack. OSINT is often used for reconnaissance (see Sect. 14.7.2).
2. **Weaponization:** The threat actor constructs exploit malware based on e.g. zero-day vulnerabilities, packaged in an appropriate format that can be delivered to the victim. The format can e.g. be a corrupt PDF document or a media file.

3. **Delivery:** For example, the malware can be sent as an attachment with a phishing email intended to trick the recipient into opening the attachment or installing the malware. This is usually the step where for the first time the attacker's interaction with the victim can be observed in the victim's infrastructure, and where detection possibilities will increase.
4. **Exploitation:** Assuming that the recipient has been tricked into opening/installing the malware, the exploit malware will start executing.
5. **Installation:** The direct effect of the exploit is usually to download a backdoor in the form of a C&C program from a remote server. After this step, the attackers have remote access to the infected system.
6. **Command & Control:** Attackers can explore the computer network around the infected system, propagate to other systems, delete traces, and identify assets that can be stolen or sabotaged.
7. **Actions on Objective:** Data is collected, exfiltrated and sent out by the network to servers controlled by the attackers. If sabotage is the threat actor's goal, destructive actions are taken.

An important step in preventing targeted cyberattacks that follow the pattern of the CKC is through cyber threat intelligence, which is described in the next section.

16.2 CTI: Cyber Threat Intelligence

Cyber threat intelligence, commonly abbreviated *CTI* is threat intelligence for cybersecurity. Below is a definition of threat intelligence[1] which is often used.

> *Threat intelligence is evidence-based knowledge, including context, mechanisms, indicators, implications and actionable advice, about an existing or emerging menace or hazard to assets that can be used to inform decisions regarding the subject's response to that menace or hazard.*

The importance of "evidence-based knowledge", is that CTI must be reliable and credible to form the basis for decisions and actions. It is also important that this type of knowledge actually has practical utility in identifying and protecting the business against real threats, which means that the knowledge must be relevant, precise and sufficiently fresh.

CTI is valuable because information about threat actors and cyberattacks can be produced and shared among collaborating entities, and that this information can be easily analysed and integrated into processes and security solutions within each entity. Through the sharing and enrichment of details and context by CTI, collaborating entities will understand the threat landscape much more clearly and quickly than if each had worked alone. Without a clear and up-to-date understanding of the threat landscape, entities would often have to make guesses and assumptions instead

---

[1] https://www.gartner.com/en/documents/2487216

16.2 CTI: Cyber Threat Intelligence

of being able to make evidence-based decisions about defending against cyber threats and responding to incidents.

16.2.1 Categories and Levels of Digital Threat Intelligence

CTI can be divided into different categories, where Fig. 16.2 illustrates a common model with the three hierarchical categories (1) technical, (2) tactical/operational, and (3) strategic CTI. At the same time, the figure shows that the three intelligence categories represent different levels of what the cyber agency is able to detect, which is described in the DML (Detection Maturity Level) model of Stillions.[2] The DML model says that it is relatively easy to detect technical indicators such as IP addresses, domains, and file names of malware, because it can be read directly out of the logs, meaning it only requires low maturity in incident response and CTI. Greater maturity is required of an agency to be able to detect what tactics, techniques, and procedures a threat actor has used in an attack, because these aspects cannot be read directly out of the logs. Even greater maturity is required to understand the attribute attacks to a specific grouping and to understand its high-level strategies.

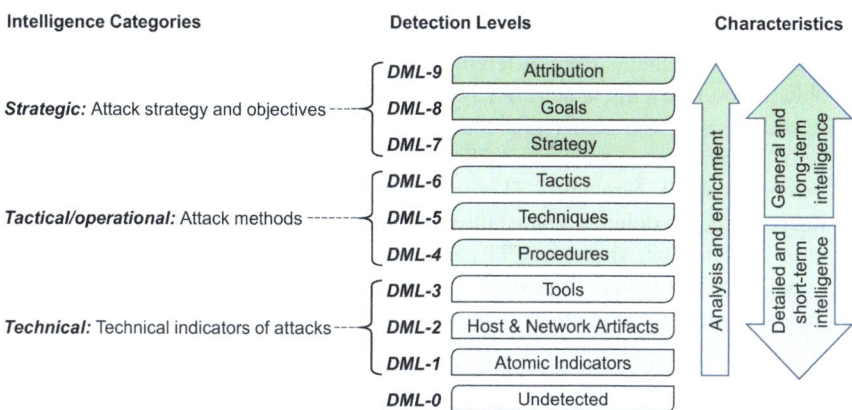

Fig. 16.2 DML model and categories of CTI

- **Technical CTI**
  This type of CTI consists of technical indicators of attack and compromise, which are typically collected through monitoring and logging of activities in systems and networks. Indicators of compromise consist e.g. of IP addresses, domain names, signatures (hash values) of observed malware and observed use of specific tools. It is very common to exchange technical CTI between collaborating entities. Technical CTI is sometimes called "tactical CTI" which is illogical because CTI about TTPs belongs to tactical/operational CTI.

---

[2] http://ryanstillions.blogspot.com/2014/04/the-dml-model_21.html

- **Tactical/operational CTI**

    This is specific information about *Tactics, Techniques and Procedures* (TTP) used by a threat actor in an attack. A tactic is an operational objective, such as gaining initial access to a system. A technique is a method used to achieve the tactical objective. A procedure is an "attack recipe" consisting of a set and sequence of techniques that the threat actor employs to achieve the tactical objective. A threat actor usually has a selection of TTPs that they master and are familiar with. Exchanging tactical/operational CTI between entities is becoming increasingly common and is considered important for strengthening defenses against cyber threats. MITRE ATT&CK, described in the next section, is a catalogue of various TTPs observed and used as part of cyberattacks in the wild. Tactical/operational CTI can be specific to an organization or a sector. Because attackers often conduct campaigns against a specific sector, entities within the same sector can benefit greatly from exchanging CTI on campaigns that can be used to warn each other about a likely impending attack. Sectoral and national cyber agencies (described in Sect. 17.2) have as part of their role the responsibility to coordinate the exchange of tactical/operational CTI.

- **Strategic CTI**

    This is general information about an APT or about some type of threat actor. Strategic CTI is produced over time by compiling and analysing many different information elements from different sources. The focus of strategic CTI is to be able to describe a threat actor/APT based on observed activities and other intelligence, as well as to understand the overall objectives and strategies of the threat actor. *Attribution* means that a CTI agency is able to identify which grouping is behind an attack. Strategic CTI is important for better positioning and prioritising resources in defence against threats, and for being able to do long-term planning. In addition, strategic CTI can form the basis for political response to cyberattacks. It happens regularly that the government of a state that has been attacked goes public to attribute the attack to a specific APT/state.

Large amounts of technical data are collected, such as logs from systems, networks and firewalls or alarms from IDS (Intrusion Detection System). Figure 16.2 shows that technical CTI data can be used in analysis and enrichment to produce tactical/operational and strategic CTI.

When a specific type of attack is observed and becomes known, measures are typically taken to stop that type of attack. That means threat actors are constantly forced to change their attack method to succeed with their next attacks. Some things can be easily changed by the attacker, while other things require more work and restructuring, indicating whether the related CTI is short-term or long-term, that is, whether it has short or long usefulness for the defender.

Technical CTI is short-term because a threat actor can change IP addresses, domain names, and malware with relative ease. Technical CTI therefore has limited value over time. It is relatively easy to obtain this type of CTI, but at the same time it has a limited shelf life and short usefulness in the future.

16.2 CTI: Cyber Threat Intelligence

Tactical/operational is CTI is more long-term because a threat actor or APT cannot easily modify their TTPs, as it takes time to establish expertise and to develop tools to carry out certain types of attacks. Strategic CTI is the most long-term because a threat actor does not typically change "what it is" or what overall objectives it has. It is relatively challenging to produce this type of CTI, but because it is long-term, it can be of great value over a relatively long period of time.

The model in Fig. 16.2 illustrates that technical CTI is detailed, which is because IP addresses, domain names, and signatures can be accurately described. CTI about TTPs is necessarily more general because it is challenging to create a precise description of a threat actor's TTP.

The figure also illustrates that strategic CTI is general. This must be understood in the sense that this type of CTI cannot be directly used to create rules in a firewall or IDS to stop a certain type of attack. The value of this type of CTI lies in being able to be generally better prepared for attacks from certain groups and be able to adjust levels of protection and preparedness accordingly.

16.2.2 MITRE ATT&CK

MITRE ATT&CK is a framework and knowledge base of TTPs (tactics, techniques, and procedures) that threat actors use throughout the lifecycle of cyberattacks. The framework is intended to be a tool for understanding and describing how threat actors operate, thereby strengthening organizations' readiness against cyberattacks. As described above, a tactic is an operational objective, which may be to carry out reconnaissance. For each tactic, a threat actor can choose to execute a set of techniques, in which Fig. 16.3 shows that the tactic *reconnaissance* includes 10 different techniques. The general MITRE ATT&CK knowledge base describes 14 tactics, with each tactic encompassing a greater or lesser number of techniques.

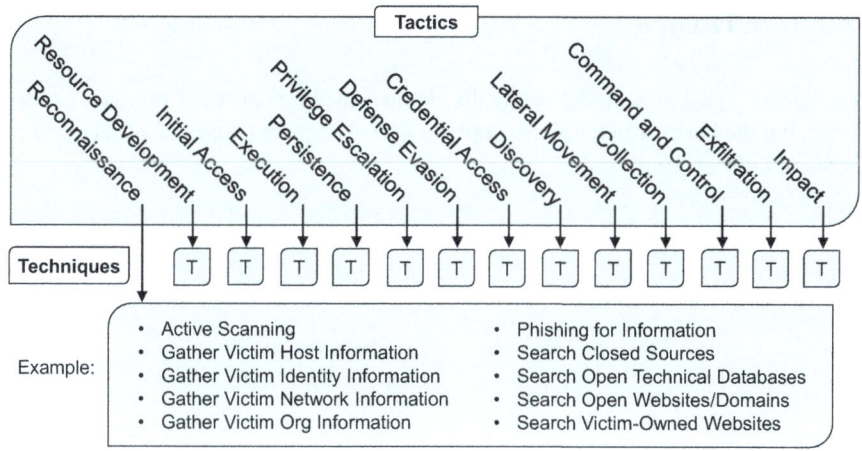

Fig. 16.3 MITRE ATT&CK

MITRE ATT&CK describes different types of cyberattacks from the threat actor's perspective. By detecting a sequence of specific attack techniques in its systems and networks, the organization can infer the tactical objective of the threat actor. This is very useful for businesses to be prepared to stop the next steps that the threat actor plans as part of the attack procedure.

MITRE continually maintains and updates the ATT&CK knowledge base by incorporating attack tactics and techniques from publicly available threat intelligence and cyberattack reporting, as well as from research into emerging attack techniques.

MITRE is a nonprofit organization created to provide technical guidance to the United States government, but it also provides valuable support to private organizations. MITRE originally developed the framework for use in a MITRE research project in 2013 and was named after the data it collects, which is Adversarial Tactics, Techniques & Common Knowledge—or, in acronym form, ATT&CK. Today, MITRE ATT&CK is used by organizations around the world as part of cyber readiness and threat intelligence.

During a parallel activity, MITRE maintains an overview of various threat groups, so-called APTs (Advanced Persistent Threats). As of 2024, MITRE has information on 163 APTs, which are described, among other things, by which TTPs in the AT&CK matrix they have been observed to use. When a TTP typical of an APT is observed during an attack, it can help attribute the attack to that APT. In addition to the ATT&CK knowledge base and APT mapping, MITRE runs several associated projects that are used, among other things, as a basis for threat modelling.

Note that the MITRE ATT&CK and APT mapping have a US/Western perspective, so it does not mention the US NSA, which is very active in offensive cyber operations. Many companies would like a neutral (politics-agnostic) mapping of TTPs and APTs, since they basically cannot know which group is behind observed attacks or spying activities against their own systems.

16.2.3 CTI Cycle

Intelligence cycle is a model originally defined in the context of military intelligence, but the basic principles also apply to CTI. The main phases of a CTI cycle is illustrated in Fig. 16.4.

16.2 CTI: Cyber Threat Intelligence

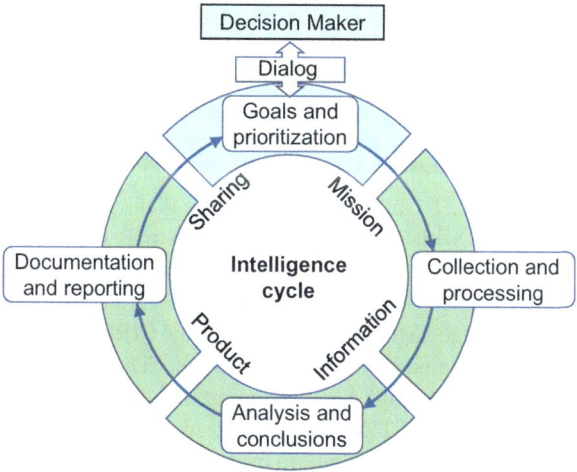

Fig. 16.4 Cycle for CTI

The phases of the CTI cycle in Fig. 16.4 is described below.

- **Goals and prioritization:** Threats can be considered from different perspectives depending on whether it is a business, a sector or an entire state. Which assets are important to protect also depends on the perspective. This phase defines the focus of CTI efforts with a view to mapping general threat categories and suspected threat actors that are relevant, as well as what assets they are typically trying to target.
- **Collection and processing:** There are many different sources for collecting data for CTI. For example, sources can be manual or automated, and they can come from your own organization or from external sources. Processes must be implemented to efficiently collect and store data for CTI. The data collected must be quality assured before they can form the basis for analysis. There are three main considerations to make: (i) the reliability of the source, (ii) whether the data is relevant to the purpose of the business, and (iii) whether the data is of sufficient quality. Poor-quality data, or unreliable or irrelevant, can be discarded. Cleaning the data makes analysis easier because it removes noise and reduces the workload. As larger and larger amounts of data are collected, CTI eventually becomes "big data" and must be processed accordingly.
- **Analysis and conclusions:** Analysis of collected data is necessary to make CTI a product that is useful for the organization as well as for external parties. Enrichment consists of identifying semantic relationships between data of different types and from different sources. Enrichment of data has great potential to see connections between indicators and thus provides greater situational awareness. The analysis and assessment can produce information about general threats, threats to a sector, to specific applications or information about imminent or ongoing attacks. Analysis can be done with AI (artificial intelligence) or logical-semantic technologies.

- **Reporting and dissemination:** It is easier said than done to say that CTI should be shared among the various entities mentioned above. Traditionally, CTI is articulated as plain text and shared as pdf documents with other entities. This way of sharing CTI is patently inefficient, and scales poorly in comparison to the rapid increase in both the number and speed of cyberattacks. The only way to achieve sufficient scalability and speed is by automated sharing based on machine-readable CTI. To make CTI machine-readable, it must be expressed in formal language, i.e. with standardized attributes and well-defined semantics. There are several such languages, including MISP which traditionally is the most prevalent in cyber agencies. MISP is the name of both the language and the platform for CTI exchange. MISP is maintained by CIRCL (Computer Incident Response Center Luxembourg). STIX (Structured Threat Information eXpression) is a standardized language for exchanging threat information that is not tied to a specific platform. As STIX is standardized, different tools and platforms from different manufacturers can talk to each other and exchange CTI.

CTI can be used manually by analysts working on incident response or by using tools for *Security Information and Event Management (SIEM)*. Such tools and products can integrate CTI to make automated decisions and take automated action fully or partially, and control system and network appliances based on the Collaborative Automated Course of Actions Operations (CACAO) standard. Automated sharing and use of CTI can enhance the ability to detect and analyse a threat before and during an attack, thereby reducing the likelihood of it resulting in a cyber incident. As mentioned earlier, current SIEM tools are still not fully able to process tactical/operational and strategic CTI according to Fig. 16.2.

CTI can be prepared for different audiences. Purely technical, tactical/operational and strategic CTI is most often used for incident management and for sharing with other entities, but special types of CTI can also be created for management and decision-makers for assessment of prioritization of handling incidents, for implementing new security controls, etc.

16.2.4 Sharing CTI

National or sectorial hubs in cyber-organizational structures play an important role not only in incident response within their sectors, but also in collecting, producing, analysing and sharing CTI. They share CTI among themselves, with businesses within their respective sectors and with entities in other countries. There are also many commercial MSSPs (Managed Security Service Providers). These may be providers of SOC (Security Operations Centre) or CSIRT (Computer Security and Incident Response Team) services. In addition to conducting incident response, these entities are also active in the collection, production, analysis and exchange of CTI.

16.2 CTI: Cyber Threat Intelligence

ISAC (Information Sharing and Analysis Centre) is the term for an entity that focuses primarily on the collection, production, analysis, and sharing of CTI. At European level, various ISACs have been created in specific sectors, such as EE-ISAC[3] in the energy sector. At the same time, it may be mentioned that the United States has more experience in establishing ISACs than the EU. The ISAC that perhaps most people in the security community have heard of is FS-ISAC (Financial Services ISAC) in the US.[4]

Sources of CTI can be free or paid, which is not necessarily equivalent to open or closed. To receive CTI from other agencies, an organization needs knowledge of open sources, money to pay for closed sources and role/position to be allowed to participate in free yet closed forums/sources of CTI. For an analyst, understanding the CTI landscape is important to know which CTI is practically possible to obtain. OSINT (Open-Source Intelligence) is a type of intelligence in which information is collected from sources that are available to the public. This contrasts with closed-source intelligence, where information is obtained from sources that are not available to the public.

The Traffic-Light Protocol (TLP)[5] is a scheme specified by FIRST for labelling CTI that indicates how widely the sender wants the information to be circulated beyond the nearest recipient. It is important that anyone handling TLP-labelled CTI understands and complies with the rules of the protocol. Only then can trust be established and the benefits of information sharing be realized. The recipient must consult the source of the information if there is a need for wider dissemination. The four levels of the TLP protocol are described below.

- **TLP:RED:** Personal, for named recipients only. For example, during meetings between professionals, RED information is limited to those present at the meeting. The distribution of RED CTI will usually take place via a defined list, and in extreme circumstances it can only be communicated verbally or in person.
- **TLP:AMBER:** Limited distribution. The recipient may share AMBER information with others in the organization and its clients, but only on a need-to-know basis. The sender can specify the intended recipients in a detailed way if needed. **TLP:AMBER + STRICT** restricts sharing to the organization only.
- **TLP:GREEN:** Within the community. Information in this category can be widely circulated within a particular sector. However, the information may not be published or posted publicly on the Internet, nor may it be forwarded outside the environment.
- **TLP:CLEAR:** Unlimited. According to standard practice, CLEAR information can be distributed freely, without restriction. This label was formerly called WHITE.

---

[3] https://www.ee-isac.eu/

[4] https://www.fsisac.com/

[5] https://www.first.org/tlp/

The classification of a document shall always be indicated with the abbreviation "TLP", followed by a colon and classification level, such as "TLP:RED".

The incident response process (described in Sect. 14.5) can benefit greatly from CTI for easier analysis of what has happened in connection with a security incident, and to be able to make faster and better decisions about what actions should be taken as part of incident management.

16.2.5 Representation and Use of CTI

It is challenging to describe TTPs uniformly, even with MITRE ATT&CK. This means that two CTI specialists will typically describe the same attack in slightly different ways. Thus, it is difficult to compare and analyse instances of the same TTP from different CTI sources with automated tools. Since similar aspects of a TTP can be described in different ways, analysts will not be able to identify and analyse similarities sufficiently, making it difficult to achieve automatic exchange and processing of this type of CTI. Describing tactical/operational CTI in terms of TTPs means describing general behavior as opposed to purely technical aspects. As an analogy, tactical/operational threat intelligence for detecting illegal smuggling and trafficking could be a description that states the following:" *They first drive several small cars with illegal goods across the border, then a lorry with legal good across the border, then they all stop at a rest area, and exchange legal and illegal goods."* This contrasts with technical threat intelligence which states the following:" *A small car with this registration plate number was used by the smugglers last time."* Over time, it is relatively difficult for smugglers to avoid detection based on tactical/operational CTI, because it is difficult for them to change routines, while a car with a specific registration number can easily be replaced.

Note that technical indicators are useful, as it is an advantage to have detailed technical information to be able to introduce controls during a specific attack. It enables a defender to block specific IP addresses or create signatures that can identify and stop steps in the execution of an attack.

16.3 Cyber Warfare

The term *cyber warfare* is a relatively imprecise term with varying meanings. Below is a simple definition used in this book:

> *Cyber warfare is all forms of cyber operations between states to carry out sabotage and espionage, as well as defense against these.*

Cyber operations is a subcategory of *information operations* (NATO term), also called *information warfare* (US term).

Information operations include physical bombing of communications infrastructure, electronic jamming of radio communications, influence operations (PsyOps: psychological operations) such as propaganda, and last but not least, *cyber operations*

16.3 Cyber Warfare

(NATO term: Computer Network Operations; American term: Cyber Operations). The cyber operations category in turn consists of the following three subcategories:

- *Cyber espionage* (NATO term: CNE, Computer Network Espionage)
- *Cyberattacks* (NATO term: CNA, Computer Network Attack), also called *Cyber sabotage*
- *Cyber defence* (NATO term: CND, Computer Network Defense)

The connection between these concepts is shown in Fig. 16.5.

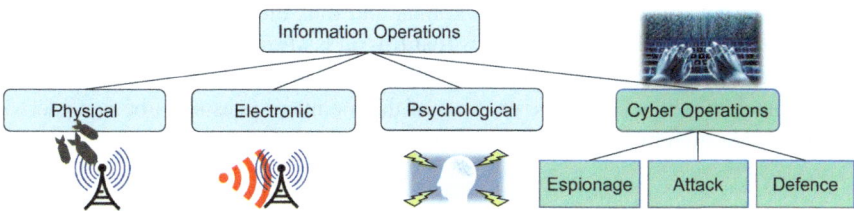

Fig. 16.5 Information operations and subcategories

Most countries, and especially the industrialized countries, practice all three types of cyber operations. Of these three, cyber espionage is the most useful and potent. Compared to sending a person to a foreign country to conduct espionage, with all the costs and risks it entails, cyber espionage is cheap and effective.

Having a good defense against cyberattacks with defensive cyber operations is of course absolutely essential for states. In contrast to traditional military defence where the defence sector has the primary responsibility in the event of war, all sectors are in principle involved in cyber defence and have a role in defensive cyber operations, as explained in Sect. 17.2.

The fact that a country practices offensive cyber operations in peacetime can be controversial, and hence most countries do not speak openly about it. A prominent exception is the United States, which has an official policy around offensive cyber operations, with a separate DoD division called *US Cyber Command*. The White House regularly publishes press releases e.g. to inform that the president has ordered the Cyber Command to conduct offensive cyber operations, e.g. in 2016 to carry out cyberattacks against ISIS,[6] and in 2019 to install backdoors in the Russian power sector.[7] The latter was obviously carried out as cyber deterrence, as described in Sect. 16.3.2.

In isolation, offensive cyber operations have a limited effect because organizations gradually get better at cyber continuity planning so that they relatively quickly eradicate the threat actor and restore systems and processes after an incident. It is assumed that offensive cyber operations have their greatest potential in combination with traditional military operations. If offensive cyber operations can paralyze enemy IT systems for a short period of time, while conducting physical operations, the enemy may lose the ability to coordinate its defenses just when it matters. In this

---

[6] ISIS Targeted by Cyberattacks in a New U.S. Line of Combat. New York Times, April 24, 2016.

[7] U.S. Escalates Online Attacks on Russia's Power Grid. New York Times, June 15, 2019.

way, offensive cyber operations can amplify the effect of traditional military operations. This was clearly the purpose of Russian cyberattacks against Ukraine in the initial phase of the Feb. 24, 2022 invasion. However, the intended effect of these cyberattacks was not achieved because Ukraine had relatively good cyber readiness, and also because they received help from, among others, Microsoft, to neutralize malware by quickly removing vulnerabilities, as described in Sect. 16.3.3.

Directly following the Russian invasion of Ukraine on February 24, 2022, Western countries noticed a decline in the number of cyberattacks from Russia. Possible explanations for this are, firstly, that the Russian APT actors were busy attacking targets directly related to Ukraine, and thus did not have time to attack others. Another possible explanation is that Russia is afraid of the possibility that a cyberattack against a NATO country could trigger Article 5 of the NATO Treaty, which states that attacks on one or more of the member states shall be considered attacks on all, thereby triggering a military counterattack.

16.3.1 Comparison of Weapons

Cyber weapons consists of the different types of malware and attack vectors described in Chap. 2. In terms of warfare, it is interesting to compare characteristics of cyber weapons and traditional kinetic weapons (bombs and missiles). Table 16.1 describes and compares some characteristics of cyber weapons and kinetic weapons.

Table 16.1 Comparison of cyber weapons and kinetic weapons

Property	Cyber weapons	Kinetic weapons
Harmful effect	Harms CIA in IT systems and affects processes that depend on IT systems. May affect far more actors and sectors than the attack was aimed at. May cause indirect physical damage to cyber-physical systems.	Kinetic weapons cause direct physical damage to infrastructure. Only affects the area where the weapon hits.
Reuse	Can be reused/recycled through malware reverse-engineering, so that the victim can reuse the cyberweapon to attack others.	Kinetic ammunition is destroyed in the attack, can only be used once.
Openness	Cyber weapons are immaterial, and thus easy to hide. Cyber weapons are usually kept hidden because transparency would make it possible to develop countermeasures that render the weapon useless.	Kinetic weapons are often physically large, and thus visible, e.g. from satellites. Transparency does not necessarily weaken the weapon.
Attribution	It is often challenging to identify which group or state is behind cyberattacks. The threat actor can leave misleading traces to frustrate attribution.	It is usually relatively easy to see where a kinetic attack is coming from, thus attributing to an actor.
Durability	Cyber weapons are often based on zero-day vulnerabilities that have a limited shelf life, often less than a year. CyberOps actors therefore need a continuous supply of new cyber weapons, see below.	Kinetic weapons have a long shelf life, typically decades. Large arsenals of weapons can be stored for a long time.

Zero-day vulnerabilities are the main ingredient in the development of cyberweapons consisting of exploit code, which exploits the vulnerability. To find zero-day vulnerabilities, many actors use automated tools and techniques, such as fuzzing (see Sect. 11.3.5). Zero-day vulnerabilities and associated exploit code are bought and sold in closed marketplaces, where both state and private actors participate. Assuming that a state cyber agency has obtained a highly potent zero-day vulnerability with associated exploitation code, the authorities in the country face the following dilemma: On the one hand, they can protect the country's IT infrastructure by reporting vulnerabilities to the software manufacturer and MITRE CVE, as described in Sect. 3.3, but on the other hand, they can use the zero-day vulnerability and exploit code to spy on enemies who threaten the country's IT infrastructure, thus helping to protect the country's IT infrastructure. It is also risky to keep zero-days and corresponding exploits, when considering that they can be stolen and misused by others. This happened when the Shadow Brokers reportedly stole and leaked exploits from NSA in 2016.[8]

16.3.2 Cyber Deterrence and Cyber Privateering

The theory behind military deterrence is that it must be based on the exchange of signals of credible capability and intent to retaliate against attacks. This requires that the parties know each other's assessments and priorities, know each other's capacities and political will for destruction, and actually fear destruction.

A balance of terror based on deterrence is intended to prevent war by making it pointless to wage war. It is believed that nuclear deterrence has contributed to preventing the use of nuclear weapons after World War II.

Deterrence has already spilled over to cyberspace. Through actual incidents, including sabotage against Ukraine in December 2015, it is known that Russia has been compromising power grids in Western countries since 2014. Reports of compromise and spying indicate that Russia is actively compromising power grids in the United States.

In return, since 2018, the United States has begun to compromise power grids in Russia. These facts are not based on intelligence reports, but on press releases from the US government.[9] The purpose of such press releases is obvious: This is cyber deterrence. The logic is that if Russia tries to sabotage U.S. power grids, the U.S. will retaliate by sabotaging Russian power grids.

While there are similarities between nuclear deterrence and cyber deterrence, there are also fundamental differences. First, a cyberattack is far less destructive than a nuclear attack, which means that the threshold for carrying out a cyberattack is much lower. Second, cyberattacks are often carried out by non-state groups, while we must assume that only states can be behind a possible nuclear attack.

---

[8] https://en.wikipedia.org/wiki/The_Shadow_Brokers
[9] https://www.nytimes.com/2019/06/15/us/politics/trump-cyber-russia-grid.html

When a non-state group carries out cyberattacks, it is essential for them to know if they can operate with impunity in their country of residence. In that context, Russia's President Putin has stated that Russian groups carrying out cyberattacks against other countries are not considered criminals, "because they do not violate Russian law."[10] This is in practice a modern form of privateering, which can be called *cyber privateering*.

From 1600 to 1850, piracy was run by private groups with the permission of the country's authorities, and was used as a tool in warfare at sea. The aim of privateering was to take control of the enemy's merchant ships and bring them and their cargo to the pirates' port as *a prize*. Pirates were typically rewarded with half the value of the prize. While the privateering activity was going on, a kind of balance prevailed in that all countries recognized this activity as legitimate.

Cyber privateering, on the other hand, is asymmetrical in that certain states, such as Russia, allow private groups to carry out cyberattacks against other countries, while states in the Western world typically do not. By allowing private groups to attack targets in other countries for profit, a state can gain a strategic advantage in cyber warfare. In May 2021, US company Colonial Pipeline suffered a ransomware cyberattack that crippled gasoline supplies in southeastern parts of the United States. The investigation pointed out that the attack was carried out by the Russian group DarkSide, which demanded $4.4 million to decrypt the data. After this attack, a discussion about retaliation started in earnest in the United States (and in other Western countries).[11]

It would violate the principles of the rule of law if Western countries were to allow cyber privateering, i.e. allowing private groups to attack targets in other countries for profit without prosecution in their own country. Accepting cyber privateering as a legal tool in international relations would most likely destabilize the entire global economy, and therefore seems unthinkable. The logical alternative for Western countries is for retaliation to be done in the form of state cyber operations. However, there are factors that limit the types of cyber operations that can be performed. For example, it is probably unacceptable both legally and ethically for Western state cyber operations to be financed through ransomware extortion. One could therefore conclude that the expenditure of conducting cyber operations puts a relatively greater strain on state budgets in Western countries than in Russia. This is because private and state groups in Russia actually profit from their cyber operations and thus may not need state funding. Other countries, such as North Korea, are also known to make large amounts of money from offensive cyber operations. The lack of internationally recognised rules for cyber operations between states is very unfortunate for the entire global community, because it provides incentives for increasing levels of lawlessness in cyberspace.

The Geneva Conventions define what constitutes lawful targets of attack in warfare. In that sense there is a clear distinction between soldiers and civilians, where

---

[10] https://www.nytimes.com/2021/05/29/world/europe/ransomware-russia-darkside.html

[11] https://edition.cnn.com/2021/06/07/politics/president-joe-biden-cyber-attacks-russia-putin-trump-economy/

civilians are defined as illegal targets. Nevertheless, if private actors conduct offensive cyber operations against a belligerent, they are considered legitimate targets of the belligerent. Offensive cyber operations conducted by private individuals can legitimize cyber warfare against civilian targets, which we do not want to happen. An important principle is therefore that offensive cyber operations should only be carried out by state authorities under a well-defined policy.

The Paris Call for Trust and Security in Cyberspace[12] (2018) was an attempt to bring about an international convention on cyber operations and the use of cyber weapons. Small countries supported the initiative, but the great powers were opposed. First, the major powers see that they benefit greatly from cyber operations, and second, they believe, perhaps rightly, that it would be almost impossible to monitor compliance with such an agreement. With the proverb *"All Is Fair in Love and War"*, we can add that this also applies in cyber warfare.

16.3.3 The Role of Big Tech in Cyber Warfare

On Wednesday, February 23, 2022, a few hours before Russian tanks began rolling into Ukraine the following morning, Microsoft's Threat Intelligence Center (MSTIC) found indicators of a never-before-seen "wiper" malware that was observed in Windows servers, placed there by Russian cyber operators in attempts to attack Ukraine's government ministries and financial institutions.

MSTIC quickly picked apart the malware, calling it "FoxBlade", and alerting Ukraine's top cyber defense authority. The FoxBlade malware was programmed to wipe all data on computers accessed on a computer network. Within three hours, Microsoft's virus detection systems on Windows servers, both in Ukraine and around the world, had been updated to block FoxBlade. This example shows how players in the IT industry, especially Big Tech, have enormous potential to implement controls with immediate effect in large parts of the global IT infrastructure. This and other similar controls supported by Western IT companies and governments allowed the IT systems of Ukraine's government ministries to continue to function during the invasion, allowing them to coordinate their defense operations. This robbed Russia of the planned effect of crippling Ukraine's IT systems for a short period while the invasion was ongoing.

A simple review of similar scenarios reveals that partnering with Big Tech can provide significant advantages in cyber warfare. For example, each Windows operating system connects to a Microsoft server at least weekly to check for software and configuration updates. From a technical perspective, this gives Microsoft full control over all computers around the world that run Windows and that are connected to the Internet. It would be relatively easy to identify the computers of individuals, as well as the computers of a particular organization or in a specific geographical region. Technically, Microsoft is able to remotely control each computer, for example for espionage or for sabotage. Similar types of control will also be possible for other

---

[12] https://pariscall.international/en/

Big-Tech companies. It is plausible that a small selection of Big-Tech companies have the technical ability to "switch off" the Internet in large parts of the world. As the Snowden revelations have shown us, private companies are willing, or legally obligated, to cooperate with their national governments, so the potential of global control of core functions in systems, applications, and the entire Internet through collaboration with Big-Tech companies is realistic. A natural conclusion to such reflections is that having an advantage in cyber warfare depends on having Big Tech on your side. Since Big-Tech companies are mostly based in the US, this means that a country's strength in cyber warfare largely depends on having the US as an ally.

When large IT manufacturers become partners in cyber operations, a natural consequence is that the market share of their products and services will be affected by global political divides. For example, if it were the case that a US-based provider implements a backdoor for the NSA in their products, and this became public knowledge, a likely result is that markets in countries that are largely U.S. antagonistic would stop using these products for fear of being subject to scrutiny and surveillance. The fear that China's Huawei could install backdoors in the 5G network was precisely the reason why Western governments did not allow telecom operators to build the 5G network with technology from Huawei. These events are likely just the tip of the iceberg in global ICT and cyberwarfare politics, reflecting a trend toward fragmentation of the technology market along geopolitical divides.

16.4 Tasks

1. Cyber threat intelligence

    (a) Why is sharing CTI important?
    (b) What type of CTI is short-term and what type is long-term?
    (c) Why is it important to be able to (partially) automate CTI?
    (d) Mention challenges for (partially) automating CTI.

2. Cyber warfare

    (a) What is the connection/difference between information operations and cyber operations in a military context?
    (b) Describe the usefulness of various forms of military cyber operations.
    (c) Describe aspects and characteristics of cyber weapons.
    (d) What is cyber deterrence and why can it be useful?
    (e) Why is it difficult to achieve an international convention on cyber warfare?

3. MITRE ATT&CK

    (a) What is the similarity between the DML (Detection Maturity Level) model and the MITRE ATT&CK framework?
    (b) Look at Fig. 16.3 or visit the *ATT&CK Matrix for Enterprise* on https://attack.mitre.org/.
    For each tactic, indicate which stage of the Cyber Kill Chain it corresponds to.

Chapter 17
Cyber Organizational Structures and Regulation

> *Good people do not need laws to tell them to act responsibly, while bad people will find a way around the laws.*
>
> Plato

Cyberspace is defined by NATO and other countries as one of the five domains of military operations together with land, sea, air and space. At the same time, cyberspace is a fundamental element in our modern society. Hence, every country has established organizational structures and regulation to try to enforce jurisdiction over cyberspace and cybersecurity. Given the extremely dynamic nature of digital technology and consequences for cybersecurity, the organizational structures and regulations are currently "work in progress", meaning that it is challenging for governments to follow the rapid innovation in IT by establishing adequate organizational structures and regulations for cyber. In the EU in particular, there are multiple initiatives to establish new cyber organizations and cyber acts.

In this chapter you learn why cybersecurity regulations are important, and what it means to be accountable for compliance with regulations. Then you learn about how cybersecurity is organised at national and regional levels in USA, Europe, China and Russia, as well as their main cybersecurity laws and regulations.

17.1 The Importance of Cybersecurity Regulations

Laws and regulations have always been necessary to maintain a well-functioning society. Through digitisation, society has become completely dependent on digital systems to function. An increase in attacks against digital systems with serious consequences thus poses a serious threat, so the security of digital systems must be regulated to maintain a well-functioning society.

Technology innovation is moving at breathtaking speed, while the development and revision of regulations is typically a slow and laborious process. It is therefore very challenging to create laws that can effectively regulate the use of new IT (information technology). This is all the more problematic when new IT brings disruptive changes in society, as the introduction of the Internet did, and which is now happening with the large-scale introduction of AI and ML (machine learning).

Cybersecurity regulations are an essential source of requirements for cybersecurity, as described in Sect. 1.7. Businesses have a duty to comply with laws and regulations, where the executive and the board are responsible for ensuring that the business complies with these. For an enterprise, it can be more challenging to comply with laws requiring that cybersecurity governance must be based on risk assessments, than it is to comply with requirements to implement specific technical security controls. The organization needs a certain maturity in cybersecurity governance to manage cybersecurity based on risk assessments as described in Chap. 18.

Efforts to comply with cybersecurity requirements specified in regulations naturally entail a cost that businesses must integrate into their budgets. Not complying with, or only partially complying with, certain requirements can therefore result in cost savings. On the other hand, there will be a risk that the enterprise, or individuals, will be subject to penalties, or other negative consequences if e.g. an audit reveals that the enterprise does not comply with laws and regulations. From a purely economic point of view, businesses may view this as *compliance risk*, which is the risk that the possible costs of non-compliance are greater than savings through reduced costs through reduced compliance. The responsibility for such assessments lies with senior management which ultimately is accountable.

17.1.1 Responsibility and Accountability

Top management and owners of a business are accountable for ensuring that the operation of the business takes place in accordance with applicable regulations. In the event of a breach of compliance with applicable regulations, the business may be fined, which are a *corporate penalty*. Alternatively, or in addition, individuals may be punished with fines or imprisonment, which is a *personal penalty*. Who in the business can be given a personal penalty for an offense depends on circumstances. In principle, it is the person who has the highest authority in the enterprise and who at the same time should have known about, and could have prevented, the violation of the law. It is a misconception to believe that the employee who clicked on a malicious link in a phishing email is liable for the resulting security breach and negative impacts. Although everyone in the company is "responsible" for cybersecurity, the "accountability" always lies with senior management.

Many cybersecurity laws express the principle of accountability very clearly. GDPR Art. 5-2, (see Sect. 10.7.1) points out that *the data controller* (in principle the company's management) is accountable for complying with the GDPR vis-à-vis the data subjects and the authorities. In the event of a breach of compliance, the enterprise may, pursuant to the GDPR or other relevant security regulations, be sanctioned with a business penalty.

GDPR Art. 84 specifies that each member state shall determine the maximum penalties imposed on individuals in the event of serious violations of the GDPR. As a national adaptation, Norway, through section 48 of the Personal Data Act, has set the maximum penalty to fines and/or imprisonment of up to 1 year (3 years in particularly aggravating circumstances).

17.1.2 Hierarchy of Regulation

Regulations are ranked according to a hierarchy, which means that regulations further up in the hierarchy are more general than regulations further down. It also means that regulations further up override regulations further down if they were to somehow contradict each other. Figure 17.1 shows the hierarchy of regulations divided into three categories.

Fig. 17.1 Hierarchy of regulations

The top category consists of laws and regulations determined by the government. In this category, of course, a country's constitution comes out on top, followed by ordinary laws, parliamentary resolutions, presidential/royal decrees, administrative regulations and other decisions from various authorities. For an enterprise, this category constitutes external regulatory requirements.

The middle category consists of external requirements from parties other than authorities. Contracts with external parties will primarily set requirements that are agreed between the business and external parties. In many sectors, there are formal norms that set requirements for businesses to participate within the sector. An example of such requirements is the International *Payment Card Industry Data Security Standard (PCI DSS)*, which sets security requirements for all businesses that participate in payment transactions with debit cards and credit cards. There are also less formal norms that businesses are expected to follow, without necessarily having any direct consequences if they do not. For example, it is considered good practice for businesses to provide public information when they have been subjected to serious cyberattacks.

The last category of requirements are those that the business sets for itself through policies and internal rules for procedures, business processes, and governance. For example, company policies can set requirements for password strength or employee use of personal devices.

17.2 Cyber-Organizational Structures

Cyberspace is not only a domain for conducting crime, but also considered as the 5th dimension in warfare in addition to land, sea, air, and space. Major nations, regions and alliances have established structures for combatting cybercrime and for conducting military cyber operations. To strengthen these operations there is great potential in sharing information and coordinating activities. This is relevant for organizations in all sectors, both private and public. Countries and regions have therefore established cyber-organizational structures that can harness and share the cyber resources and capabilities to make the organizations stronger together. The following sections describe such structures in the US, in the EU, in Russia and China, as well as their major cybersecurity laws. Note that the descriptions are based on information from the web which might be inaccurate or outdated. Hence, these descriptions must be seen as general and approximative.

17.3 USA

The organizational structure of agencies dealing with cybersecurity in the US is vast and complex where the diagram in Fig. 17.2 gives a simplified overview.[1]

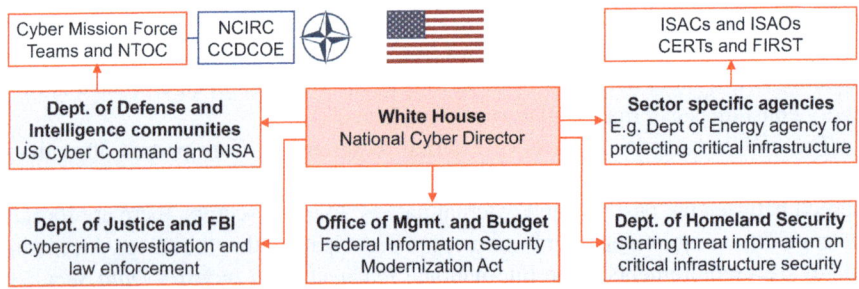

Fig. 17.2 US structural organization of cyber agencies

---

[1] https://www.defensepriorities.org/explainers/cyber-as-statecraft-not-war

17.3 USA

As shown on the left of Fig. 17.2, the US Cyber Command (USCYBERCOM) is one of the eleven unified combatant commands of the US Department of Defense (DoD). It cooperates with NSA and has been concurrently headed by the director of the NSA since its inception in 2009. There are more than 100 different cyber mission force teams under USCYBERCOM.

The NSA Cybersecurity Threat Operations Center (NCTOC) represents the hub for the agency's 24/7/365 cybersecurity operations mission.[2] NCTOC collaborates with USCYBERCOM to protect the unclassified DoD Information Network (DoDIN), a global network encompassing 3 million users from Washington DC to military missions abroad. NTOC and the cyber mission force teams collaborate with NATO Computer Incident Response Capability (NCIRC) for CTI exchange, and with the NATO CCDCOE (Collaborative Cyber Defence Centre Of Excellence) for training.

As shown on the right of Fig. 17.2, many industry sectors have established an Information Sharing and Analysis Centre (ISAC) that gathers and shares information on cyber threats to the sector's critical infrastructure. ISACs also have the role of sharing of data between public and private sector groups. ISACs are typically financed by membership fees and have infrastructure and SOCs for monitoring threats on a global scale. The National Council of ISACs is the umbrella organization for the various sector ISACs.[3]

In relation to the CISA act described in Sect. 17.3.3, Information Sharing and Analysis Organizations (ISAOs) were initiated by a White House directive in 2015 to promote voluntary cyber threat information sharing within industry sectors. Organizations are protected by limited liability when sharing CTI within an ISAO, which precisely has the purpose of encouraging the sharing of CTI. The evolution of ISACs has been somewhat exclusionary, hence the goal in promoting ISAOs was to make it easier for all organizations to share CTO and not just those belonging to ISACs.

In the 1980s the Internet consisted of a few interconnected mainframe computers at universities and research institutions, mainly in the US but also in Europe. The Morris worm was created in 1988 by MIT student Robert Morris who released it to show that computer networks of the time had many weaknesses (and they still do). To his surprise, the worm turned out to be more damaging and infectious than originally planned, and it basically brought the Internet to a halt. Realising the disastrous impact of attacks like that of the Morris worm, the US DoD sponsored Carnegie Mellon University to establish the first CERT (Computer Emergency Response Team) in 1988. Other CERTs soon followed, and by 1990 the Forum of Incident Responders and Security Teams (FIRST) was established as a non-profit organization for CERTs and other organizations with response teams in the cybersecurity

---

[2] https://www.nsa.gov/portals/75/documents/resources/cybersecurity-professionals/top-5-soc-principles.pdf

[3] https://www.nationalisacs.org/

industry. As of 2024 there are 717 FIRST team members in 107 countries around the world.[4]

There exist many laws and regulations related to cybersecurity in the US, both at federal and state level. Some of these are briefly described in the section below. Many other laws and regulations relate to cybersecurity, such as laws regarding surveillance including FISA (Foreign Intelligence Surveillance Act) described in Sect. 10.7.5.

17.3.1 CFAA (Computer Fraud and Abuse Act)

Enacted as early as 1986, the CFAA is the primary federal law addressing cybercrimes and cyberhacking. Key provisions of the CFAA are:

- **Unauthorized Access:** It is illegal to access a computer without authorization or exceeding authorized access. This covers e.g. hacking into computer systems or networks without permission.
- **Obtaining National Security Information:** It is illegal to access a computer system to obtain national security information without authorization.
- **Computer Fraud:** It is illegal obtaining information from a computer to defraud or to commit identity theft.
- **Unauthorized Damaging:** It is illegal to knowingly cause damage to a computer system without authorization. This includes the transmission of malware or viruses.
- **Trafficking in Passwords:** It is illegal to involve in trafficking of passwords or other access credentials that facilitate unauthorized access to computer systems.
- **Extortion Involving Computers:** It is illegal to use a computer to extort others, such as by threatening to damage or impair a computer system unless certain demands are met. This covers ransomware attacks.
- **Conspiracy and Attempt:** The CFAA also criminalises conspiracy to commit offenses under the Act and attempts to violate its provisions, even if the actual offense is not completed.

Note the prominence of the term "authorization" in the provisions above, which reflects the importance of correct interpretation of authorization, as explained in Sect. 1.11.

---

[4] https://www.first.org/

17.3.2 FISMA (Federal Information Security Management Act)

Enacted in 2002 as part of the E-Government Act, FISMA prescribes a framework for improving the security of information systems and data within federal government agencies. The primary objectives of FISMA are:

- **Establishment of Security Standards:** Federal agencies are required to use security standards and guidelines for their information systems.
- **Risk Management Framework:** Agencies are required to adopt a risk-management approach to information security, i.e. to assess the risks to their information systems and implement appropriate security controls to mitigate those risks.
- **Security Assessments and Authorization:** Agencies are required to conduct regular security assessments of their information systems, to evaluate the effectiveness of security controls.
- **Continuous Monitoring:** Agencies are required to perform continuous monitoring of information systems to detect and respond to security threats in real-time.
- **Incident Response and Reporting:** Agencies are required to establish incident response capabilities to respond to security incidents promptly. Agencies are also required to report significant security incidents to appropriate authorities and affected individuals.
- **Annual Reporting Requirements:** Agencies are required to submit annual reports to the Office of Management and Budget (OMB) and Congress on their information security programs and compliance with FISMA requirements.

FISMA assigns responsibilities to federal agencies, in particular to NIST (National Institute of Standards and Technology) and OMB (Office of Management and Budget), to oversee and enforce compliance. NIST's role is to develop security standards and guidelines that federal agencies must follow under FISMA, such as the CSF described in Sect. 18.2, and SP 800-63 described in Sect. 8.7.1.

17.3.3 CISA (Cybersecurity Information Sharing Act)

Enacted in 2015, CISA has provision for sharing CTI between the government and private sector entities in order to enhance overall cybersecurity defenses and incident response capabilities. Key provisions are:

- **Facilitation of Information Sharing:** CISA provides a basis for voluntary sharing of CTI, vulnerabilities and cybersecurity issues between the private sector and federal agencies. This includes liability protections for private sector entities for sharing CTI in good faith.

- **Privacy and Civil Liberties Protections:** CISA protects privacy and civil liberties by requiring the removal of personal information that is not directly related to a cybersecurity threat before sharing information with the government.
- **Use of Shared Information:** CISA gives federal agencies the right to use shared CTI for cybersecurity purposes as well as for investigating and prosecuting cybercrime.

17.3.4 HIPAA (Health Insurance Portability and Accountability Act)

Enacted in 1996, HIPAA primarily focuses on protecting the privacy and security of patients' health information. Indirectly this sets requirements for cybersecurity within the healthcare sector based on the following provisions.

- **Privacy Rule:** HIPAA mandates the safeguard of patients' "protected health information (PHI), thereby necessitating robust cybersecurity measures to safeguard electronic PHI (EPHI) from unauthorized access, disclosure, alteration, or destruction.
- **Security Rule:** HIPAA establishes national standards for safeguarding EPHI with technical, administrative, and physical safeguards.
- **Data Breach Notification:** HIPAA requires health care entities to notify affected individuals, the Department of Health and Human Services (HHS), and sometimes the media in the event of a data breach of EPHI.
- **Penalties for Non-Compliance:** HIPAA violations can result in significant penalties, including fines, civil monetary penalties, and even criminal charges in cases of wilful neglect.

Compliance with HIPAA not only forces healthcare organizations to mitigate risks to health information, but to their operations in general, which is a positive side-effect.

17.4 Europe

The landscape of cybersecurity agencies in the EU is much less coherent than e.g. in the US. Each member state has its own structure of cybersecurity agencies, both civilian and military. Hence, the coordination and collaboration across the EU has been limited. Figure 17.3 gives an overview of the main players in the European cybersecurity landscape.

17.4 Europe

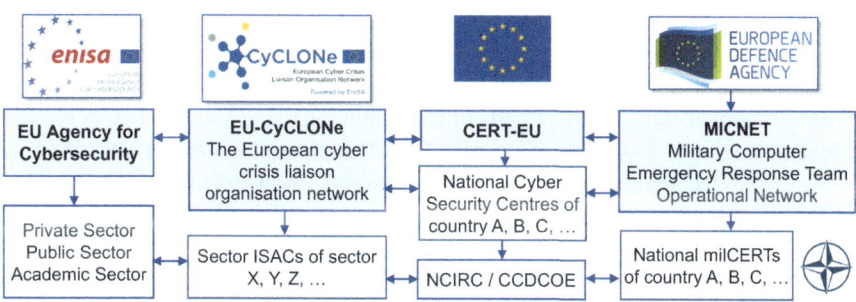

Fig. 17.3 Landscape of cybersecurity organizations in the EU

CERT-EU acts as a coordination centre for many national and sector CERTS in the EU, by acting as a hub for CTI exchange and incident response coordination. CERT-EU cooperates with ENISA, with NATO Computer Incident Response Capability (NCIRC) for CTI exchange, and with the NATO CCDCOE (Collaborative Cyber Defence Centre Of Excellence) for blue-teaming and red-teaming exercises. Each member state in the EU typically has a NCSC (National Cybersecurity Centre) or national SOC (Security Operations Centre). Through the draft EU Cyber Solidarity Act (see Sect. 17.4.5) the exchange of CTI between EU NCSCs will be much more formalised than it is today.

MICNET was established in 2023 to act as a hub for the national milCERTs, to strengthen cooperation in the cyber domain at EU level. MICNET is managed by the European Defence Agency (EDA) and represents a significant element in the EU Policy on Cyber Defence.

ENISA was created in 2004 under the name of European Network and Information Security Agency. In 2019 the agency changed its name to the European Union Agency for Cybersecurity, but kept its acronym. ENISA's general mission is to ensure a high common level of cybersecurity across the Union, e.g. through knowledge sharing, capacity building and awareness raising. In particular ENISA manages EU's cybersecurity certification scheme for IT products.

EU-CyCLONe (European cyber crisis liaison organisation network) is a cooperation network for member states national authorities in charge of cyber crisis management. The main task of EU-CyCLONe is to support the coordinated management of large-scale cybersecurity incidents and crises affecting member states and to ensure the regular exchange of relevant information among member states and Union institutions, with support from sector ISACs (Information Sharing and Analysis Centres). EU-CyCLONe shall also support decision-making at political level in relation to such incidents and crises.

17.4.1 EU Regulation

The European Union has a comprehensive regulatory framework consisting different types of "acts". The main three types of acts are:

- A "regulation" is a binding legislative act which must be applied without modification across the EU, only translated into the national language. The national law is then a referral law. However, a regulation may contain provisions that allow member states to adapt parts of the regulation in the national law. A typical example of this is GDPR (General Data Protection Regulation) described in Sect. 10.5, where member states can make national adjustments, such as to specify different sanctions in case of non-compliance.
- A "directive" is a legislative act that defines a regulatory goal that each EU member state must achieve. However, it is up to each member state to define their own laws on how to reach these goals. A typical example is the NIS2 directive described in Sect. 17.4.2.
- "Implementing acts" and "delegated acts" are non-legislative acts typically used to specify detailed rules related to other legislative acts. Situations where such acts are useful are when there is an urgent need for measures to control e.g. epidemics and exceptional situations, or to specify detailed rules on the exchange of data or on prices for agricultural products. To develop implementing acts, the EC must involve the European Parliament or Council, whereas to develop delegated acts, the EC can do the work alone, but the European Parliament or Council can veto the act. The EC's power to develop delegated acts must be based on a provision in the overlaying legislative act (regulation or directive).

The EEA countries Iceland, Lichtenstein and Norway decide together whether an EU act shall apply in the EEA countries; hence they can perfectly well decide to not implement an EU act. However, they must decide together, all or none. The main cybersecurity-related acts are briefly described below, but GDPR is described in Sect. 10.5, and eIDAS is described in Sect. 8.7.2.

17.4.2 NIS2 Directive

The NIS1 directive (Network and Information Security) was enacted in 2016 but was quite limited in scope. For this reason, EU revised it as the NIS2 directive, which was came into force in 2022. By 24 October 2024, the member states had to have implemented the directive into national law. From this point on, the old NIS1 directive was repealed. Elements of NIS2 include:

- Cooperation on a European infrastructure for dealing with cyber crises through EU-CyCLONe (European Cyber Crises Liaison Organisation Network)
- Harmonization of security requirements and reporting obligations
- Ensuring that member states' cybersecurity strategies cover supply chain security, vulnerability management, core network security, and general cyber security culture and cyber hygiene

17.4.3 Cybersecurity Act

The EU Cybersecurity Act is an EU regulation that came into force in 2019, repealing the previous EU Cybersecurity Act from 2013. The title sounds like a "flagship" cybersecurity regulation, but disappointingly the title is a misnomer. The regulation should have been called "the ENISA Act" because it mainly concerns ENISA which through the regulation is granted a permanent status with a specific mandate. ENISA was established in 2004 as *European Network and Information Security Agency*, but had no permanent status and was dependent on annual budget allocations from the EC. The Cybersecurity act of 2019 gives ENISA permanent status and a strengthened mandate. ENISA shall, upon request, be able to assist member states with cross-border incident management, including consulting, analysis, and technical investigations. ENISA shall also assist and facilitate member states' capacity building, operational cooperation and research and development. The regulation also establishes a common European framework for voluntary certification of ICT products, services, and processes.

On 18 April 2023, the EC proposed an amendment to the EU Cybersecurity Act to establish an EU certification schemes for "managed security services" (MSS) that e.g. provide incident response, penetration testing, security audits and consultancy.

17.4.4 Cyber Resilience Act (in Progress)

The proposed Cyber Resilience Act focuses on specifying security requirements for digital products, both hardware and software. This regulation is linked to the certification of ICT products as part of the EU Cybersecurity Act described above. Two main objectives with the Cyber Resilience Act are:

- To set requirements for the development of secure digital products with fewer vulnerabilities, and to ensure that manufacturers follow the principles of "secure by design" as described in Chap. 11.
- To create market conditions which allow users to take cybersecurity into account when selecting and using digital products.

17.4.5 Cyber Solidarity Act (in Progress)

The EU Cyber Solidarity Act aims to ensure adequate capacities in the EU to detect, prepare for and respond to serious cybersecurity threats and attacks. The Act proposes an EU Cybersecurity Alert System based on a network of SOCs across the EU, and a Cybersecurity Emergency Mechanism to improve the EU's cyber readiness. This includes an EU Cybersecurity Reserve that can be deployed at the request of member states or EU institutions, bodies, and agencies to assist with responding to significant or large-scale cybersecurity incidents. As a pilot, launched in November 2022, three consortia of cross-border SOCs were selected as part of the ECCC described in Sect. 17.4, bringing together cybersecurity centres from 17 member states and the EEA state Iceland.

17.5 Russia

FSB (Federal Security Service), SVR (Foreign Intelligence Service) and GRU (Military Intelligence) are the main Russian government organizations engaging in cyber operations around the globe, for which the government spends significant resources. However, it can also be assumed that these operations are partially self-funded since some of them also engage in criminal activities, as described in Sect. 16.3.2.

The UK NCSC has publicly attributed malign cyber activity to Russian APT groups, where Fig. 17.4 shows how these groups are organized in the structure of Russian cyber agencies.[5]

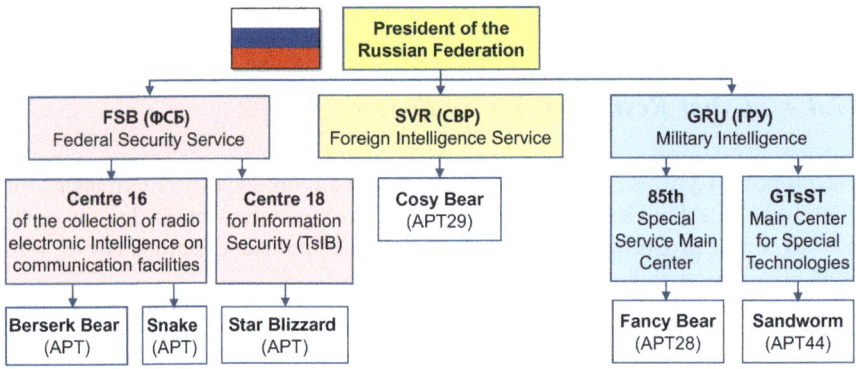

Fig. 17.4 Russian structural organization of cyber agencies

FSB Centre 16 conducts cyber operations including the intercepting, cryptanalysis and processing of electronic messages, and engage in technical penetration of foreign targets. Berserk Bear and Snake are two APT groups under Centre 16. Several other names are being used for the same APT groups. Snake is also the name of an attack tool which has been used in espionage operations conducted by Centre 16 for two decades. Infestation with the Snake malware has been identified in more than 50 countries across the world.

FSB Centre 18 for Information Security sits within the Counter-Intelligence Service of the FSB. According to the UK NCSC, the APT group *Star Blizzard* is organized under Center 18 and conducts cyber espionage and interference with

---

[5] https://www.gov.uk/government/publications/russias-fsb-malign-cyber-activity-factsheet/russias-fsb-malign-activity-factsheet

democratic and political processes in Western countries. The group has also selectively leaked information to discredit governments in other countries to undermine trust in their political and democratic processes.

SVR RF is Russia's foreign intelligence agency, focusing mainly on civilian affairs. The APT group *Cozy Bear* is suspected of being organized under SVR. Cozy Bear runs cyber operations which focus on military, government, energy, diplomatic and telecom sectors.

GRU is the foreign military intelligence agency of the Russian armed forces and maintains its own cyber force units. GRU is reputedly the largest of Russian foreign-intelligence agencies and has a reputation for conducting risky and complicated high-stakes operations.

Fancy Bear is a cyber espionage group sponsored by GRU and is organised under GRU's 85th Unit Special Services Main Centre. Fancy Bear is known for targeting government, military, and security organizations, especially Transcaucasian and NATO-aligned states. Among attacks attributed to Fancy Bear are the hacking of emails of the Democratic National Committee to influence the outcome of the United States 2016 presidential elections, the hacking of German government ministries in 2015 and 2018, as well as the hacking of the Norwegian Parliament in 2020.

Sandworm is a cyberwarfare unit of the GRU, and is organised under GRU's GTsST Main Centre for Special Technologies. Among attacks attributed to Sandworm are the 2015 Ukraine power grid cyberattack, the 2017 cyberattacks on Ukraine using the NotPetya malware, interference efforts in the 2017 French presidential election, and the cyberattack on the 2018 Winter Olympics opening ceremony in South Korea.

Cybersecurity regulation in Russia is governed by several laws and regulations aimed at protecting the country's information infrastructure, ensuring data privacy, and combating cyber threats. Some key aspects of cybersecurity regulation in Russia are described in the sections below.

17.5.1 *Federal Law on Personal Data*

First enacted in 2006 the Federal law No. 152 FZ on Personal Data was amended in 2022 and 2023.[6] The law is in many ways similar to GDPR described in Sect. 10.5.

- The Personal Data Law defines what constitutes personal data and special categories of personal data such as health data, union memberships and religious faith. The law defines legal bases for processing, rights of data subjects, and security requirements to protect personal data.

---

[6] https://www.dataguidance.com/notes/russia-data-protection-overview

- Transfer of personal data outside of Russia is restricted, but not a complete ban. The Russian data protection authority (Roskomnadzor) can issue adequacy decisions for specific countries if their data protection laws offer equivalent protection to those in Russia. Without an adequacy decision, organizations need to ensure they comply with other requirements such as contractual clauses or binding corporate rules for multinational organizations, before transferring data across borders.
- There are requirements for notification to data subjects and to Roskomnadzor in case of data breaches.
- Personal data must be deleted in case of the data subject's request or consent withdrawal.
- Fines for breaching the law can range from 10,000 roubles (approximately $130) to 18 million roubles (approximately USD 230,000).

Similarly to the how the EU GDPR provides law enforcement agencies with a special basis for processing personal data, law enforcement agencies in Russia have their own basis for collection and processing of personal data.

17.5.2 Criminal Code of the Russian Federation

The Criminal Code's Chapter 28 from 1996 contains clauses related to computer crime, where the main articles are listed below.

- Article 272 penalises unauthorized access to computer information that is legally protected. Computer information means any message or data which is presented in electronic form.
- Article 273 penalises the creation, dissemination, and usage of harmful computer programs through a computer with malicious intent.
- Article 274 penalises the violation of the rules of operation of computers, computer systems, and networks.

As mentioned in Sect. 16.3.2, the Russian government is typically turning a blind eye to Russian hacking groups attacking targets outside of Russia. After the invasion of Ukraine in 2022, tolerating attacks on foreign organizations has started to become official policy, where the Russian Duma plans to issue an order for absolving Russian hackers from criminal liability when the attacks are useful to the government, because this can be seen as a contribution to the war effort.[7]

---

[7] https://cybernews.com/news/russia-cybercrime-for-homeland/

17.5.3 Sovereign Internet Law

The Sovereign Internet Law is a nickname for a set of 2019 amendments to the existing "Law on Communication" and "Law on Information, Information Technologies, and Information Protection." The main amendments are:

- The compulsory installation of technical equipment at ISPs for inspecting and controlling Internet traffic.
- Centralized management of telecommunication networks and control mechanism for Internet traffic in and out of Russia.
- The implementation of a separate Russian national Domain Name System (DNS).

The amendments mandate Internet surveillance and grants the Russian government powers to isolate Russia from the rest of the Internet with a national fork of the Domain Name System.

17.5.4 FSB Law

As mentioned in Sect. 17.5, FSB is the Russian Federal Security Bureau. The FSB Law Article 11.2 establishes FSB as the authority in the information security field, including government encryption technology, and for enforcing restrictions on encryption in the private sector. Article 13 covers the FSB's general authorities. According to Article 13, the FSB is entitled to:

- Conduct operational-search methods (defined in another law) to fight espionage, organized crime, corruption, illicit arms and drug smuggling and threats to Russia's national security;
- Penetrate foreign intelligence services, criminal groups, and organizations;
- Conducting espionage and other activities damaging Russia's security;
- Ensure secrecy of cryptographic material in cryptographic entities in state bodies, enterprises, institutions and organizations irrespective of ownership;
- Assist businesses, institutions and organizations irrespective of ownership in developing measures to protect trade secrets.

17.6 China

China enforces a very strict Internet policy compared to most other countries in the world. On the one hand there is the "Great Firewall" which blocks access to selected foreign websites, and enforces internal censorship of social media and web content in general. On the other hand, China makes a strong effort to crack down on cybercrime and activities considered illegal. For example, from being one of the major countries for Bitcoin mining around 2015, the mining activity is reduced to a small proportion globally thanks to a policy which makes Bitcoin mining illegal combined with a government crackdown on the industry.

China is also very active in cyber operations, where most cyber units are organised under the Ministry of State Security and under the People's Liberation Army Strategic Support Force, as shown in Fig. 17.5.

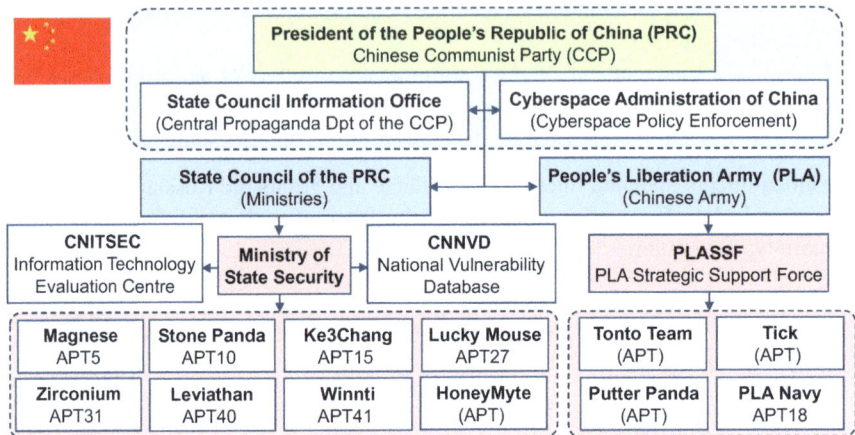

Fig. 17.5 Chinese structural organization of cyber agencies

Figure 17.5 is a simplified version of the diagram provided by Sekoia in their 2023 Threat Detection Research Report "My Tea's not cold" [52].

The Chinese Ministry of State Security is the principal civilian intelligence, security and secret police agency of the PRC, responsible for foreign intelligence, counterintelligence, and the political security of the CCP. The ministry controls semi-autonomous branches at the provincial, city, municipality and township levels throughout China. Under the ministry are numerous cyber operations teams that are typically assigned APT numbers and colorful names by Western cybersecurity organizations. Numerous cyberattacks have been attributed to these APTs.

Established in 2015, the PLA Strategic Support Force (PLASSF) is the space, cyber, political, and electronic warfare force of the PLA. Its focus is to provide capacity to fight what China calls "informationized conflicts" and strengthen military power in space and cyberspace. Responsibilities of the PLASSF include target acquisition, space and cyberspace reconnaissance, operation of the BeiDou navigation satellites, offensive electronic and cyber warfare as well as defensive countermeasures. The PLASSF operates cyber operations teams that are typically assigned names and APT numbers by Western cybersecurity organizations. Several cyberattacks have been attributed to these military APTs of China.

Cybersecurity regulation in China is grounded in three established pillars of law: the Cybersecurity Law (CSL), Data Security Law (DSL), and Personal Information Protection Law (PIPL). In addition, the National Intelligence Law regulates Chinas cyber intelligence activities. These laws are briefly described below.

17.6.1 Cybersecurity Law (CSL)

The Cybersecurity Law of China, first implemented in 2017, is the primary legislation governing cybersecurity. In 2022 the law was amended by the CAC (Cyberspace Administration of China). The main purpose of the law is to ensure state control over the Internet in China, which naturally has broad implications for technology companies. The most important aspects of the Cybersecurity Law are briefly described below.[8]

- **Real-name requirements:** Anonymity online is not tolerated, meaning that every messaging service and social network in China must verify users' identity. Users must always use their real names online, and service providers are required to deny service to anyone who refuses to comply.
- **Data localization:** Article 31 requires that citizens' personal information must be stored within China borders. This clause is applied to "critical information infrastructure operators" which, in effect, means any network provider with large user database. Companies that use overseas cloud services for processing must apply for a special permission. Some foreign tech companies such as Apple, were forced to store their user data in China which caused backlash from some users.
- **Prohibited content:** Communication and social network operators are required to censor content and remove any illegal material. This is aimed at enforcing the prohibition for circulating online content relating to: "activities harming national security, propagating of terrorism and extremism, inciting ethnic hatred and ethnic discrimination, dissemination of obscene and sexual information, slandering or defame others, upsetting social order, harming the public interest, infringing of other persons' intellectual property or other lawful rights and interests".
- **Technology "backdoors":** Article 23 stipulates that "for the needs of national security and criminal investigation, investigating organs may request network operators provide necessary technological support and assistance in accordance with laws and regulations." This part has caused concerns to foreign technology companies as the requirement could also mean providing encryption backdoors or other surveillance assistance to the government. The requirement of source code disclosure in an earlier draft version of the 2022 amendment was removed amid protests from US and other countries.
- **Critical information infrastructure sectors:** The law requires "critical information infrastructure" industries to implement additional checks and measures for establishing security safeguards. Entities covered under this provision include telecommunications, energy, transportation, information services, finance, public services, military and government networks as well as "networks and systems owned or managed by network service providers with massive numbers of users".

---

[8] https://sampi.co/china-cybersecurity-law/

17.6.2 Data Security Law (DSL)

The Data Security Law (DSL) defines a data classification scheme based on the data's potential impact on Chinese national security, and regulates its storage and transfer depending on the data's classification level. The law can be seen as a response to the U.S. Clarifying Lawful Overseas Use of Data Act (CLOUD Act), which gives U.S. law enforcement agencies the authority to compel companies falling under U.S. jurisdiction to produce requested data regardless of where the data is stored.[9]

17.6.3 Personal Information Protection Law (PIPL)

The PIPL is similar to EU's GDPR in that it gives Chinese consumers the right to access, correct and delete their personal data gathered by businesses. Fines can be as much as RMB50 million or up to 5% of a company's turnover from the previous financial year. Businesses may be forced to suspend operations until they can demonstrate compliance.

"Personal Information" is broadly defined to cover "any information related to identified or identifiable natural persons stored in electronic or any other format." The PIPL generally applies to all types of data activities (e.g., collection, storage, usage, reorganization, transmission, provision, disclosure and deletion) involving the personal information of data subjects in China, as well as activities outside China that are aimed at providing products or services to individuals in China or analysing their behavior. The PIPL imposes the following key obligations on data handlers:

- **Consent Requirements:** The legal basis for processing can be legal consent, but prior consent is not required for example in case of performing a contractual or statutory duty, responding to an emergency involving life and property, news reporting on a matter of public concern and where the information is already found in the public domain. This is similar to EU's GDPR.
- **Data Deletion Requirements:** Data handlers are required to delete the collected personal data when the purpose of the collection has been achieved, when the information no longer serves its disclosed purposes, when the service is no longer being provided, when the retention period has expired, when the user rescinds consent or when the processing activities contravene relevant laws and regulations. This is also similar to EU's GDPR.
- **Restrictions on Transfer of Personal Information Overseas:** For cross-border transfers, the data handler must ensure that the foreign recipient of the data has

---

[9] https://www.skadden.com/Insights/Publications/2021/11/Chinas-New-Data-Security-and-Personal-Information-Protection-Laws

in place data protection requirements at least as stringent as those that PIPL specifies for domestic companies. Companies in possession of a large volume of personal data must complete a mandatory security review led by the Cyberspace Administration of China before transmitting any data overseas.
- **General Compliance Requirements:** The PIPL requires companies handling personal data to conduct regular self-audits to assess their information security risks and implement corresponding policies and safeguards. Companies that make use of AI to analyze data subjects' personal information must abide by certain "transparency" and "fairness" principles that prohibit certain types of discriminatory pricing and marketing activities based on the data subject's personal status and protected characteristics.

Despite the PIPL law requiring a legal basis for processing personal data, government agencies in China perform widespread collection and processing of personal data as part of the Social Credit System (SCS). It is reasonable to assume that the collection and processing personal data for SCS is a legal basis according to PIPL. SCS is an extension to the existing legal and financial credit rating system in China, and is managed by the National Development and Reform Commission (NDRC).

17.6.4 National Intelligence Law

The National Intelligence Law of China, enacted in 2017, grants authorities broad powers to compel organizations and individuals to support and cooperate with intelligence work. This law has implications for cybersecurity regulation and data privacy. The resources that China dispose of are substantial and integrate human with technical components[10]:

- **Global reach of Chinese companies:** Both the human and technical reach of Chinese companies now give the intelligence services opportunities to gain direct access to many governments within the developing world as well as many allied and European countries with which China maintains economic and cultural collaboration.
- **Everyone can be compelled to assist state security**. As long as national intelligence institutions are operating within their proper authorities, they may, according to Article 14, "request relevant organs, organizations, and citizens to provide necessary support, assistance, and cooperation". According to Article 16, intelligence officials "may enter relevant restricted areas and venues; may learn from and question relevant institutions, organizations, and individuals; and may read or collect relevant files, materials or items".

---

[10] https://www.canada.ca/en/security-intelligence-service/corporate/publications/china-and-the-age-of-strategic-rivalry/chinas-intelligence-law-and-the-countrys-future-intelligence-competitions.html

These aspects are the reason for why most Western countries banned Huawei from delivering 5G networks. Hence, every branch of a Chinese company located in China or overseas must be considered as an instrument for Chinese state intelligence.

17.7 General Remarks on Cyber-Organizational Structures

The reader might have noticed the different focus of the structures illustrated and described in the previous sections. For the US and EU, the illustrated structures mainly focus on defensive entities, whereas for Russia and China there is also a strong focus on offensive entities. In reality, all these countries and regions have both defensive and offensive entities. The reason for the difference in focus is mainly due to the skewed availability of public information on which the descriptions are based. For example, Western cyber organizations typically give APT numbers and names to adversarial offensive groups, but not to Western offensive groups. Based on publicly available information it is unclear how Russian and China call adversarial offensive groups from their perspective.

17.8 Tasks

1. Accountability

    Imagine that Company A and Company B of comparable size become victims of cyberattacks, and that both companies suffer significant and comparable losses that negatively affect customers and shareholders. Both companies are subject to legal requirements to have cyber governance/ISMS in place. During the investigation of the incidents, it is found that company A has implemented an ISMS (information security management system) and good cyber maturity, while company B has inadequate ISMS and substandard cyber maturity. If one assumes that the loss for both companies is of the same magnitude, explain possible differences between the companies regarding management accountability and sanctions against the companies or their management.
2. Cyberlaws and agencies

    (a) What are similar cyberlaws in the US, Europe, Russia and China respectively?
    (b) What are cyber agencies with similar roles in the US, Europe, Russia and China?

17.8 Tasks

3. <u>EU legislation and cyber organizations</u>
   (a) What are EU regulations, directives, and acts, and how are they implemented in the member states?
   (b) Which EU acts are aimed at strengthening defensive cyber capacities of the EU?
   (c) Does the EU have an offensive cyber capacity? Discuss this question.

Chapter 18
Governance and Information Security Management

> *Management is efficiency in climbing the ladder of success; leadership is to make sure the ladder is leaning against the right wall.*
>
> Stephen Covey, American educator, author and businessman

Cybersecurity must be integrated in the governance and management of any organization. At the same time, there are several unique aspects of cybersecurity that require special competencies, which means that cybersecurity in many ways is a separate area within governance and management. Cybersecurity governance ensures that cybersecurity activities are aligned with business objectives, regulatory requirements, and ethical values. Information security management ensures that the organization has a systematic approach to protecting information assets from harm. It encompasses the policies, procedures, technologies, and practices that organizations implement to safeguard their information assets. In the end it is about managing cyber risks as part of enterprise risk management in general.

In this chapter, you learn about key principles, standards, and frameworks for cybersecurity governance and information security management. Furthermore, you learn about the required processes and work tasks that go under governance and management. Finally, you learn about assessing an organization's maturity regarding its information security posture.

18.1 Information Security Management Levels

Organizational structures are hierarchical. All levels of an organization are in their own way responsible for performing tasks related to information security, but the board and top executive management are legally accountable for ensuring that cybersecurity is well managed.

Levels of information security management can be divided in different ways. A common division of information security management is shown in Fig. 18.1. This can also be called cyber governance levels.

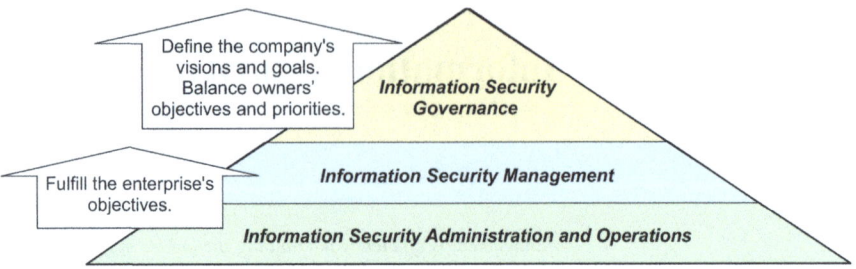

Fig. 18.1 Levels of information security management

Below is a short description of concepts related to the levels of governance in Fig. 18.1. Subsequent sections give a more detailed description.

Corporate governance is the highest level of governance in an organization, which is the responsibility of the board of directors and senior management. Information security governance is part of corporate governance, and focuses on setting goals for information security and verifying that goals are achieved. Priorities for information security are typically balanced with other goals that the board and top management want to achieve.

Information security management is about managing controls that an organization needs to implement to ensure a reasonable level of security. Information risk management is a central component which forms the basis for selecting security controls. As part of information security management, an organization may implement an information security management system (ISMS) as described in ISO/IEC 27001, NIST CSF and other frameworks on information security.

Information security administration and operations consist of the daily work to ensure security in the operation of systems and networks. This work includes, for example, monitoring and configuring systems and networks, handling security incidents and reporting.

In the sections below we describe in more detail the three management levels illustrated in Fig. 18.1. Note that there is often inconsistency in the usage of terms related to information security management. For example, the terms "cybersecurity governance" and "information security management" should be considered equivalent, and this book does not express any preference for one of them, because both are correct in their contexts. However, in this section we make a clear separation between governance, management, and administration of information security, to emphasise that there are different layers in the broader management spectrum.

18.1.1 Information Security Governance

The board of directors and senior management of an enterprise are responsible for defining the company's objectives, visions, and value propositions. To the extent that a balance must be struck between different objectives, it is the responsibility of the board and senior management to define the right trade-off. This also applies to information security management.

18.1 Information Security Management Levels

Businesses are under constant pressure to increase efficiency and quality and to be competitive in the market. A common means of achieving these things is to reduce costs and increase efficiency through the digitization of business processes. Along with the increasing dependence on digital business processes, threats to these same processes are also increasing through online exposure. When digital business processes have vulnerabilities that can be exploited by threat actors from around the world, information security risks arise. To the extent that important business processes and services can be attacked and be disrupted as a result of cyber incidents, risks arise for the company's objectives and perhaps for its entire existence. For the digitization of business processes to be sustainable, these security risks must be managed responsibly.

The board of directors and management team also have the overall responsibility for ensuring that laws and regulations are complied with at all times. Many organizations are also required to manage security according to an ISMS (Information Security Management System), which simply means that processes and activities around information security must be systematically structured with policies in place and follow a recognized framework.

Complying with the requirements described above can generally be described as information security governance. A more concise definition of information security management is given by ISACA [53].

> *Information security governance is a subset of enterprise governance that provides strategic direction, ensures that objectives are achieved, manages risks appropriately, uses organisational resources responsibly, and monitors the success or failure of the enterprise security programme.*

ISACA (Information Systems Audit and Control Association) is an international association for IT auditors and professionals working with IT management. ISACA publishes frameworks and guidance material for IT governance in general, including information security governance. In addition to the definition above, ISACA defines the following five main objectives for information security governance:

- **Strategic Alignment.** The usefulness of activities around information security is inevitably related to how well they support the objectives of the organization and at what cost. Without organizational goals as a reference point, other principles such as best practice can become too comprehensive, inadequate or misdirected in other ways. From a business perspective, an adequate practice that is proportional to requirements is usually more cost-effective and appropriate than always following best practice. Anything else will initially be difficult for the board and senior management to defend, especially in commercial companies. From a governance perspective, information security is not an end in itself – it should contribute to achieving the company's goals.

- **Risk Management.** Optimal risk management is a fundamental goal – not just for information security activities, but really for all security activities in an organization. The benefits of risk management cannot be measured directly, but must be measured using indicators that correlate indirectly with the benefits of risk management. Put in a relatively abstract way, risk management is beneficial if it meets the expectations of achieving defined security objectives in a consistent and cost-effective manner.
- **Resource Management.** As with any other organizational resource, resources related to information security must be used sensibly and efficiently. For example, it is important to map already implemented solutions and consider whether a solution can be reused in other areas or departments. Competence and human resources are very important. Knowledge must be captured, preserved, disseminated and made available when needed. Expertise must be built, used and safeguarded in the best possible way. Management systems and processes must be standardised as far as possible to reduce administration and training costs. Everything must be well documented, referenced and accessible.
- **Value Delivery.** Some may find it paradoxical that work with information security creates value, because it is primarily a major cost item in organizations. However, good information security management creates value on the bottom line through reduced losses, increased reputation and market confidence, balanced risk management and more efficient organization and use of resources. Optimal value delivery occurs when strategic information security objectives are achieved, legal requirements are complied with and security threats are balanced with an acceptable risk, all at the lowest possible cost.
- **Performance Measurement.** A simple rule of governance can be expressed like this: "You can't manage what you can't measure." Measurement, monitoring and reporting of activities related to information security is therefore a prerequisite for being able to assess whether organizational goals are being achieved. Methods for measuring activities and events related to information security across the organization must be developed, and analytical models that provide an indication of the effect of activities and processes around information security must be defined. There will typically be a spectrum of such measurements and analytical models that cannot be directly translated into a single metric for how "secure" the organization is. The degree of information security in an organization must therefore be measured indirectly through the assessment of quality in processes, and can ultimately be expressed as the degree of maturity in the management of information security. A relevant standard and guide for measuring, evaluating and analysing information security is ISO/IEC 27004 [54] or NIST CSF [9].

18.1.2 Questions that the Board and Executive Should Ask Themselves

For many board members and senior executives, cybersecurity challenges can seem overwhelming. On the one hand, they have responsibilities and are accountable for governance of information security in the organization, on the other hand, they may have insufficient expertise to take this responsibility in a good way. As most business processes are digitized and exposed to cyber threats, which often result in significant cyber risk, expertise in this area is a prerequisite for the board and top management to be able to do a good job.

The list of questions below can be used as a simple checklist for directors and senior managers to test their control and awareness of information security governance in the organization they are accountable for.

1. How does the organization keep up-to-date with cyber threats in general, and for its own sector and organization in particular?
2. Does the organization have a good overview and understanding of risks related to cybersecurity?
3. Does the management have sufficient competence in cybersecurity?
4. How well is cyber risk integrated into enterprise risk management?
5. Should cyber insurance be included in the organization's insurance policies?
6. How mature is the company's security governance/ISMS?
7. How well is security governance/ISMS supported by management?
8. Is there sufficient focus on security culture as part of the organizational culture?
9. Do the emergency and contingency plans also cover cyber security incidents, and have the plans been tested?
10. What is the status of compliance with laws and regulations on cybersecurity and privacy?
11. Is information adequately protected when it is transferred to third parties?
12. To what extent should the organization outsource security services, such as e.g. through purchasing MDR (Managed Detection and Response)?

18.1.3 Information Security Management

Management of information security is the creation, operation, and maintenance of a set of processes and controls that protect the organization's information assets from security threats, thereby helping to manage and reduce cybersecurity risk. Risk management involves assessing relevant security risks that an organization is exposed to, and how these can best be managed and reduced to an acceptable level. Risk management is described in Chap. 19.

The activities of information security management should be described by governance policies such as in the form of an *ISMS (Information Security Management System)*. An ISMS is a set of information security processes and activities based on a selection of standards and frameworks which are complemented with the organization's own policies. There are multiple standards and frameworks to choose from.

In the US, the most commonly used framework is *NIST CSF (Cybersecurity Framework)* [3] combined with related standards and frameworks such as those in the NIST SP 800 series of standards, frameworks and guidelines.[1] NIST CSF is described in Sect. 18.2. The NIST documents are generally of high quality, and can be downloaded free of charge.

In other parts of the world (Europe, Africa, Asia, Australia/Oceania, and South America), the most commonly used framework is *ISO/IEC 27001 ISMS (Information Security Management System) – Requirements* [5] combined with other ISO/IEC standards in the 27000-series. The ISO/IEC 27000 series is described in Sect. 18.5. These standards are developed in collaboration between ISO (International Organization for Standardization) and IEC (International Electrotechnical Commission). An important feature of ISO/IEC 27001 is that it has the same structure as several other ISO/IEC management system standards such as ISO 31000 Risk Management and ISO 9000 Quality Management. This provides a better basis for organizations to establish a comprehensive management system (which encompasses more than just information security). One disadvantage of standards from ISO/IEC is that you have to pay for them. Note that NIST CSF is becoming increasingly popular also outside of the US.

We can briefly mention some additional management frameworks that are relevant for cybersecurity. ISACA publishes the *COBIT framework*[2] which is widely used in the management and audit of IT. The Centre for Internet Security (CIS) publishes the *CIS Controls and Resources*[3] which provides a comprehensive description of controls and associated resources for information security, as described in Sect. 18.4. Axelos publishes the ITIL (Information Technology Infrastructure Library) which includes information security and continuity management [8], as described in Sect. 1.6.

The above set of standards and frameworks is far from exhaustive, and it can almost be too much of a good thing in the sense that it is confusing with so many different ones. A general advice is to select the frameworks/standards that are typically used by other organizations in the same sector, to best facilitate learning and exchange of experience.

---

[1] https://csrc.nist.gov/publications/sp800
[2] https://www.isaca.org/resources/cobit
[3] https://www.cisecurity.org/

18.1.4 Administration and Operation of Cybersecurity

Administration and operation of cybersecurity consist of a portfolio of practical activities. An organization can choose to handle these activities themselves or outsource them to an *MSSP (Managed Security Service Provider)*. There is a growing market for MSSPs with an increasing number of companies that specialize in providing these types of services. The large consulting firms also offer services in this market.

Small businesses typically have no cybersecurity staff, while larger businesses may have a *SOC team* for security monitoring and incident response.

A SOC (Security Operations Centre), CSIRT (Computer Security Incident Response Team) or CERT (Computer Emergency Response Team) is a team of experts who have the expertise, tools and resources to handle security incidents when they are discovered. CERT is a trademark owned by Carnegie Mellon University, while CSIRT is a term that can be used freely. Since the activities of a SOC, CSIRT and CERT are largely overlapping, the term SOC is increasingly being used as a collective term, which we also do in the following. A SOC detects and handles incidents as described in Sect. 14.5.

The necessary competence and complexity to perform SOC activities may entail significant costs. It may therefore be cost-effective to purchase such services. As previously mentioned, there is a rapidly growing market for SOC and CSIRT services. These topics are described in more detail in Chap. 16.

Note that there are many other security activities in addition to those that are performed by the SOC team. An organization that expects the SOC team to do all the security work will never achieve a reasonable level of security. For example, most activities related to preventive security controls are performed by other parts of the organization, such as the IT department, HR department and the legal/GRC department. This is described in Sect. 18.3.

18.2 NIST Cybersecurity Framework

The NIST Cybersecurity Framework (CSF) describes a set of guidelines aimed at helping organizations reducing their cybersecurity risks and strengthening their security posture. NIST CSF 2.0 was published in 2024, precisely 10 years after version 1.0 was published in 2014. The main change in version 2.0 is to include GOVERN as a new high-level function to complement the previous five functions IDENTIFY, PROTECT, DETECT, RESPOND and RECOVER. NIST CSF 2.0 is typically illustrated in the form of a wheel as in Fig. 18.2.

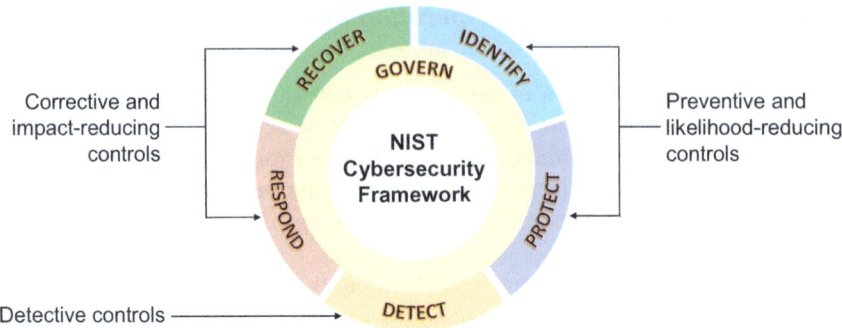

Fig. 18.2 NIST Cybersecurity Framework

Although the meaning of the functions might seem obvious from their names, it is useful to give a brief description of each function.

- GOVERN covers activities for defining the organization's risk management strategy, expectations, defining roles and responsibilities for cybersecurity, articulating policies and procedures, communication and dissemination, and measuring and monitoring the maturity of the cybersecurity programme. Contingency planning (incident and recovery plans) also belongs in this function.
- IDENTIFY covers everything that has to do with understanding the organization's current cybersecurity risks. This involves asset management, identifying threats, discovering weaknesses and performing vulnerability scanning, as well as performing risk assessments.
- PROTECT consists of implementing controls to manage and reduce the organization's cybersecurity risks to an acceptable level. Controls can apply to all asset classes such as devices, networks, applications, data and users.
- DETECT consists of detecting cybersecurity attacks and compromises, communicate incidents, and analyse what happened to be informed about how best to eradicate the attacker from the organization's systems.
- RESPOND consists of actions regarding a detected cybersecurity incident, in order to minimize impact, eradicate the attacker, and communicate with stakeholders. The incident response plan should be followed.
- RECOVER consists of restoring assets and operations affected by a cybersecurity incident. The recovery plan should be followed.

The rationale for representing the functions as a wheel is because all of the functions relate to one another. For example, DETECT, RESPOND and RECOVER which address cyber incidents will IDENTIFY unknown vulnerabilities and the need to better PROTECT the assets that have been affected by incidents. GOVERN is in the centre of the wheel because it informs how the organization should implement the other five functions.

All the functions should be performed concurrently. Activities related to GOVERN, IDENTIFY, PROTECT, and DETECT happen continuously, and activities related to RESPOND and RECOVER should be ready at all times and be performed as soon as cyber incidents are detected.

18.2 NIST Cybersecurity Framework

With reference to the security-control diagram of Fig. 1.6, the functions IDENTIFY, and PROTECT cover protective security controls, whereas the functions RESPOND and RECOVER cover corrective security controls. The function DETECT is of course detective.

With reference to the bow-tie diagram of Fig. 14.1, the functions IDENTIFY and PROTECT cover likelihood-reducing controls, whereas the functions DETECT, RESPOND and RECOVER cover impact-reducing controls.

NIST CSF is divided into three parts, which are Core, Profiles and Tiers. These are briefly described in the following sections.

18.2.1 NIST CSF Core

CSF Core is structured around the six security functions GOVERN, IDENTIFY, PROTECT, DETECT, RESPOND, and RECOVER, as shown in Fig. 18.3. For each function, there is a set of categories that each consists of a set of subcategories. There are in total 22 categories and 106 subcategories. Each subcategory provides informative references to external sources that describe relevant security controls and profiles for that particular subcategory. CSF core is provided in the form of a huge spreadsheet with all the information specified in rows and columns.

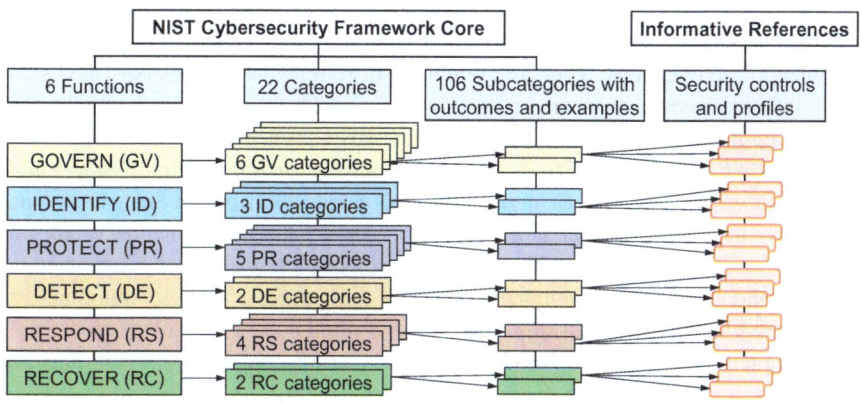

Fig. 18.3 Structure of NIST CSF Core

NIST CSF uses the term "outcome" to denote a security goal. Each category represents a high-level outcome (high level security goal), and each subcategory represents a specific outcome (specific security goal) that supports the high-level outcome of the category. This is analogous to the way MITRE ATT&CK describes high-level "attack tactics" that are supported by specific "attack techniques", as described in Sect. 16.2.2. The 6 functions with their respective 22 categories are listed in Table 18.1.

Table 18.1 Functions and categories of NIST CSF Core

Functions	Categories (high-level outcomes)
GOVERN (GV)	Organizational Context (GV.OC) Risk Management Strategy (GV.RM) Roles, Responsibilities, and Authorities (GV.RR) Policy (GV.PO) Oversight (GV.OV) Cybersecurity Supply Chain Risk Management (GV.SC)
IDENTIFY (ID)	Asset Management (ID.AM) Risk Assessment (ID.RA) Improvement (ID.IM)
PROTECT (PR)	Identity Management, Authentication, and Access Control (PR.AA) Awareness and Training (PR.AT) Data Security (PR.DS) Platform Security (PR.PS) Technology Infrastructure Resilience (PR.IR)
DETECT (DE)	Continuous Monitoring (DE.CM) Adverse Event Analysis (DE.AE)
RESPOND (RS)	Incident Management (RS.MA) Incident Analysis (RS.AN) Incident Response Reporting and Communication (RS.CO) Incident Mitigation (RS.MI)
RECOVER (RC)	Incident Recovery Plan Execution (RC.RP) Incident Recovery Communication (RC.CO)

The informative references to external sources provide guidance on how the organization may achieve a specific outcome, or to relevant profiles (see next section). Informative references can be sector- or technology-specific, and are not necessarily published by NIST. Typical informative references are security controls from *NIST SP 800-53 Security and Privacy Controls for Information Systems and Organizations*, or from *CIS Critical Security Controls* described in Sect. 18.4. One could say that CSF Core describes "what is needed" (security outcome), but not "how to do it" (security control). Anyway, there is no clear distinction between an outcome and a control, hence it is probably OK to interpret a "security outcome" as a "security control". In addition to informative references, each subcategory provides implementation examples with concise, action-oriented steps for how to achieve the outcome, which in fact are example security controls. For example, Table 18.2 describes outcome PR.DS-11 with corresponding examples.

Table 18.2 Example security outcome in NIST CSF Core

PR.DS-11: Backups of data are created, protected, maintained, and tested
Ex1: Continuously back up critical data in near-real-time, and back up other data frequently at agreed-upon schedules.
Ex2: Test backups and restores for all types of data sources at least annually.
Ex3: Securely store some backups offline and offsite so that an incident or disaster will not damage them.
Ex4: Enforce geographic separation and geolocation restrictions for data backup storage.

18.2 NIST Cybersecurity Framework

A security outcome in NIST CSF Core is equivalent to a security control in ISO/IEC 27002:2022. In that sense, outcome PR.DS-11 in the table above corresponds to security control 8.13 Information backup as described in Sect. 18.5.4.

18.2.2 CSF Profiles

CSF Profile is an approach for organizations to describe its current and/or target *cybersecurity posture* in terms of security outcomes from CSF Core. Hence, the organization can list currently achieved outcomes, and call that its current profile. Then the organization can complement the list of outcomes with additional desired outcomes, and call that its target profile.

Organizational profiles are typically sector specific and are adjusted to the organization's size, mission objectives, stakeholder expectations, threat landscape, and regulatory requirements. The organization can then prioritize its efforts to achieve the outcomes of the target profile for better managing cyber risk, and can inform relevant stakeholders about its current and target profiles. Auditors can also assess to what extent an organization achieves a given profile.

18.2.3 CSF Tiers

CSF Tiers describe levels of maturity in cybersecurity governance, for which NIST CSF defines four levels. A similar model is the CMMI model of maturity in information security management with five levels, which is described in Sect. 18.6.

A given tier characterizes the rigor of an organization's cybersecurity governance and risk management practices. The tiers are defined as Partial (Tier 1), Risk Informed (Tier 2), Repeatable (Tier 3), and Adaptive (Tier 4). An organization which has ambitions for improving its cybersecurity maturity and posture can use tiers as progressive targets from informal, ad hoc governance with holes in the profiles, to mature governance where the organization's current profile and security posture is reasonable and continuously improving. Selecting tiers gives a clear signal for how an organization intends to manage its cybersecurity risks. An approximate mapping between tiers defined in NIST CSF and maturity levels in CMMI is shown in Table 18.3.

Table 18.3 Correlation between maturity levels in NIST CSF and CMMI

NIST CSF Tiers	CMMI maturity levels
Tier 4 (Adaptive)	Level 4: Systematized process and Level 5: Optimized process
Tier 3 (Repeatable)	Level 3: Formalized process
Tier 2 (Risk Informed)	Level 2: Fragmented process
Tier 1 (Partial)	Level 1: Random Process

18.3 The Cyber Defense Matrix

The Cyber Defense Matrix (CDM) provides a two-dimensional view on cybersecurity governance which thereby expands NIST CSF's perspective on cybersecurity [10]. The first dimension of CDM is represented by the original five functions of NIST CSF which are IDENTIFY, PROTECT, DETECT, RESPOND, and RECOVER, as described in the previous section. The second dimension of CDM is represented by the five main asset classes which are DEVICES, APPS, NETWORKS, DATA, and USERS.

Although the meaning of the asset classes might seem obvious from their names, it is instructive to give a brief description of each asset class.

- DEVICES are of course hardware devices, but also the commodity software that runs on them. That means that firmware, operating systems and standard applications from vendors belong to this class. There is of course a wide spectre of devices, from handheld, to servers. It is also useful to distinguish between devices owned by different parties, such as the organization, staff (BYOD), or cloud providers. Networking devices like switches and routers are also devices because they must be patched and protected in the same way as other devices.
- APPS are business applications and processes specially developed by or for the organization itself. The APPs are typically producing the services and products which deliver the organization's value and business functions. Hence, the continuity and integrity of APPS are typically critical.
- NETWORKS are the communication channels, routings, protocols, and firewall filterings that enable traffic to flow efficiently and securely. Note that this is different from the infrastructure of devices with which networks are built. Hence, by networks, we typically mean the *configuration* of networking devices which then becomes DNS (Domain Name System), and PKI (Public Key Infrastructure), and which provides network routing, firewall filtering, email filtering, and VPN (Virtual Private Networks).
- DATA is all types of business data whether residing on (data-at-rest), passing through (data-in-motion), or being processed by (data-in-use) by the asset classes listed above. This class includes databases, file systems, and data residing in computer memory, microprocessor registers, on disk, and in the cloud.
- USERS are the people and AI agents using the resources listed above, with their associated identities and credentials. An AI agent could perhaps be considered to be a device, but if it carries the characteristics of a user with its own ID, authentication credentials, access authorizations, and intelligent decision logic, then it should be considered as a user.

There are several use cases for CDM, where the list above provides some examples:

- Resource allocation
- Organizational handoffs
- Measurement and metrics
- Cybersecurity portfolio gap analysis
- Structural and situational awareness
- Cybersecurity vendor product classification

18.3 The Cyber Defense Matrix

Take as example the current and target profiles from NIST CSF described in Sect. 18.2.2. We can use CDM to do a gap analysis as illustrated in Fig. 18.4.

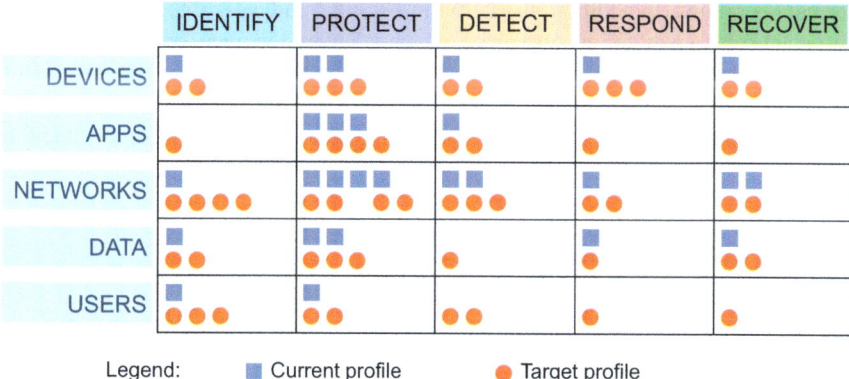

Fig. 18.4 Example gap analysis of cybersecurity profiles using the Cyber Defense Matrix

The blue squares and red dots can of course be replaced by more descriptive symbols. Anyway, the main idea is that CDM allows the cybersecurity practitioners to build a clear picture of the status, makes it easier to prioritize and allocate resources, to monitor progress, and to communicate to management and other stakeholders. Note that the CIS controls described in Sect. 18.4 also refers to asset types when they describe specific safeguards as part of each security control.

Another example is to define cybersecurity handoffs, i.e. responsibilities for activities related to cybersecurity. An unfair and unreasonable expectation that the security team will look after everything cybersecurity in an organization. However, many of the activities covered by the NIST CSF should be performed by other departments, i.e. by non-security staff. The CDM is perfect for assigning reasonable security handoffs and for understanding the division of security responsibilities in the organization, as shown by the example in Fig. 18.5.

	IDENTIFY	PROTECT	DETECT	RESPOND	RECOVER
DEVICES	IT services, Owners		SOC/CERT/CSIRT Device Monitoring		IT services, Owners
APPS	DevOps, Owners		SOC/CERT/CSIRT App monitoring		DevOps Owners
NETWORKS	Network Administration		SOC/CERT/CSIRT Network Monitoring		Network Admin
DATA	Chief Data Officer, Data Protection Officer, Owners		IT Operations Data Loss Prevention		CDO, DPO Owners
USERS	Human Resources		Insider Threat Monitoring	Human Resources Physical Security	

Fig. 18.5 Cybersecurity handoffs, example

In particular, the example CDM above illustrates that the security team (SOC/CERT/CSIRT) should only be responsible for incident detection and response, and that all other security related activites are handled by other departments as part of their normal operations. In addition, the CDM in Fig. 18.5 indicates that HR also has an important role to play for cybersecurity, which is ignored in many organizations. For example, cybersecurity awareness training should be the domain of HR, instead of being done by the IT department. Basic security skills and awareness is important for everyone in the organization; hence it is something that HR should be responsible for.

18.4 CIS Critical Security Controls

The CIS Critical Security Controls (CIS Controls) describe actions for cyber defense across 18 operational domains, as shown in Fig. 18.6. The CIS Controls are meant to provide a relatively short list of effective defensive actions that make it easy for enterprise to priorities efforts to strengthen their cybersecurity posture.

The development of the CIS Controls started in 2008 as an international, grassroots consortium of security experts from all types of sectors. The expert volunteers who develop the CIS Controls use their first-hand experience to specify well-tested actions for cyber defense. As of 2024, the CIS Controls are in their 8th edition, and they can be downloaded free of charge.[4]

There exist many mappings between CIS Controls and other cybersecurity frameworks such as the NIST Cybersecurity Framework described in Sect. 18.2, NIST 800-53 [7], the ISO/IEC 27000 series describe in Sect. 18.5 as well as regulations such as PCI DSS, HIPAA, NERC CIP, and FISMA. Mappings to the CIS Controls have been defined to provide a practical basis for implementation. In particular, NIST CSF outcomes refer to the CIS Controls as "informative references" to help implementation.

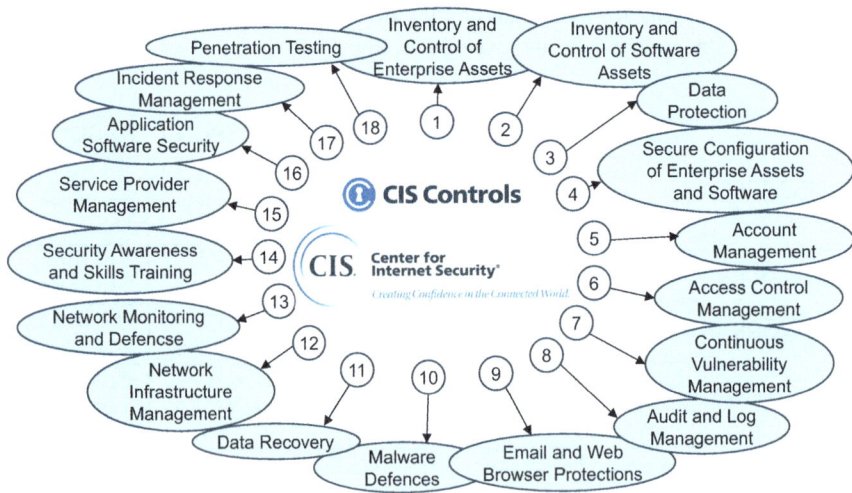

Fig. 18.6 CIS Critical Security Controls

---

[4] https://www.cisecurity.org/controls

18.4 CIS Critical Security Controls

Each of the 18 CIS control consists of a set of safeguards, which in fact are specific detailed controls. The safeguards are labelled according to priority in three *Implementation Groups* called IG1, IG2, and IG3. The IGs are intended to help organizations select the safeguards that best fit their size and context. The text below is copied from CIS Controls version 8.

- *An IG1 enterprise is small to medium-sized with limited IT and cybersecurity expertise to dedicate towards protecting IT assets and personnel. The principal concern of these enterprises is to keep the business operational, as they have a limited tolerance for downtime. The sensitivity of the data that they are trying to protect is low and principally surrounds employee and financial information. Safeguards selected for IG1 should be implementable with limited cybersecurity expertise and aimed to thwart general, non-targeted attacks. These Safeguards will also typically be designed to work in conjunction with small or home office commercial off-the-shelf (COTS) hardware and software.*
- *An IG2 enterprise employs individuals responsible for managing and protecting IT infrastructure. These enterprises support multiple departments with differing risk profiles based on job function and mission. Small enterprise units may have regulatory compliance burdens. IG2 enterprises often store and process sensitive client or enterprise information and can withstand short interruptions of service. A major concern is loss of public confidence if a breach occurs. Safeguards selected for IG2 help security teams cope with increased operational complexity. Some Safeguards will depend on enterprise-grade technology and specialized expertise to properly install and configure.*
- *An IG3 enterprise employs security experts that specialize in the different facets of cybersecurity (e.g., risk management, penetration testing, application security). IG3 assets and data contain sensitive information or functions that are subject to regulatory and compliance oversight. An IG3 enterprise must address availability of services and the confidentiality and integrity of sensitive data. Successful attacks can cause significant harm to the public welfare. Safeguards selected for IG3 must abate targeted attacks from a sophisticated adversary and reduce the impact of zero-day attacks.*

Each safeguard is structured as shown in Fig. 18.7 which is an example showing the safeguard for data backup.

NR.	TITLE/DESCRIPTION	ASSET TYPE	SECURITY FUNCTION	IG1	IG2	IG3
11.1	**Establish and Maintain a Data Recovery Process**	DATA	RECOVER	●	●	●
	Establish and maintain a data recovery process. In the process, address the scope of data recovery activities, recovery prioritization, and the security of backup data. Review and update documentation annually, or when significant enterprise changes occur that could impact this safeguard					

Fig. 18.7 CIS Controls safeguard for backup

Note that the safeguard of Fig. 18.7 refers to the DATA asset type of the Cyber Defence Matrix described in Sect. 18.3, as well as the RECOVER function of NIST CSF described in Sect. 18.2. It can also be seen that this safeguard is part of all implementation groups, meaning that all organizations, no matter their size or context, should implement this safeguard. This particular safeguard corresponds to outcome PR.DS-11 of NIST CSF, and to security control 8.13 of ISO/IEC 27002 described in Sect. 18.5.4.

18.5 ISO/IEC 27000 Series of Security Standards

ISO and IEC collaborate through JTC1 (Joint Technical Committee 1) to develop standards for information security. One set of standards that JTC1 is responsible for is the 27000 series, which focuses on information security management. There are many different standards in the 27000 series, the two most important of which are the standard *ISO/IEC 27001 Information Security Management System (ISMS) – Requirements* [5], and the standard *ISO/IEC 27002 Information security controls* [4], described in the following sections. Another important standard is *ISO/IEC 27005 Risk Management* for information security, which is described in Chap. 19.

18.5.1 The History of ISO/IEC 27001 and 27002

Most people in the information security community have heard about ISO/IEC 27001, the international standard that describes requirements for an ISMS, and ISO/IEC 27002, which is accompanying to 27001. Fewer people are aware that it was originally the other way around, i.e. that ISO/IEC 27001 was an accompanying standard to 27002. This needs to be explained.

In the early 1990s, information security began to gain importance. At that time, a major problem was that IT professionals had little or no expertise in information security. At the same time, there was a lack of a common understanding of terminology, technology and controls related to information security.

The oil company Shell was relatively early in structuring information security activities within its own organization. They created a separate guide for information security, which was picked up by *BSI (British Standards Institution)* in the early 1990s. BSI developed this guide further, and published it in 1995 as a norm for information security, called *BS 7799: Code of Practice for Information Security Management*. The standard focused on describing a range of information security controls, which was well received by the IT community – not only in the UK but also in other countries.

18.5 ISO/IEC 27000 Series of Security Standards

However, it quickly became clear that BS 7799 did not cover governance aspects of information security. During the 1990s, IT governance became an important field, and thus it was also important to define practice norms for governance of information security. Against this background, in 1999 BSI published an additional standard called *BS 7799-2: Information Security Management System (ISMS) – Specification with guidance for use.* This standard was also well received and quickly adopted, both in the UK and in many other countries.

ISO and IEC, which publish international standards, saw the need to publish standards precisely in the field of information security. Due to this urgency, they decided in 2001 to publish the standards from BSI, simply by adding the digit "1" in front, and otherwise keeping the same titles, so that the standards were numbered ISO/IEC 17799 and ISO/IEC 17799-2.

ISO (and IEC) has established a series of standards for management/management systems within specific disciplines, where the list below gives some examples:

- ISO 9001 Quality Management
- ISO/IEC 20000 Service Management
- ISO 22300 Security and Resilience
- ISO/IEC 27001 ISMS: Information Security Management System – Requirements
- ISO 31000 Risk Management
- ISO 39001 Road Traffic Safety Management Systems (RTSMS)
- ISO 22000 Food Safety Management System

All these management standards follow the harmonized approach and structure for management system standards as specified in Annex SL of the ISO Directives.[5] It means that all these standards contain 10 clauses (sections) with fixed titles as described in the next section.

Because the 27000 series was defined to be a separate series of standards for the management of information security, it was necessary to give new numbers to the already published standards for information security, and here something interesting happened: The standard for ISMS (17799-2), which was originally an accompanying standard, was considered to be the fundamental standard for information security management, and thus was assigned the number 27001, i.e. number 1. The Code of Practice in Information Security Management (17799), which was originally considered to be the fundamental standard, was "demoted" to number 27002, i.e. number 2. This development is illustrated in Fig. 18.8.

---

[5] https://www.iso.org/sites/directives/current/consolidated/

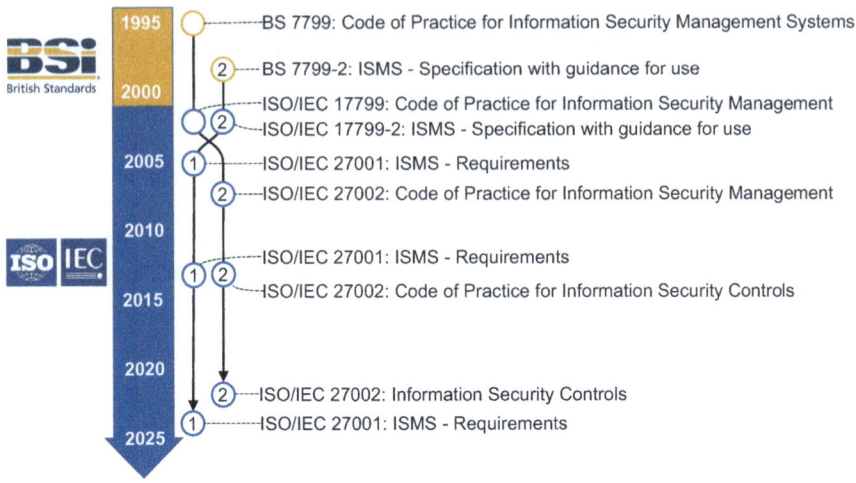

Fig. 18.8 Evolution of the ISO/IEC 27001 and 27002 standards

In addition to changing numbers, ISO/IEC 27002 has undergone several name changes. From having the confusing name *Code of Practice for Information Security Management Systems* in 1995, the 2022 edition was given the simple and intuitive name *Information Security Controls*.

18.5.2 ISO/IEC 27001 ISMS: Requirements

ISO/IEC 27001:2022 ISMS—Requirements is the fundamental standard of the ISO/IEC 27000 series of standards for the management of information security. ISO/IEC 27001 describes the requirements for establishing, implementing, maintaining, and continuously improving an ISMS. In other words, it is an overview of activities, processes and documentation that an organization should maintain in order to achieve maturity in security governance and comply with typical regulations that require an information security management system.

> An ISMS is a systematic approach for establishing, implementing, maintaining and improving an organization's information security to achieve business objectives.

Although ISO/IEC 27001 does not even mention the word "governance", an ISMS is in reality a governance program similar to the GOVERN function of NIST CSF.

The scope of an organization's ISMS should be proportional to the size of the organization, and the bulk of information security requirements should be planned based on risk assessments and regulatory requirements. Scope refers to the types and number of documents, processes and activities. Small organizations typically need a simple ISMS with few policies, while large organizations with significant processing of sensitive information need a more comprehensive ISMS with multi-level policies and multiple roles for information security. ISO/IEC 27001 defines seven categories of requirements which are briefly described below. The standard also includes Annex A, which lists all the security controls from ISO/IEC 27002.

The requirements are succinctly worded. Organizations can use them as guidance for establishing policies and processes for information security management. Auditors also use the standard for conformity auditing and certification. The requirements are presented as seven separate clauses with the same titles as those in other ISO management standards. We use the original clause numbers 4–10 as described in ISO/IEC 27001. Clauses 1–3 are: "(1) Scope", "(2) Normative references", and "(3) Terms and definitions", which we do not describe here. The next seven clauses are summarised below.

4. **Context of the organization.** A fundamental requirement is for the organization to establish, implement, maintain and continuously improve its ISMS, which includes policies, roles and processes for information security. The scope of the ISMS shall be determined based on needs and the size of the organization. The organization shall identify and identify external and internal constraints that are relevant for its purpose and that affect information security. For example, regulatory compliance requirements and industry norms are important external constraints. Furthermore, the organization must understand the information security needs and expectations of relevant third parties and stakeholders.
5. **Leadership.** Senior management shall demonstrate leadership by, among other things, ensuring that information security policies and objectives are established and aligned with the organization's overall strategy, providing adequate resources for information security work, communicating the importance of information security to the entire organization, and monitoring that the ISMS achieves the expected results. Senior management shall ensure that responsibilities and authority for roles relevant to information security are assigned and communicated internally in the organization.
6. **Planning.** The organization shall select the approach and method of risk assessment and define criteria for risk acceptance. Furthermore, the organization shall establish a process for managing identified risks. When selecting security controls to reduce risks to an acceptable level or to satisfy regulatory requirements, the organization shall take into account the list of security controls in Annex A (described in detail in ISO/IEC 27002). In addition, other security controls may be considered.

A SoA (Statement of Applicability) is a document that lists all the security controls in Annex A and which for each control briefly explains whether, and why, the control has been implemented or not. When a risk assessment is carried out (as part of point Clause 8. Operation) to select security controls, and the enterprise wishes to prepare documentation as a basis for audit or certification, an SoA must be prepared which shall contain, at a minimum:

- A table with the security controls from Annex A
- Rationale for the applicability of security controls
- information on whether the necessary security controls has been implemented or not
- the grounds for excluding some of the security controls in Annex A

Organizations may choose security controls from a framework other than ISO/IEC 27002, but audit documentation must then provide an SoA with a mapping from the controls of Annex A to the controls in the other framework.

7. **Support.** The organization shall allocate adequate resources needed for the ISMS. The organization must specify requirements and ensure that relevant employees actually have that competence. The organization must ensure that employees have awareness of security policies, the importance of following policies and the consequences of not following policies.

   Documentation is a fundamental requirement. The organization's ISMS shall include documented information explicitly required, and documented information necessary for the effectiveness of the entire ISMS, including for the various processes involved.

8. **Operation.** The organization shall designate relevant domains, functions, and business processes to undergo information security risk assessment, and conduct risk assessments according to the approach described under Clause 6. Planning.

   Based on the risk assessment and risk acceptance criteria, as well as regulatory compliance requirements, the organization shall determine and implement the various risks treatment strategies, which may include reducing risk by implementing security controls, transferring risk for example by purchasing cyber insurance, accepting risk according to the level of risk acceptance or avoiding risk by stopping the business process resulting in risk. The organization shall retain documented information about risk assessments and how the risks have been managed.

   This clause implicitly also covers incident detection, response, and recovery as security controls to be implemented and operated.

9. **Performance evaluation.** For monitoring, measuring, analysing, and evaluating the ISMS, the organization shall determine what needs to be monitored and measured, as well as methods for this. The methods chosen should yield comparable and reproducible results to be considered valid. The organization will conduct internal audits at scheduled intervals to provide information on whether the ISMS is functioning as planned. Senior management shall review the organization's ISMS at scheduled intervals to ensure ongoing suitability, adequacy, and effectiveness.

   The results of the management review shall include decisions relating to continuous improvement opportunities and any need for changes in the ISMS. The management review must be documented.

10. **Improvement.** The organization shall continually improve the effectiveness of the ISMS and make modifications as appropriate. When weaknesses in the ISMS are identified, the organization shall take steps to identify the cause of the weakness and correct it.

18.5.3 ISMS Process Cycle

The seven clauses of ISMS requirements described in the previous section do not directly follow logical phases that organizations typically follow for security management. In contrast to the logically connected functions of NIST CSF described in Sect. 18.2, the clauses of ISO/IEC are not designed to be logically connected, but to provide complete and complementary set of requirements for managing information security. For better understanding, it can be helpful to place the clauses of from ISO/IEC 27001 in a cycle of separate information security functions. Figure 18.9 is a typical interpretation of the ISMS requirements in terms of a process cycle of four phases. The activities of the different phases are intended to run concurrently in parallel.

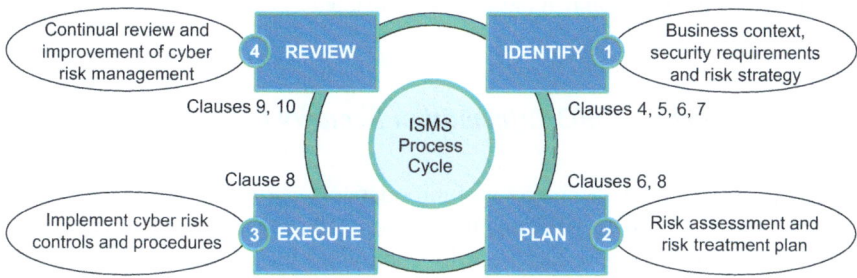

Fig. 18.9 Interpretation of the clauses of requirements from ISO/IEC 27001 in terms of an ISMS process cycle

References to clauses 4–10 are placed adjacent to the phase of the ISMS cycle in Fig. 18.9 where they most naturally belong. It is difficult to try to make the placement of clauses perfectly logical. For example, clause "8. Operation" covers both risk assessments as well as the implementation and operation of security controls. The reason for this relative misfit is that ISO/IEC 27001 is a so-called "management system standard" where the structure of the clauses described in Sect. 18.5.2 is mandatory for all ISO management standards mentioned in Sect. 18.5.1. That structure of clauses has been designed to be generic for all management system standards, but might not be the most intuitive for information security management. The advantage of a standardized structure of clauses is that it simplifies an organization's efforts when implementing multiple management standards.

The benefit of working according to an ISMS is that the organization achieves a consistent, predictable, and measurable information security posture, which is documented and can be audited. The more comprehensive the ISMS, the more maturity the organization has achieved in information security governance.

An organization that meets the requirements of the standard may choose to be audited by an accredited certification body, which after successful audit can certify the organization according to ISO/IEC 27001. The fact that an organization is 27001 certified gives an indication of good information security management, which may provide market advantages or may even be required by other organizations for participation in business partnerships and competitive bidding.

Note that an organization with a mature ISMS, and which may even be 27001 certified, may still suffer serious security incidents. The difference is that an organization with high maturity probably experiences fewer serious security incidents, and that the organization is better prepared when an incident occurs, than an organization with low maturity. If, after a serious security incident, there are questions about compensation in the event of losses, or sanctions due to non-compliance of laws and regulations, the organization with high maturity in security governance will be in a better position than the organization with low maturity. In extreme cases of serious security incidents resulting from lack of due care for information security, senior management may be charged with negligence and punished by paying compensation or with imprisonment, as described in Chap. 17.

18.5.4 ISO/IEC 27002 Information Security Controls

ISO/IEC 27002:2022 describes a set of information security controls that an enterprise can use as a basis for selecting those controls that best reduce its cyber risks. The standard explains how each security control works, what its purpose is, and how it can be implemented. The standard is intended to be used in conjunction with an ISMS based on ISO/IEC 27001. Not all controls are relevant for every organization, controls are selected according to need for managing and reducing cyber risk.

The standard has changed over the years. ISO/IEC 27002:2022 describes 93 security controls across four domains: organizational (37 controls), people (8 controls), physical (14 controls), and technological (34 controls), as shown in Fig. 18.10. The previous edition from 2013 had a completely different structure which followed operational categories in the same way as *CIS Critical Security Controls* described in Sect. 18.4.

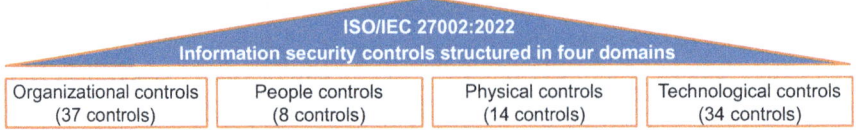

Fig. 18.10 Domains of security controls specified in ISO/IEC 27002:2022

18.5 ISO/IEC 27000 Series of Security Standards

All the controls in ISO/IEC 27002:2022 are described in a uniform way for simple reading. A useful aspect of each control description is *attributes,* which make it easy to select security controls based on specific needs. The control domain in ISO/IEC 27002:2022 do not follow the natural phases of information security management, such as *Identify, Protect, Detect, Respond* and *Recover* functions in NIST CSF. The purpose of the attributes is precisely to make it easy to find security controls that are related to specific aspects. By sorting security controls by specific attributes, one can create one's own categorization of the controls, such as the security functions used in NIST CSF described in Sect. 18.2 or to the operational categories of CIS Controls described in Sect. 18.4.

As an example, attributes are shown for security control *8.13 Information backup* from ISO/IEC 27002:2022. Attributes (market with #) are specified in a table for each control, where Table 18.4 shows attributes for the particular security control 8.13. The text under "Control" and "Purpose" below also provides a brief explanation of the control and its purpose.

Table 18.4 Attributes of control 8.13: Information backup

8.13 Information backup				
Control type	Information security properties	Cybersecurity concepts	Operational capabilities	Security domains
#Corrective	#Integrity #Availability	#Recover	#Continuity	#Protection

Control
Backup copies of information, software and systems should be maintained and regularly tested in accordance with the agreed topic-specific policy on backup.

Purpose
To enable recovery from loss of data or systems.

The text under the terms "Control" and "Purpose" corresponds to "outcome" in the jargon of NIST CSF Core. Hence, security control 8:13 above corresponds to security outcome PR.DS-11 of NIST CSF shown in Table 18.2. Control 8:13 also corresponds to safeguard 11.1 of CIS Controls shown in Fig. 18.7. Each security control in ISO/IEC 27002 comes with *Guidance* which provides suggestions for how the control can be implemented. In comparison, NIST CSF typically provides guidance in the form of specific examples.

18.5.5 27000 Family of Standards

The 27000 series of ISO/IEC standards for information security has gradually grown large and varied. The strong dynamics of international standardization in information security reflect the importance of the field of information security in the world. Figure 18.11 shows a selection of the standards in the 27000 series as of 2024. ISO27001security[6] is a website that provides a description of all the security standards of the 27000 series.

---
[6] https://www.iso27001security.com/

Fig. 18.11 The family (selection) of standards of the 27000 series

The series of standards on the bottom right side of Fig. 18.11 addresses the implementation of ISMS and information security in various industrial sectors. The standards in the lower left corner of the figure focus on auditing and certifying businesses with respect to ISMS and other security-related activities. The standard *ISO/IEC 27005 Information security risk management* is described in Chap. 19.

It typically takes several years to develop an international ISO/IEC standard, and it is a relatively bureaucratic process in which representatives from all over the world participate. Although the process is thorough, standards can have weaknesses. As technology also changes rapidly, there is regulaarly a need to revise the standards. When an organization considers to include a standard in its governance and management of information security, it is important to consider alternatives. Alternatives to the 27000 series are especially the CSF and other guidelines in the SP 800 series from NIST.

18.6 Maturity in Information Security Management

The information security posture of an organization cannot be measured directly. In practice, the security posture can only be measured indirectly by assessing the quality of management processes of information security. This is called "tiers" in the NIST CSF described in Sect. 18.2.3.

Capability Maturity Model Integration (CMMI) is a method for measuring the quality of processes. CMMI was originally developed by Carnegie Mellon University (CMU), and is managed by the CMMI Institute, which is a subsidiary of ISACA. An adaptation of CMMI for information security management is shown in Fig. 18.12.

18.6 Maturity in Information Security Management

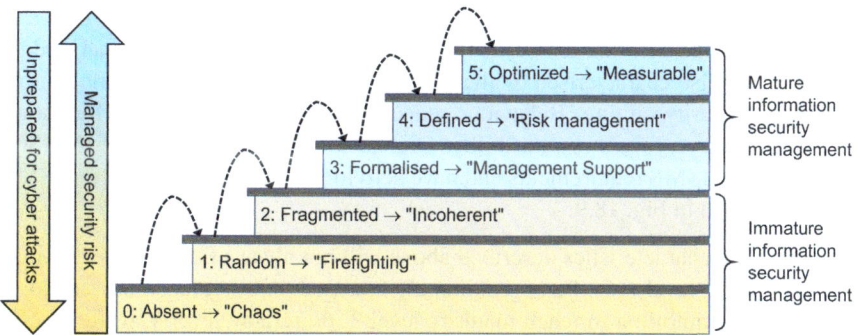

Fig. 18.12 Maturity in the management of information security

The levels of the maturity model in Fig. 18.12 is explained below.

- **Level 0: Absent information security management process.** Characterized by chaos. The organization is defenceless against most security threats.
- **Level 1: Random process.** Characterized by firefighting in the aftermath of incidents. Organizations with maturity level 1 may have some information security expertise, but the approach is characterized as ad hoc where little or nothing is planned. There is no documentation.
- **Level 2: Fragmented process.** Characterized by being incoherent. Organizations with maturity level 2 generally have a reactive approach, but may have systematic requirements for information security within isolated areas and routines for incident management. Some documentation exists.
- **Level 3: Formalized process.** Characterized by having management support and some form of Information Security Management System (ISMS). Organizations with maturity level 3 have committed leaders and clearly defined roles and organizational structure related to information security. Information security requirements are mapped relatively well, incident management is organised and processes are documented.
- **Level 4: Systematized process.** Characterized by being based on risk management in the sense that risk assessments form the basis for information security requirements. Organizations with maturity level 4 have well-established processes for monitoring threats and carrying out risk assessments. Prioritisation and implementation of security controls are followed up with adequate budgets. Contingency planning is established, and incident management is organized accordingly. The board and senior management are informed and committed.
- **Level 5: Optimized process.** Characterized by the fact that the enterprise ensures continuous improvement of the information security management system (ISMS). For example, the benefits of security controls can be measured with

approaches as e.g. described in ISO/IEC 27004 [54], and the quality of documentation and information security processes assured through internal audits. Level 5 is also characterised by the fact that the organization has a strong information security culture. Behaviour reflects an understanding that information security is everyone's responsibility. Organizations at this level meet the ISO/IEC 27001 ISMS requirements and have activities within all phases of the ISMS cycle shown in Fig. 18.9.

Naturally, the characteristics described above will not always fit with a given organization, so its level must be approximately assessed. All organizations should at least have the ambition to reach maturity level 3, preferably level 4, while level 5 can be an ideal to work towards, even if that goal might never be fully reached.

One reason for at least achieving maturity level 3 with management support is to establish a formal link to the board and executive that are legally accountable for ensuring due care in information security management. Many laws require organizations to have an information security management system, as described in Chap. 17. about cybersecurity regulations. For the Information Security Manager (CISO), it should be a high priority in a change of leadership that a new CEO is allowed to quickly audit and sign the most important security policies, so that this is not just something that remains left behind after the previous CEO for months and years. If a business is subject to laws that require security controls to be selected on the basis of risk assessments, as is the case with the GDPR in the EU, the organization should be at maturity level 4.

18.7 Tasks

1. Information Security Standards

    Access the portals for security standards/guidelines from NIST and ISO

    - NIST SP800 (Special Publications) series: https://csrc.nist.gov/publications/sp
    - ISO 27000 Series: https://www.iso27001security.com

    (a) Try to identify similar standards in NIST SP800 series and the ISO/IEC 27000 series.
    (b) What are reason why different organizations develop separate sets of similar standards?

2. NIST Cybersecurity Framework

    (a) What is an outcome in the terminology of NIST CSF?
    (b) What are profiles?
    (c) What are tiers?
    (d) How are NIST CSF, the Cyber Defense Matrix and CIS Controls related?

3. ISMS
    (a) How are ISO/IEC 27001 and ISO/IEC 27002 related?
    (b) What is meant by "system" in the term "Information Security Management System"?
    (c) Which of the above standards forms the basis for certification and why?
    (d) How should an organization decide what security controls to implement?

Chapter 19
Cyber Risk Management

Risk comes from not knowing what you're doing.[1]

Warren Buffett, American business leader

Risk management is a key element in information security management, where the goal is to reduce the likelihood of adverse impacts resulting from cyber incidents. Cybersecurity risk management is the process of identifying, assessing, prioritizing, and mitigating risks resulting from exposure to cyber threats that can compromise an organization's information assets. It involves a systematic approach to identifying and understanding threats, vulnerabilities and risks, followed by managing the risks by reducing their likelihood and impact. In addition, all this needs to be done in a cost-effective way, meaning that the cost of reducing a risk should never be greater than the value of the risk reduction it achieves. Hence, cyber risk management requires both technical and business skills to be well aligned with business objectives.

In this chapter, you learn how cybersecurity risk management is a process of multiple steps, where *risk assessment* is the step of mapping relevant risks and their magnitude, and *risk treatment* is the step of managing those risks. You learn how an organization can perform risk assessments, and how this forms the basis for managing and reducing the risk to an acceptable level.

19.1 Interpretation of Risk and Risk Management

In everyday speech the word "risk" is used with different meanings. We intuitively understand the different meanings of the word by interpreting it from the context. Below are some examples.

(a) *"There's significant risk in buying bitcoin."*
(b) *"There is little risk in putting the money in the bank."*
(c) *"There's a risk of rain for our garden party on Saturday."*
(d) *"Ransomware poses a significant risk."*

To clarify the meaning of "risk" in each statement above, the statements can be paraphrased by replacing the word "risk" with a more precise term, as shown below.

(a) *"There is significant uncertainty about profits or losses of buying bitcoin."*
(b) *"There is little uncertainty about the profits or losses of putting the money in the bank."*
(c) *"There's a likelihood of rain for our garden party on Saturday."*
(d) *"Ransomware creates a significant likelihood of negative impacts."*

The statements above reflect that "risk" can be interpreted *as uncertainty* (about objectives) or as *likeihood* (of negative impacts). These two interpretations can be found in definitions of risk in various standards.

19.1.1 Definition of Information Security Risk

The most general standard for risk management is *ISO 31000:2018 Risk Management* [55]. This international standard defines *risk* as follows:

> *Risk is the effect of uncertainty on objectives.*

This definition is so abstract that it can almost seem meaningless. However, a general definition of risk must necessarily be sufficiently abstract to cover all interpretations of risk in different contexts. In the financial sector, uncertainty about future profits or losses from investments is a key aspect of risk, where the definition above fits very well.

In the context of information security, the standard *ISO/IEC 27005 Guidance on managing information security risks* [56] provides several interpretations of risk that are more precise than the definition above. The interpretation below from ISO/IEC 27005 is a good working definition of *information security risk*:

> *Information security risks can be associated with the potential that threats will exploit vulnerabilities of an information asset or group of information assets and thereby cause harm to an organization.*

The interpretation of risk above is illustrated in Fig. 19.1. Here, risk consists of a relevant threat that exploits a vulnerability with the consequence that an asset gets harmed.

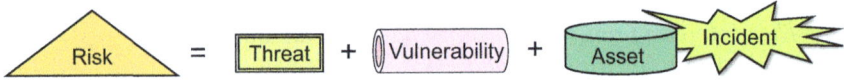

Fig. 19.1 Risk consisting of a threat that exploits a vulnerability and harms an asset

19.1 Interpretation of Risk and Risk Management

Be aware that there are many risks, each of which consists of a relevant combination of threat, vulnerability(ies) and impact on asset(s). An important part of the risk assessment is precisely to identify relevant risks in this way, as described in Sect. 19.4.

The ISO/IEC 27005 standard also provides a definition of *information security risk level*:

> *Risk level is the significance of a risk, expressed in terms of the combination of consequences and their likelihood.*

The interpretation of this definition of risk is illustrated in Fig. 19.2. Here, a risk level is calculated based on the likelihood of a security incident and the incident's impacts. The term *risk exposure* is often used as a synonym for risk level.

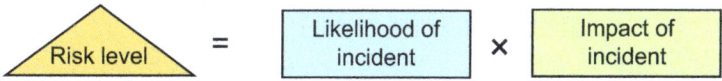

Fig. 19.2 Risk as a combination of the likelihood of an event and the impact of the incident

This interpretation of information security risk is consistent with the US standard *NIST SP 800-39 Managing Information Security Risk* [57], which provides the following definition of information security risk:

> *Risk is a measure of the extent to which an entity is threatened by a potential circumstance or event, and typically a function of: (i) the adverse impacts that would arise if the circumstance or event occurs; and (ii) the likelihood of occurrence.*

During the risk assessment, the assessment of risk level is precisely based on the combination of likelihood and impact of a security incident, where the assessment can be made in a qualitative, quantitative or relative manner, as described in Sect. 19.5.

Models for the interpretation and definition of information security risk above are described in the next section.

19.1.2 Information Security Risk Models

Information security risk models are useful for explaining and visualising what risk means, and for analysing risk. The interpretation of the risk triangle in Fig. 1.9 is that security risks arise when there are vulnerabilities that can be exploited by threats and thus harm assets. The figure also shows that the risk level (size of the triangle) increases proportionally as a function of (i) the quantity/magnitude of the assets, (ii) the severity of the vulnerabilities and (iii) the amountr/strength of the threats. The terms "threat" and "vulnerability" are described in Sect. 1.4.

A more detailed visualization of the relationship between security risks and related concepts is shown in Fig. 19.3. This visualisation provides a deeper understanding of security risk dynamics and is suitable as support for practical assessment and analysis to determine risk levels.

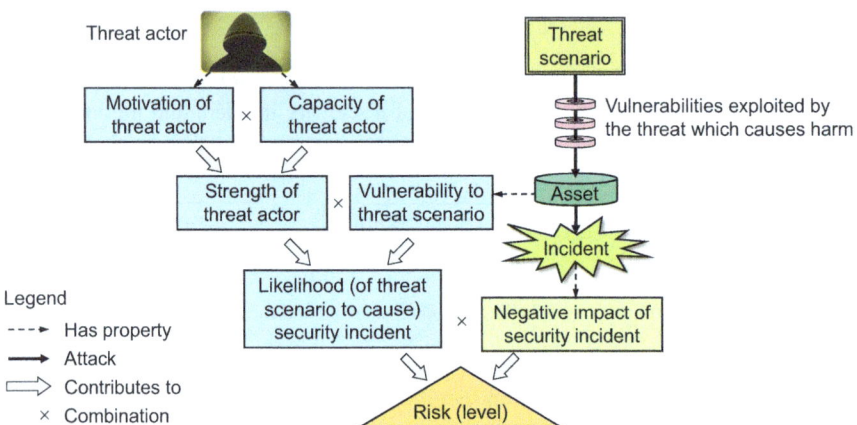

Fig. 19.3 Detailed security risk model

The interpretation of Fig. 19.3 starts with the fact that there is a *threat actor* which has a certain *motivation* and a certain *capacity* to attack an *asset* based on a *threat scenario*. Be aware that there is not always a clear threat actor with an intended attack objective. For example, data center fires can be caused by errors and mishaps. In the context of information security risk management, however, great emphasis is placed on adverse incidents resulting from intentional actions by a threat actor. The threat actor's strength can be defined as the combination of motivation and capacity. There can be a degree of vulnerability to the threat scenario that allows the threat actor to harm assets and thus cause an adverse security incident. The likelihood of a security incident can be defined as the combination of the threat actor's strength and the degree of vulnerability to the threat scenario. At this stage, the severity of the impacts resulting from the incident must be estimated. Finally, the risk level can be determined as a combination of the likelihood of the incident and its impact.

It is interesting to reflect on the elements of Fig. 19.3. Through digitization, more and more business processes, and at the same time their inherent vulnerabilities, are exposed online, which in turn helps increase threat actors' motivation.

Threat actors' motivation is largely influenced by a high expected return on investment (ROI). Cyberattacks are relatively low cost, can easily be carried out beyond the reach of the law and can yield great gains. Capacity can largely be based on increasing attacker expertise and freely available advanced tools or attack services for hire.

The threat actor's advantageous ROI in executing attacks is not necessarily reflected by a similarly advantageous ROI of organizations wanting to remove vulnerabilities to counter attacks. One example is supply chain attacks, where the threat actor is motivated by the value of an asset for the end user, while the vulnerability exploited in the attack may lie with a subcontractor that does not see the value of the asset for the end user and therefore does not have sufficient incentives to remove the vulnerabilities. This is in fact a cybersecurity market failure.

The challenge here is that the subcontractor often has a poor prerequisite, or lack of interest, for assessing the value that the subcontractor's end-user customers represent for a threat actor.

Another motivation for a threat actor is to use the company's assets or activity as a means to attack an entirely different party. An example would be attacking the navigation system of a large ship so that it runs aground in a narrow and important trade route such as the Suez Canal.

Figure 19.3 shows where measures can be taken to mitigate risks. The most common way to lower risk levels is to remove vulnerabilities to threat scenarios. Even if a threat actor is motivated and has significant capacity, attack attempts will fail if systems and networks do not have vulnerabilities that can be exploited.

In principle, the risk level can be lowered if it is possible to reduce the motivation of threat actors. Initially, it seems difficult to control a threat actor's motivation, but it may still be relevant. Political conflicts and tensions between states and political groups can contribute to an increase in politically motivated *cyber operations*. Unethical actions carried out by businesses can provoke *cyberhacktivism*. Although it may theoretically be possible to influence the motivation of threat actors, the effect of such controls will usually be unpredictable. *Cyber deterrence* is a control that is actually used to lower the motivation for cyberattacks among opponents. Cyber deterrence is typically part of the cyber strategy of prominent cyber powers such as the United States, as described in Sect. 16.3.2.

Finally, it can be mentioned that good readiness and incident response help to reduce the impacts of a security incident. Even if an incident does occur, businesses with good incident response are better able to recover from the damage and restore business processes quickly. Incident response is described in Chap. 14.

19.1.3 Definition of Information Security Risk Management

There are several different standards and frameworks for risk management, each with different definitions. The standard *ISO 31000:2018 Risk Management* defines *risk management* as follows:

> *Risk management consists of coordinated activities to direct and control an organization with regard to risk.*

Since ISO 31000 is a general standard, the above definition is necessarily relatively abstract. NIST SP 800-39 [57] provides the following definition, which is more specific and relevant to information security:

> *Risk management is a comprehensive process that requires organizations to: (i) frame risk (i.e., establish the context for risk-based decisions); (ii) assess risk; (iii) respond to risk once determined; and (iv) monitor risk on an ongoing basis using effective organizational communications and a feedback loop for continuous improvement in the risk-related activities of organizations.*

There is a general consensus on what risk management is and what should be included in the risk management process. The entire risk management process is relatively extensive, as described in the next section.

19.2 Risk Management Process

An information security risk management process is described in the standard ISO/IEC 27005 [56]. This process consists of phases and decision points, as shown in Fig. 19.4.

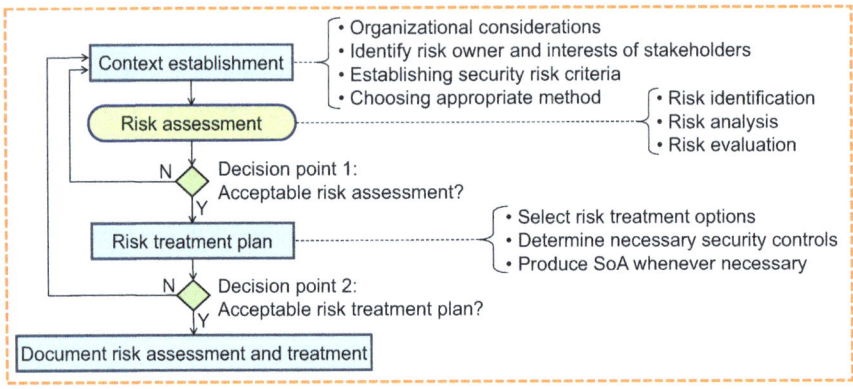

Fig. 19.4 Information security risk management process (ISO/IEC 27005)

It is not always necessary or appropriate for a business to carry out the entire process in Fig. 19.4. It may be sufficient to carry out parts of the process if, for example, the company has only one, or a selection, of the following objectives:

- Stay updated on cyberthreats
- Get an overview of vulnerabilities
- Get an overview of existing risks
- Point out relevant security controls

- Assess residual risk by assuming new proposed security controls
- Provide a basis for buying cyber insurance
- Provide a basis for specifying a budget for risk management

By carrying out the entire process, all of the above objectives can be achieved. The different phases of the risk management process in Fig. 19.4 are described in the sections below.

19.2.1 Context Establishment

Activities in the risk management process naturally need to be planned. Elements of the planning include

- Put together teams responsible for carrying out risk assessment and other activities
- Specify who is the risk owner and the focus area for risk assessment
- Identify requirements from standards, legislation, sector-specific frameworks and policies
- Specify a threshold level and/or a set of acceptable risk criteria
- Choose a method or approach for risk assessment
- Select a risk analysis model to determine risk levels
- Indicate the level of detail for identifying assets and resources

A risk management team should have a representative composition. For the risk assessment, owners of assets and resources can help identify these and specify the importance of security goals (CIA) for each asset/resource. Users and security experts can help identify threats and vulnerabilities and assess incident likelihoods. Experts in risk assessment and analysis can help guide the process. Security experts can help advise on security controls. Management must review risk assessment reports and proposed security controls, and approve the budget for the controls. An expensive security control is, for example, a top model of a firewall that can cost more than US$100,000. A security control with impact on usability may be that all employees must use two-factor authentication. As a rule, there will be a discussion about whether such controls are worth the cost or the reduced usability.

A threshold level for acceptable risk means that risks below the threshold value can be accepted, while risks above the threshold value should be managed so that the risk is modified to become acceptable.

In order to protect assets, one must know their values. Identification of assets and resources is done as part of the risk assessment, which we describe in Sect. 19.3. However, already during planning it is useful to hint at how detailed asset identification should be done. Experience shows that asset identification can be both time-consuming and frustrating. If the team tries to make asset identification very detailed, they risk never reaching the goal. Asset identification should therefore be done relatively coarse-grained.

19.2.2 Risk Assessment

The risk assessment phase is the main element of the risk management process. This phase is relatively extensive, and is described in detail in Sect. 19.3. We mention here only that risk assessment consists of the following steps:

- **Risk identification** – mapping of all relevant risks
- **Risk analysis** – calculation of levels for each risk
- **Risk evaluation** – ranking and prioritisation of risks in relation to threshold value and criteria

After the risk assessment, decision point 1 follows. This decision point considers whether the result of the risk assessment is satisfactory. It is possible that the assessed risk levels have a very skewed distribution, for example that most of the risks are at the same level (e.g. all have only very low levels or only very high levels), which makes it quite difficult to rank and prioritise the risks. If necessary, the calibration and estimation of likelihoods and impact levels can be changed, or the model for risk analysis can be changed, to achieve a better spread of risk levels. Keep in mind that qualitative or relative risk levels have no absolute significance. They should only be a basis for ranking and prioritising risks and for being able to compare with a threshold level.

Another reason to consider the report from the risk assessment to be unsatisfactory is if there is significant uncertainty about one or more high risks. In principle, a high risk must be prioritised in order to be managed, which will typically entail a cost. Uncertainty about a high risk thus means that there is doubt as to whether management of the risk is necessary. It can be difficult to reduce uncertainty around risks, but if possible, an attempt can be made to make a more thorough assessment of some risks e.g. based on statistics from the industry sector.

Risk assessment is one of the three sources of information security requirements, as explained in Sect. 1.7. This means that the implementation of security controls is partly based on risk assessments, with the aim of managing and reducing cyber risk to an acceptable level.

19.2.3 Risk Treatment

The result of the risk assessment phase is a ranked and prioritized list of risks. Priority risks are those that are considered unacceptable. What is acceptable or unacceptable may be based on comparison with a predetermined threshold level or on the basis of a separate assessment of risks that have a relatively high level. The risk treatment phase consists of proposing and planning how priority risks should be treated. The four relevant treatments are: (1) reduce risk, (2) share or transfer risk, (3) retain risk and (4) avoid risk, which are described below.

1. Reduce risk

Reducing risk is normally done by implementing security controls, where relevant controls are described in various guidelines and standards described in Chap. 18. Introducing a security control will always entail a cost, which in a business context is an investment. To determine whether the investment pays off, it is initially necessary to estimate the gain from introducing the security control. If we assume that it is possible to estimate both the cost and the benefit of introducing a security control, we can calculate its ROI (Return On Investment). The formula for calculating ROI is given below:

$$ROI = \frac{\text{Gain} - \text{Investment}}{\text{Investment}}$$

In our case, investment is the same as the cost of introducing a security control, while gain is the same as risk reduction, where both are measured in money. Figure 19.5 illustrates what it means for a security control to have positive, negative, or neutral ROI.

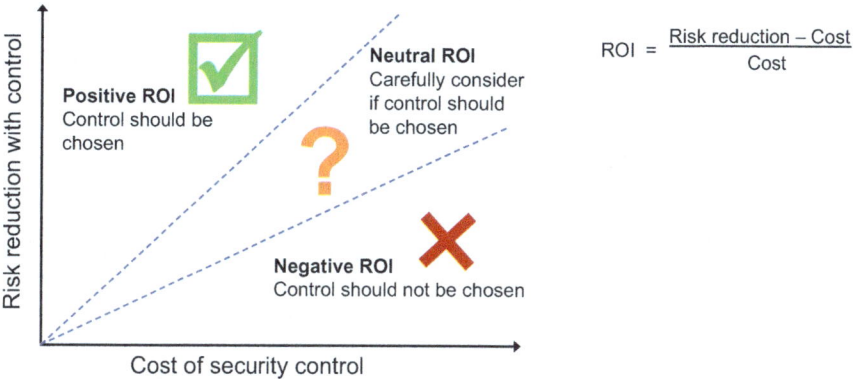

Fig. 19.5 ROI (return on investment) when introducing security controls

The ROI model for assessing security controls illustrated in Fig. 19.5 is relatively theoretical, and not directly applicable in practice. This model assumes that it is possible to estimate the expected nominal value of a risk reduction measured in money, but in practice this may be difficult other than with very rough estimates. Figure 19.5 is therefore intended to illustrate the basic principle for choosing security controls, namely that expensive controls can be chosen if they can be assumed to have a significant effect in reducing security risk. If only a moderate effect is expected from a security control, it should not entail a particularly large cost to implement. However, there are controls where estimating ROI can be done relatively accurately. Controls that reduce the time it takes to detect and handle an unwanted incident can, in principle, be measured, both through exercises and red-teaming and actual incidents, and provide a basis for good estimates.

When considering a technical security control for reducing risk, it is useful to assess the overall costs, not just the nominal cost of buying a product. It is useful to consider the technology from the dimensions of the four P's (People – Product – Partner – Process) described in Sect. 1.6. More specifically, when buying a product, the organization needs people with the skills to operate it, needs support from the vendor or other partner, and must have a process for operating the product.

In principle, controls for each risk are assessed separately, but it is often possible that a single control will be able to remove several vulnerabilities and thus affect several different risks at the same time. Upgrading underlying system components to remove vulnerabilities can also reduce *technical debt* in systems, although technical debt does not necessarily pose a security risk. Technical debt is expected future additional costs as a result of choosing easy, short-term solutions instead of good, long-term solutions.

2. **Share or transfer risk to others**

This risk treatment option is intended to share or shift risk to external entities or organizations. It can often be a sensible choice to transfer the risks associated with the operation of complex systems to another organization with greater expertise in managing these risks. Buying cloud services is an example where part of the risk of information security is transferred to the cloud provider. As a rule, a cloud provider will be able to manage the risk in a far more cost-effective way than the organization that buys the cloud service.

Alternatively, your organization can purchase cyber insurance against potential losses caused by security incidents. The purpose of cyber insurance is to protect the business against risks related to information security and privacy in the operation and management of IT systems and services. Risks of this type are usually excluded from traditional general insurance policies. The coverage offered through cyber insurance may include direct losses resulting from data corruption, extortion, theft, hacking, and denial-of-service attacks. It can also cover fines and legal costs for breaches of legal compliance, for costs associated with covering losses incurred by third parties caused by errors and deficiencies in the organization's infrastructure, for damages in the event of defamation, for security audits and for investigation costs after security incidents.

Cyber insurance is becoming increasingly common. The insurance company typically conducts an audit of a potential customer prior to entering into an agreement to assess the maturity of the governance and management of information security in the business. The premium for a cyber insurance policy will depend on the assessed maturity, in the sense that low maturity will incur a higher premium. Here, a positive side effect is that the organization gets an incentive to improve its cybersecurity governance, where reducing the insurance premium is a contribution to the gain from the improvement.

3. **Retaining risk**

Retaining a specific risk means that the business does nothing further to treat the risk. Accepting a risk as it is must be justified on the business understanding the risk and being aware of the potential impacts of cyberattacks.

This alternative is justifiable only when a risk is below the threshold level for acceptable risk, or if the costs associated with introducing controls or other treatment would otherwise be disproportionately high in relation to the level of risk.

Here there will also be cases where the organization chooses to keep risk temporarily, i.e. that a plan exists for fixing the vulnerability in the future, e.g. when a system upgrade is planned anyway. It will then be important to follow up the risk while it exists, e.g. with temporary mitigations.

Risk appetite describes the extent to which the organization is willing to accept risk as a trade-off in relation to the costs of reducing the risk or passing it on to others.

4. **Avoiding risks**

Avoiding a specific risk normally means stopping the activity or business process that is causing the risk. Any business activity carries some risk, but the benefits and profits of the activity normally outweigh the risks.

However, in special situations, it may be necessary to stop the business activity that entails risks. This will be the case if the risk is disproportionately high compared to the benefits and profits, and at the same time it is not possible to reduce, mitigate or transfer the risk in a cost-effective manner.

From a holistic perspective, there is also a risk associated with not conducting a business activity, because it takes away all opportunity for the benefit and profit of the activity. This risk should also be taken into account before the business chooses to stop a business activity to avoid risk.

Discussions of this type typically lie outside the area of information security risk treatment itself, and belong to the overall area of enterprise risk management. This is important to be aware of in the dialogue with business management if the issue of avoiding risk becomes relevant.

19.2.4 Risk Treatment Plan and Accepted Risk

A plan is drawn up for treating identified risks that are above the threshold level, or that are proposed to be addressed on the basis of other criteria. This plan will also include a proposed budget. When submitting the plan to management, we come to decision point 2 in Fig. 19.4.

At decision point 2, management decides whether the risk treatment plan is satisfactory. It is possible that the proposed plan for treating risks is unacceptable, for example, in the following cases:

- The risk treatment plan is too expensive or too ambitious. Management may decide to raise the acceptable risk level, so that it will be sufficient to introduce fewer and less ambitious security controls.
- The levels of some residual risks are considered to be too high to be accepted. If management considers the residual risk to be too high, additional security controls may be proposed that will typically require a larger budget.

- The estimated cost of the risk treatment plan is considered to be misleading due to misguided risk assessment or miscalculations of control costs such that risk assessment and/or risk treatment plan must be revised.

If the treatment plan cannot be accepted in the first instance, the management and the responsible representative for the risk assessment must jointly agree on ways to improve the plan.

Once the risk treatment plan has been accepted, the plan can be implemented. *Residual risks* are those risks that remain and are accepted after treatment.

19.3 Risk Assessment Process

The risk assessment within a focus area should be carried out regularly (e.g. once a year, as needed or requested, etc.). A risk assessment provides a temporary overview of identified risks, more or less fully described, including the calculated risk level.

Risk assessment can be carried out in several iterations – the first at a high level to identify high risks, while the others detail the analysis of the major risks and other risks of particular interest. The risk assessment process is illustrated in Fig. 19.6. The yellow oval is an expanded version of the yellow oval in Fig. 19.4.

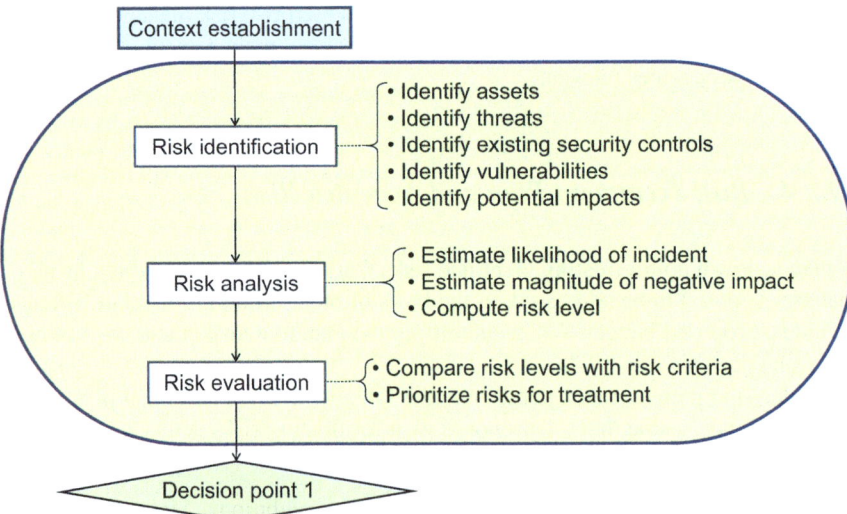

Fig. 19.6 Risk assessment process

The steps in the risk assessment process are described in the next three sections.

19.4 Risk Identification

A simple risk consists of a relevant threat that can exploit a vulnerability to causes a security incident on an asset, as illustrated in Fig. 19.1. A security incident may consist of a breach of one or more CIA security objectives (confidentiality, integrity and availability) or some other adverse impact on an asset.

In order to identify risks, it is therefore necessary to be able to identify relevant threats, vulnerabilities as well as the assets that can be harmed by the threats. If we had lists of all relevant threats, all relevant vulnerabilities and all relevant assets that can be harmed, we could identify risks by mapping correlations, as illustrated in Fig. 19.7. However, it is not very practical to create large lists of threats, vulnerabilities and assets, and then try to find risks in the form of relevant combinations.

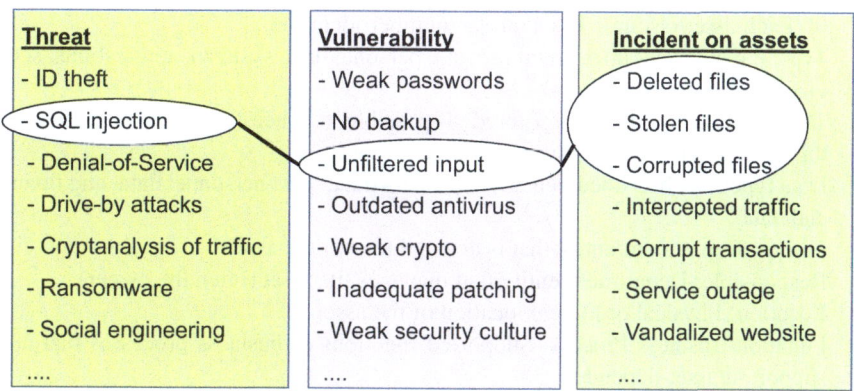

Fig. 19.7 Principle of risk identification

Experience shows that an effective approach to identifying risks may be to focus on threat scenarios first. This can be done by asking questions like *"What could go wrong here?"*, *"How can someone attack or abuse this application?"* or *"How can someone gain unauthorized access?"*. If a relevant threat scenario can be identified, the assets that get compromised will naturally stand out. To put it bluntly, we could almost say that it is unnecessary to map assets because relevant assets will stand out when we identify threat scenarios. However, it is important to have an overview of the most important assets that can be harmed, because this makes it easier to identify relevant threat scenarios where precisely these assets are exposed.

The lesson is that it is important to identify both assets and threats. This is described below, as well as the identification of impacts.

19.4.1 Identification of Assets

Identifying assets is traditionally defined as the first step of risk assessment. It can be challenging to set the right level of detail for registering assets. It should not be too detailed, otherwise the work will become overwhelming. It should also not be too general, otherwise the register will have no practical use.

It is often a good approach to map out *asset classes* as opposed to specific things. Each asset or asset class can be described with a set of *attributes*. There are standard templates and spreadsheets that can be used. Owners of assets in an organization are often best placed to identify assets and their attributes, as indicated by Fig. 18.5.

Assets that are mapped should be described systematically with a common set of attributes. The following list is an example of attributes that can be described for each asset.

- Id: Each asset/resource has a unique number/identifier.
- Type: e.g. physical infrastructure, data, personal data, systems, applications, services and employees (roles).
- Things or class: What is registered? If possible, it is helpful to register a class of things as a whole, rather than as specific things.
- Data types: such as documentation, process data, logs, personnel data, and financial data.
- Owner: Department/entity that is deemed to own the asset.
- Responsible: Department/entity that manages the asset (often the owner).
- Location: Physical or logical location of the asset.
- Function/Business Process: Supported functions or business processes that are supported/asset-dependent.
- Data classification: Sensitivity/classification of data.
- Impact assessment: It may be useful to define the degree of impact of a breach of security objectives (CIA) for the asset.

Not all attributes are relevant to all types of assets. Only relevant attributes need to be described for each identified asset.

19.4.2 Threat Modeling: Identifying Threats

Threat modelling consists of identifying, analysing, and describing potential threat scenarios, i.e. ways in which the organization's information assets could be attacked and harmed. Threat modelling for IT systems is described in Sect. 11.5 where Fig. 11.5 illustrates various interfaces of typical IT infrastructures. Each interface constitutes a possible attack channel, where the set of possible attack channels represent the threat surface.

The purpose of threat modelling is to help the owners of systems and applications see where potential weaknesses and vulnerabilities lie, so that these can be

19.4 Risk Identification

removed. Threat modelling answers questions such as: *"Where are we most vulnerable to potential attacks?"* and *"What do we need to do to protect ourselves from these threats?"*

In addition to asking questions like those above, there are concrete techniques where traditionally three approaches are often mentioned. These are briefly described below.

- **Attacker-centric threat modelling:** This approach is based on the attacker's objectives and considers options for how the attacker can achieve them. A useful question to ask is *"What is the attacker most interested in and how can they achieve it?"*. It is important to think about what threat actors are interested in, not just what is important and valuable for the organization. STRIDE (described in Sect. 11.5.4) is an example of an attacker-centric approach. A related technique is to draw an *attack tree*, which consists of a tree upside-down where the root is the attacker's objective, and the branches are (alternative) sequences of steps that the attacker must perform to reach the target.
- **Technology-centric threat modelling:** This approach is based on looking at the architecture of systems and networks, trying to follow the dynamics and logic of the technology in search of relevant attacks that can exploit vulnerabilities in various technology components. A useful question to ask is *"Where are we most vulnerable to attack?"*. This approach is used, for example, for threat modelling in *Microsoft SDL*, described in Sect. 11.1, and in *OWASP Top 10*, described in Sect. 11.6.3.
- **Asset-centric threat modelling:** This approach draws on assets considered particularly sensitive or important, and to identify how these can be harmed. A useful question is *"What assets are most important to us, and how can they be attacked?"*. Impacts of an attack against a specific asset depend on which CIA security goal(s) is/are breached in the attack.

The three approaches above can and should be combined, for none of them is considered complete more effective than another, in isolation.

A general approach is to define a set of core threat scenarios that are considered most relevant. A relatively small number, perhaps only about five, different threat scenarios can provide good coverage for the most relevant attack types one needs to consider. The attack vectors described in Chap. 2 represent general threat scenarios. In addition, inspiration can be drawn from typical events reported in the press and in reports on the current threat landscape. Governments in many countries produce threat assessments, some of which are published openly. Many private cybersecurity consulting firms also produce and publish these types of reports, which may be open or may be purchased. Open threat reports are also published by international organizations, such as ENISA.[1]

It often happens that a specific sector is subjected to a campaign of attacks over a period of time. As the threat landscape is constantly changing, it is challenging to

---

[1] https://www.enisa.europa.eu/topics/cyber-threats/threats-and-trends/enisa-threat-landscape

constantly keep up to date with the most relevant threats for your own business. Reading threat reports makes it easier to follow the changing threat landscape.

19.4.3 Identification of Impacts

An incident can lead to a breach of one or more CIA security goals for assets, where assets can be, for example, operational data, personal data, system, networks, applications, processes, services, etc.

The impact of an incident can have various aspects, where the following are typical impact aspects:

- Lost revenue and profit
- Impaired performance of service
- Breach of legal compliance
- Damaged reputation
- Costs of recovery
- Legal costs
- Litigation

The severity of each impact aspect will always vary. For a given event and asset, some impact aspects will be relevant, while others are irrelevant. Of the relevant aspects, some are more serious than others. When estimating the impact of an incident, all impact aspects shall be considered as a whole. This estimation can be challenging because it is not obvious how different aspects should be combined.

19.4.4 Risk Description

Identified risks should be systematically recorded using a common set of attributes. This can be part of a spreadsheet for risk analysis. The next step after risk identification is precisely risk analysis. The following list is an example of attributes that can be described for each risk.

- **Id:** Each risk has a unique identifier.
- **Threat:** How can anything happen? What steps are included in the scenario? Who or what triggers the incident? What capacity and motive does the threat actor have?
- **Vulnerability:** What weakness/vulnerability can be exploited? What makes the threat scenario feasible?
- **Security incident on assets:** What assets are affected? What security breaches occur? Is there a breach of one or more of the CIA security objectives for one or more assets?

- **Impacts:** Typical negative impacts are lost profits, outcome/impairment of services, legal consequences, lost reputation, cost of recovery.
- **Existing security controls:** Can the incident be prevented? Is there already something that can stop or slow down the threat at one of the steps to prevent the incident?
- **Existing detection methods:** How can the incident be detected if it occurs?
- **Likelihood:** Estimate of how likely the event is to occur.
- **Impact:** Estimate of how serious the impacts of the incident will be.
- **Risk level:** Determine the level of risk.
- **New controls:** Suggestions for how new controls can reduce likelihood and/or impact.
- **New likelihood:** Changed likelihood estimate with assumed new controls.
- **New impact:** Changed impact estimate with assumed new controls.
- **New risk level:** Changed risk level with assumed new controls.

19.5 Risk Analysis

Risk analysis can be performed once risks have been identified, where each risk is described with threat, vulnerability, and incident on assets, among other things.

Practical risk analysis typically considers two factors to determine the level of each risk. These factors are estimated *likelihood* (frequency/odds) of each type of incident and estimated *impact* (severity) on assets resulting from an incident. Figure 19.2 illustrates this principle.

There are many ways and models for representing likelihood and impact, and for determining risk levels. Below we describe examples of a qualitative, quantitative, and relative models for risk analysis. Simple spreadsheets for risk analysis are available on the website of the Digital Security Group at UiO: https://www.mn.uio.no/ifi/english/research/groups/sec/instructional

19.5.1 Qualitative Risk Analysis

Qualitative risk analysis is based on levels of likelihood and impact expressed with descriptive words, and optionally associated with ordinal numbers. The number of levels and their respective interpretations can be specified according to what seems appropriate for cyber risk, or according to what the organization typically uses in other areas of security and safety. Figure 19.8 shows an example of a qualitative likelihood range with five levels and their interpretations.

Likelihood	Interpretation
(5) Certain	There are motivated threat actors who can easily achieve attack objectives by implementing the assessed threat scenario for this risk. An incident may already have happened, or will occur shortly, probably within a week.
(4) Likely	Motivated threat actors are likely to achieve attack objectives by executing the identified threat scenario. An incident could occur within a couple of months.
(3) Possible	It may be possible for threat actors to achieve attack objectives using the identified threat scenario. An incident of this type may occur every two years.
(2) Unlikely	Threat actors have little opportunity to achieve attack objectives using the identified threat scenario. It may take decades before an incident occurs.
(1) Rare	Threat actors have no practical opportunity to achieve attack objectives using the identified threat scenario. An incident of this type will probably never occur.

Fig. 19.8 Qualitative likelihood of a security incident

The term "likelihood" may also have the meaning of probability which is statistical, or chance which is not necessarily interpreted statistically. The difficulty of interpreting "likelihood" related to risk analysis is discussed Douglas Hubbard's *"The Failure of Risk Management: Why It's Broken and How To Fix It"* [58].

Figure 19.9 provides an example of a qualitative impact range with five levels and their interpretations.

Impact	Interpretation
(5) Disastrous	Disastrous harm to assets, services paralyzed, disastrous financial loss and possible bankruptcy. Recovery requires long-term work with large resources. External functions that depend on the business may lapse for a long period.
(4) Major	Major harm to assets, serious service interruption and major financial loss. Considerable resources required to handle the incident. Functions outside the affected business may be adversely affected, but without long-term impacts.
(3) Significant	Significant harm to assets that can result in significant service interruption and significant financial loss. Recovery and service continuity require significant work. Probably little impact on functions outside the affected business.
(2) Minor	Minor harm to assets or operations, and probably little or no service interruption. Just small financial loss. Can be handled with moderate resources.
(1) Insignificant	Insignificant harm to assets, and without any service interruption. The incident is easily handled as part of routine operations. Little or no financial loss.

Fig. 19.9 Qualitative impact of a security incident

Qualitative risk analysis is typically based on a two-dimensional matrix called *heat map*, where likelihood and impact each constitute their own dimension, as shown in Fig. 19.10. High risk is "hot" and therefore colored red, whereas low risk is "cool" and therefore colored green.

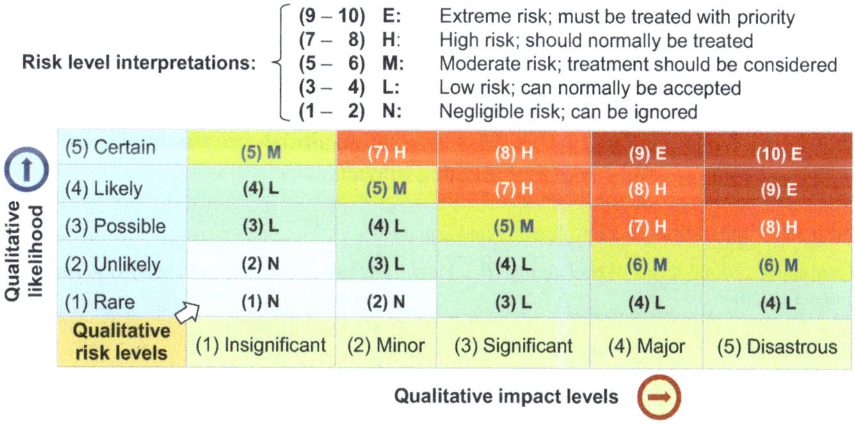

19.5 Risk Analysis

Risk level interpretations:	(9 – 10) **E:**	Extreme risk; must be treated with priority				
	(7 – 8) **H:**	High risk; should normally be treated				
	(5 – 6) **M:**	Moderate risk; treatment should be considered				
	(3 – 4) **L:**	Low risk; can normally be accepted				
	(1 – 2) **N:**	Negligible risk; can be ignored				

Qualitative likelihood						
(5) Certain	(5) M	(7) H	(8) H	(9) E	(10) E
(4) Likely	(4) L	(5) M	(7) H	(8) H	(9) E
(3) Possible	(3) L	(4) L	(5) M	(7) H	(8) H
(2) Unlikely	(2) N	(3) L	(4) L	(6) M	(6) M
(1) Rare	(1) N	(2) N	(3) L	(4) L	(4) L
Qualitative risk levels	(1) Insignificant	(2) Minor	(3) Significant	(4) Major	(5) Disastrous

Qualitative impact levels →

Fig. 19.10 Example heat map (risk matrix) for qualitative risk analysis

The risk matrix is a lookup table, where the estimated likelihood and estimated impact are used to determine the level of risk. Each organization must choose an appropriate distribution of risk levels and populate the risk matrix accordingly, where Fig. 19.10 is just an example. For multiple separate risk analyses to be comparable, it is necessary that the organization settles on a specific risk matrix that is used for all risk analyses.

Risk matrices can also be populated with ordinal numbers in the range [1, 10] as shown in Fig. 19.10. A simple approximative way of determining risk levels is to calculate the risk level as the sum of the ordinal numbers according to the formula below. Let P_{Qual} denote qualitative likelihood, let V_{Qual} denote qualitative impact, and let R_{Qual} denote qualitative risk. Then the ordinal qualitative risk level R_{Qual} is the sum of qualitative likelihood and impact:

$$R_{Qual} = P_{Qual} + V_{Qual}$$

By expressing likelihood (probability) and impact as levels on the scale 1–5, a simple calculation of risk levels can be done using the formula above, which gives risk levels on a scale of 2–10. This way of calculating qualitative risk levels is logically sound when the ordinal number for qualitative likelihood is interpreted as a function of the logarithm of probability (relative frequency), and the ordinal number for qualitative impact is interpreted as a function of the logarithm of relative impact. This is explained in the next section.

19.5.2 Relative Risk Analysis

When calculating risk levels in relative terms, each level of likelihood and impact is expressed with a relative value in the interval [0, 1]. In addition, to make it more understandable, descriptive words can be assigned to each relative level of likelihood and impact, in the same way as for qualitative risk analysis.

Let likelihood be a statistical probability which is already a relative value. In addition, impact can be expressed as a relative value in the interval [0, 1]. It is natural that both probability and impact levels are logarithmically distributed, meaning that the difference between each level amounts to one order of magnitude, i.e. a tenfold difference. Table 19.1 shows the relationship between ordinal qualitative levels and relative levels of probability and impact.

Table 19.1 Relative levels of probability and impact

Qualitative ordinal levels of probability (P_{Qual}) and impact (V_{Qual})	Relative levels of probability (P_{Rel}) and impact (V_{Rel})
5	$1.0000 = 10^0$
4	$0.1000 = 10^{-1}$
3	$0.0100 = 10^{-2}$
2	$0.0010 = 10^{-3}$
1	$0.0001 = 10^{-4}$

Relative risk calculation is multiplicative and is done according to the formula below. Let P_{Rel} denote probability (relative likelihood), let V_{Rel} denote relative impact, and let R_{Rel} denote relative risk. Then the relative risk R_{Rel} is the product of relative likelihood and impact:

$$R_{Rel} = P_{Rel} \times V_{Rel}$$

This way of calculating relative risk is logically sound and formally correct. However, relative risk is difficult to interpret intuitively. For better understanding, relative risk levels can be linked to qualitative descriptions of risk as words, so that relative risk calculations can also be interpreted qualitatively.

The ordinal qualitative risk level is a function of the logarithm of the relative risk level according to the formula below. We define the following variables:

$P_{Qual} = \log(P_{Rel}) + 5$	Qualitative likelihood
$V_{Qual} = \log(V_{Rel}) + 5$	Qualitative impact
R_{Qual}	Qualitative risk
R_{Rel}	Relative risk

Then the qualitative risk level is calculated according to:

$$R_{Qual} = \log(R_{Rel}) + 10 = \log(P_{Rel} \times V_{Rel}) + 10 = \log(P_{Rel}) + \log(V_{Rel}) + 10 = P_{Qual} + V_{Qual}$$

Table 19.2 shows how qualitative risk levels can be expressed as relative risk levels calculated by using the above formulas.

19.5 Risk Analysis

Table 19.2 Correlation between qualitative and relative risk levels

Qualitative risk levels (R_{Qual})	Relative risk levels (R_{Rel})
10	$1.00000000 = 10^0$
8	$0.01000000 = 10^{-2}$
6	$0.00010000 = 10^{-4}$
4	$0.00000100 = 10^{-6}$
2	$0.00000001 = 10^{-8}$

Impacts can have different aspects, as described in Sect. 19.4.3. Each impact aspect can be expressed as a relative value in the interval [0, 1]. In the event that a risk has several impact aspects, it is necessary to calculate an overall level of impact. Various impact aspects can be, for example, lost revenue and damaged reputation. Let V express the overall level of impact, and V_i express the level for a specific impact aspect i.

If each impact aspect is considered independent, the overall relative impact valuer can be calculated as the sum of the impact values from each aspect according to the following formula:

$$V = v_1 + v_2 + v_3 + \ldots$$

If the impact aspects are considered dependent, the overall impact value can be set equal to the greatest of the various impact values according to the following formula:

$$V = \max[v_1, v_2, v_3, \ldots]$$

For example: If lost revenue is the result of damaged reputation, these two aspects can be considered dependent, and overall impact can be set equal to the greatest of the two.

19.5.3 Quantitative Risk Analysis

Quantitative risk analysis is about representing risk in absolute monetary values. There are several different models for quantitative calculation of risk. Below we describe a simple model.

Let the likelihood be denoted as P (for probability) with likelihoods in the in the interval [0, 1]. This is interpreted as the relative frequency of a certain type of incident per year so that $P = 0.5$ means, for example, that the incident is expected to occur every two years. Let impact be denoted as V (for value) with values from zero to any monetary value.

The risk level R is calculated as expected loss (per year) in absolute monetary value. Let P denote likelihood and V denote impact value, then the risk value R is calculated as:

$$R = P \times V$$

Quantitative risk analysis is intriguing because on the one hand, quantitative risk can easily be assessed in the context of budgeting security activities in an enterprise. However, on the other hand, there is typically considerable uncertainty associated with the quantitative input values, so that the calculated quantitative risk levels are necessarily also infested with great uncertainty.

Quantitative risk calculation can be linked to qualitative risk by using ordinal levels of probability and impact from Table 19.1. If we define a specific frequency for the highest probability, then frequencies for the lower probabilities will follow naturally. Likewise, by choosing a specific dollar amount for the highest impact, the lower impacts will follow naturally.

For example, let the highest likelihood frequency (level 5) be that an event occurs 50 times per year, or said differently, probably within a week. By multiplying by the relative probability of level 4 from Table 19.1, the frequency for level 4 will be 5 times per year, meaning that an incident probably will occur within a couple of months. From this example, Table 19.3 shows how qualitative probability levels correspond to quantitative probabilities.

Table 19.3 Quantitative probability frequencies

Qualitative ordinal likelihood level	Quantitative probability frequencies based on a maximum expected number of 50 incidents per year
5	$50 \times 10^0 = 50$ times per year, i.e. approximately within a week
4	$50 \times 10^{-1} = 5$ times per year, i.e. within a couple of months
3	$50 \times 10^{-2} = 0.5$, meaning that it could happen every 2 years
2	$50 \times 10^{-3} = 0.05$, meaning that it could happen every 20 years
1	$50 \times 10^{-4} = 0.005$, meaning every 200 years, or probably never

The risk analysis team must specify the frequency corresponding to the highest likelihood level. The example in Table 19.3 uses 50 times per year as the highest likelihood, but it could be 100 times or maybe 10 times. The frequencies corresponding to the other likelihood levels then follow mathematically on a logarithmic scale.

Similarly for impacts, suppose as an example that the highest impact (level 5) is $10 million. By multiplying by relative impact level for ordinal impact level 4 from Table 19.1 the impact corresponding to level 4 will then be $1 million. As a generalization of this example, Table 19.4 shows how qualitative impact levels correspond to quantitative impact values.

Table 19.4 Quantitative impact levels

Qualitative ordinal impact levels	Quantitative impact values based on a company's maximum market valuation $10 million
5	$10 million $\times 10^0 = \$10,000,000$.
4	$10 million $\times 10^{-1} = \$1,000,000$
3	$10 million $\times 10^{-2}. = \$100,000$
2	$10 million $\times 10^{-3}. = \$10,000$
1	$10 million $\times 10^{-4}. = \$1,000$

As a simple example, assume a company with a market valuation of $10 million, and assume that the company goes bankrupt if the incident with the highest impact occurs, i.e. the share value goes to zero. Let the likelihood of an incident have qualitative ordinal number 3, and the incident's impact have qualitative ordinal likelihood 4. By multiplying the respective relative values, the quantitative risk level is calculated as:

$$\text{Risk level} = P \times V = 0.5 \times 1,000,000 = \$500,000.$$

19.6 Risk Evaluation and Reporting

Each company must assess for itself what they consider to be the highest impact. For a relatively large company, for example, the highest impact can be estimated to $10 million, which means that the lower impacts follow naturally with 1 order of magnitude less for each level. The lowest impact, i.e. with ordinal level 1, would then correspond to impact value $1,000. This is a convenient way for companies to adapt risk analysis to their size and market valuation.

Of course, if for a given risk the estimated impact V is bankruptcy, and the estimated likelihood P is once per week, it does not mean that the company will go bankrupt every week. In addition, if a lesser risk does materialize, it typically only happens once, since the company will learn from the incident and will recover to a more secure state. For such risks, the expected loss is equal to the estimated impact V, where the estimated likelihood P indicates how soon that loss is predicted to materialise. Anyway, their product indicates the severity of that risk relative to other risks.

19.6 Risk Evaluation and Reporting

Risk evaluation is based on the outcome of risk analysis to prioritise risks for later treatment. The evaluation is primarily based on ranking of calculated risk levels, where the greatest risks naturally stand out.

Visualisation of risk is important for decision-makers to understand and decide what should be prioritised. Figure 19.11 gives an example of how risk can be visualized in terms of probability and impact, and shows nine hypothetical risks as numbered circles.

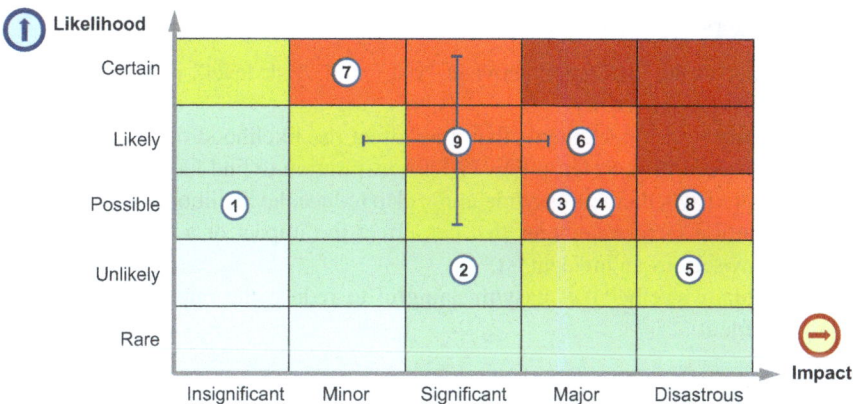

Fig. 19.11 Heatmap visualization of risks

The threshold level for acceptable risk and any other criteria should be included in the evaluation. In Fig. 19.11 the threshold level can be drawn as a diagonal from top left to bottom right. For example, risks 3 and 4 on the heatmap have equal probability and impacts, which makes it logical that they are given the same priority. Although risks 3 and 7 have the same level, they have different probability and impacts, which may be a reason for assigning different priorities.

The heatmap visualization of Fig. 19.11 can be interpreted as qualitative in the sense that each risk is placed in the middle of a "qualitative" field. If the risk analysis is quantitative, placing a risk in the figure is not bound by being in the middle of a field, but will have a free placement based on relative or quantitative probability and impact.

If it is desirable to visualise the degree of uncertainty surrounding the risks, a risk circle may also have a horizontal line to express uncertainty about impact, or a vertical line to express uncertainty about likelihood, as shown for risk 9 in Fig. 19.11. The length of a line represents the degree of uncertainty. There are many other ways of representing uncertainty, which to varying degrees make it intuitively easy to understand uncertainty. Although the representation of uncertainty brings more detail to the visualization, it is of little use if those reading the report do not understand what it means. It is better to use a simple visualization that people understand than an advanced visualization that people do not understand. Communication of risk is important for good overall risk management. It is therefore important to think about how the visualisation can best help communicate the results of the risk assessment.

The result of the risk evaluation is typically communicated with a report describing the outcome of the risk assessment and how identified and analysed risks are prioritised according to a threshold value and other criteria. Decision point 1 (described in Sect. 19.2.2) determines whether the report can be accepted, or whether parts of the risk assessment need to be revised.

19.7 Tasks

1. Risk factors

    (a) What is the difference between "risk" and "risk level" in the context of information security?
    (b) Mention factors/elements that may affect the likelihood of a security incident occurring. Refer to Fig. 19.3 for inspiration to find factors.
    (c) Explain whether it is easy/meaningful to reduce the likelihood of an incident.
    (d) Mention factors/elements that can affect the impact of a security incident (how serious an incident is).
    (e) Explain whether it is easy/meaningful to reduce the impact of a security incident.

2. Risk aspects

    (a) What is meant by an incident resulting in the breach of a security goal?
    (b) Mention relevant impact aspects of a security incident.
    (c) In the event that a security incident has different impact aspects – how should these be included in the analysis of the incident's risk?

3. Risk communication

    (a) Why is it important to communicate threat/risk assessments?
    (b) Who should threat/risk assessments be communicated to?
    (c) Mention factors that contribute to good communication of threat/risk assessments.

References

1. R. Howard, *Cybersecurity First Principles* (Wiley, 2023)
2. ISO/IEC, 27000:2018 Information security management systems, Overview and vocabulary, 2018
3. NIST, *The NIST Cybersecurity Framework (CSF) 2.0* (National Institute of Standards and Technology, 2024)
4. ISO/IEC, 27002:2022 Information security, cybersecurity and privacy protection — Information security controls, 2022
5. ISO/IEC, *27001:2022 Information Security Management System*. International Organization for Standardization, 2022
6. Center for Internet Security, *CIS Critical Security Controls v8*, 2021. [Online]. Available: https://www.cisecurity.org/controls
7. NIST, *SP 800–53 Rev. 5: Security and Privacy Controls for Information Systems and Organizations*, 2020
8. L. Hunnebeck, *ITIL Service Design (ITIL Service Lifecycle)*, 2nd edn. (TSO, The Stationery Office, Norwich, 2011)
9. NIST, Cyber Security Framework (CSF) 2.0, 2024. [Online]. Available: https://www.nist.gov/cyberframework
10. S. Yu, *Cyber Defense Matrix: The Essential Guide to Navigating the Cybersecurity Landscape* (JupiterOne Press, 2022)
11. ITU, *X.800: Security Architecture for Open Systems Interconnection* (International Telecommunication Union, 1991)
12. NIST, SP 800-161 Rev. 1: Cybersecurity Supply Chain Risk Management Practices for Systems and Organization, 2022
13. Department of Homeland Security, *Increasing Threats of Deepfake Identities—Phase 2: Mitigation Measures* (DHS, 2022)
14. W. Diffie, M.E. Hellman, New directions in cryptography. IEEE Trans. Inf. Theory **22**(6), 644–654 (1976)
15. S. Singh, *The Code Book* (Doubleday, 1999), pp. 279–292
16. R.L. Rivest, A. Shamir, L. Adleman, A method for obtaining digital signature and public-key cryptosystems. Commun. ACM **21**(2), 120–126 (1978)
17. A. Jøsang, D. Povey, A. Ho, What You See is Not Always What You Sign, in *Proceedings of the Australian UNIX and Open Systems*., Melbourne, 2002
18. A. Kerckhoffs, La cryptographie militaire. Journal des sciences militaires **IX**, 161–191 (1883)
19. D. Kahn, *The Codebreakers*, 2nd edn. (SCRIBNER, New York, 1996)
20. C. Shannon, A communication theory of secrecy systems. Bell Syst. Tech. J. **28**, 656–715 (1949)

21. P. Miranda, M. Siekkinen, TLS and energy consumption on a mobile device: A measurement study, in *011 IEEE Symposium on Computers and Communications (ISCC)*, Kerkyra, 2011
22. S. Nakamoto, *Bitcoin: A Peer-to-Peer Electronic Cash System*, 2009
23. S. Chamanara, S.A. Ghaffarizadeh, K. Madan, The environmental footprint of bitcoin mining across the globe: Call for urgent action. Earth's Fut. (2023)
24. NIST, *SP 800–57 Part 1 Rev. 5: Recommendation for Key Management: Part 1—General* (National Institute of Standards and Technology, 2020)
25. ITU, *Recommendation X.509, The Directory: Authentication Framework (also known as ISO/IEC 9594–8:2020)* (International Telecommunications Union, Telecommunication Standards Sector (ITU-T), 2019)
26. IETF, *Internet X.509 Public Key Infrastructure Certificate and Certificate Revocation List (CRL) Profile* (Internet Engineering Task Force, 2008)
27. NIST, NIST SP 800–207 Zero Trust Architecture, 2020
28. A10 Networks, Zero Trust is Incomplete Without TLS Decryption, 2020
29. NCSC, *TLS Interception: Considerations and Preconditions for the Deployment of TLS Interception (ver. 1.1, 2020)* (National Cyber Security Centre, 2017)
30. NIST, *SP 1800-16B Securing Web Transactions; TLS Server Certificate Management*, 2020
31. T. Tucker, H. Searle, K. Butler, P. Traynor, Blue's clues: Practical discovery of non-discoverable Bluetooth devices, in *2023 IEEE Symposium on Security and Privacy (SP)*, (2023)
32. NIST, *SP 800-63A: Enrollment & Identity Proofing*, 2024
33. ANSSI-BSI, *Remote Identity Proofing* (ANSSI-BSI Joint Release, 2023)
34. C.W. Munyendo, P. Mayer, A.J. Aviv, "I just stopped using one and started using the other": Motivations, Techniques, and Challenges When Switching Password Managers, in *30th ACM Conference on Computer and Communications Securit (CCS'23)*, (Copenhagen, 2023)
35. H. Flanagan, *Government-Issued Digital Credentials and the Privacy Landscape* (OpenID Foundation, 2023)
36. Ministry of Electronics and Information Technology, *Standards and Specification for e-Pramaan: Framework for e-Authentication* (Government of India, 2014)
37. Digital IID System, *Trusted Digital Identity Framework - Part 02 Overview* (The Australian Government, 2023)
38. D.A. Bell, L.J. LaPadula, *Secure Computer Systems: Mathematical Foundations and Model* (MITRE Corporation, 1973)
39. R. Binns, Tracking on the web, Mobile and the internet of things. Foundations and Trends in Web Science **8**(1–2), 1–113 (2022)
40. S. Zuboff, *The Age of Surveillance Capitalism: The Fight for a Human Future at the New Frontier of Power* (PublicAffairs, 2019)
41. ENISA, *Cybersecurity Culture in Organisations* (European Union Agency for Cybersecurity, 2018)
42. KnowBe4, *The Security Culture How-To Guide* (KnowBe4 Research, 2023)
43. P. Martin, *Rules of Security: Staying Safe in a Risky World* (Oxford University Press, 2019)
44. S. Dekker, *The Field Guide to Understanding 'Human Error'*, 3rd edn. (Taylor & Francis, 2017)
45. A. Whitten, D. Tygar, Why Johnny Can't Encrypt: A Usability Evaluation of PGP 5, in *Proceedings of the 8th USENIX Security Symposium*, (1999)
46. C. Espinosa, *The Smartest Person in the Room: The Root Cause and New Solution for Cybersecurity* (Lioncrest Publishing, 2021)
47. K. Laakso, *Organizing Civil Preparedness in FIinland, Sweden and Norway* (Tampere University, 2019)
48. NIST, *SP 800-61r2: Computer Security Incident Handling Guide* (National Institute of Standards and Technology, 2012)
49. SANS, Incident Handler's Handbook, 2012. [Online]. Available: https://www.sans.org/white-papers/33901/
50. NIST, *Adversarial Machine Learning: A Taxonomy and Terminology of Attacks and Mitigations (NIST AI 100-2e2023)* (NIST Trustworthy and Responsible AI, 2023)

51. C. Szegedy et al., Intriguing properties of neural networks, in *2nd International Conference on Learning Representations (ICLR 2014)*, (2014)
52. Sekoia, My Tea's not cold. An overview of China's cyber threat, 2023
53. IT Governance Institute, *Information Security Governance: Guidance for Boards of Directors and Executive*, 2nd edn. (ISACA, 2006)
54. ISO/IEC, *27004:2016 Monitoring, Measurement, Analysis and Evaluation* (International Organization for Standardization, 2016)
55. ISO, *31000:2018 Risk Management* (International Organization for Standardization, 2018)
56. ISO/IEC, *27005:2022 Guidance on Managing Information Security Risks* (International Organization for Standardization, 2022)
57. NIST, *SP 800–39 Managing Information Security Risk* (National Institute for Standards and Technology, 2011)
58. D. Hubbard, *The Failure of Risk Management: Why It's Broken and How to Fix It*, 2nd edn. (Wiley, 2020)
59. NIST SP 800–161 Rev. 1, Cybersecurity Supply Chain Risk Management Practices for Systems and Organizations, 2022
60. Y. Mirsky et al., The threat of offensive AI to organizations. Comput. Secur. **124** (2023)

Index

A

Access control, 22, 192, 206, 274
Accountability, 20, 166, 227, 356, 374
Accountable, 230
Address Space Layout Randomization (ASLR), 53
Advanced Encryption Standard (AES), 65, 88, 89, 91, 103
Advanced Persistent Threat (APT), 222, 338, 344, 350, 366, 370
Application Security Verification Standard (ASVS), 261
Artificial Intelligence (AI), 33, 183, 240, 274, 290, 321
Asset, 2, 388, 406, 418
Attribute-based Access Control (ABAC), 211
Attribution, 342, 350
Audit, 31, 230, 250, 301, 319, 379, 395, 414
Australia, 205
Authentication, 17, 70, 165, 191
Authorization, 21, 191, 274
Availability, 17, 37, 233, 257

B

Backdoor, 40, 331
Bell-LaPadula, 208
Biometrics, 166, 175, 179
Bitcoin, 95, 114
Blockchain, 94, 114
Bluetooth, 32, 93, 125, 150
Botnet, 35, 38
Buffer overflow, 50
Building-Security-In Maturity Model (BSIMM), 244
Business Continuity Planning (BCP), 305
Business Impact Analysis (BIA), 305
Backup, 7, 37, 305

C

CA authorization (CAA), 114
Caesar cipher, 81
Central Processing Unit (CPU), 34, 58, 130
Certificate, 77, 80, 99, 107, 129, 140, 149
Certificate Authority (CA), 107
Certificate Revocation List (CRL), 113
China, 31, 96, 217, 369
CIA security goals, 4, 14
Cipher Block Chaining (CBC), 69
CIS controls, 390
Closed-Circuit TV (CCTV), 274
Cloud security, 266
Compliance, 6, 228, 355, 398, 420
Computer Emergency Response Team (CERT), 301, 306, 359, 383, 390
Computer Fraud and Abuse Act (CFAA), 360
Computer security, 1
Computer Security Incident Response Team (CSIRT), 306, 383, 390
Confidentiality, 14, 64, 126, 134
Contingency planning, 8, 301, 303
Control, 7, 8, 12, 385, 390, 398, 413
Covert channel, 60
Cryptocurrency, 94
Cryptography, 63
Cryptovirus, 37
CTR (Counter Mode), 68
CVE/CVSS, 46
Cyber defense matrix, 11, 388

Cyber Incident Response Plan (CIRP), 306
Cyber insurance, 411, 414
Cyber kill chain, 339
CyberOps, 337, 348, 358
Cyber range, 318
Cybersecurity, 3, 269, 340, 381
Cybersecurity Framework (CSF), 9, 302, 361, 378, 383, 388, 390, 399
Cybersecurity Information Sharing Act (CISA), 244, 361
Cyber Threat Intelligence (CTI), 309, 318, 330, 337, 340, 344
Cyber warfare, 348

D
Dark web, 28, 37
Data controller, 227
Data Encryption Standard (DES), 89
Data processing agreement, 227
Data processor, 234
Data Protection Impact Assessment (DPIA), 235
Data subject, 356
Deepfake, 33, 166, 326
Defense Advanced Research Projects Agency (DARPA), 123
Demilitarized Zone (DMZ), 139
Deterrence, 351
DevOps, 268
Diffie-Hellman, 71, 80, 88, 90, 103, 128
Digital forensics, 314
Digital security, 1
Digital signature, 19, 57, 77, 94, 108, 175
Discretionary Access Control (DAC), 206
Distributed Denial of Service (DDoS), 12, 17, 34, 38, 137
Domain Name System (DNS), 114, 119, 123, 139
Domain Name System Security Extensions (DNSSEC), 112, 121
Downgrade attack, 161
Doxing, 37
Drive-by, 28, 41
Dynamic Host Configuration Protocol (DHCP), 120

E
eID, 178, 185, 199
eIDAS, 112, 187, 201, 203
Electric power supply, 279

Electronic Code Book (ECB), 66
Elliptic Curve Cryptography (ECC), 91, 103
Energy consumption, 95
Enigma, 85
Equal Error Rate (EER), 183
Espionage, 349
Ethical hacking, 246, 248, 316
EU, 141, 187, 224, 318, 362
European Union Agency for Cybersecurity (ENISA), 283, 363, 419
Exploit, 39, 50, 53, 339

F
Fail safe and fail secure, 273
Faraday cage, 144, 162, 278
Federal Information Security Management Act (FISMA), 361
FEIDE, 204
FIDO, 175
Fingerprint, 177, 180
Firewall, 126, 135
Firmware, 40, 44, 59
FMR/FNMR, 182
Forum of Incident Response and Security Teams (FIRST), 46, 347, 359
Forward secrecy, 77, 80, 150
Four P's, 10, 414
France, 204
Fuzzing, 39, 248, 351

G
General Data Protection Regulation (GDPR), 38, 141, 180, 220, 224, 367, 372
Geo-blocking, 132
Germany, 178, 204
Gnu Privacy Guard (GPG), 112
Governance, 8, 377, 381

H
Hacking, 25, 246, 316
Hash function, 69
Health Insurance Portability and Accountability Act (HIPAA), 362
Honeypot, 139
HTTP Strict Transport Security (HSTS), 130
Hypertext Transfer Protocol Secure (HTTPS), 123, 129, 139
Hypervisor, 49, 54, 267

Index 435

I
Identity, 27, 194
Identity and Access Management (IAM), 22, 191
Identity federation, 186, 196, 266
Identity Provider (IdP), 195, 199, 200, 202, 203
Identity theft, 18, 27, 33, 130, 163, 257, 292
ID-portal, 205
IEC, 382, 392
Impact, 6, 235, 259, 302, 305, 407, 420
IMSI catcher, 35, 158, 161
Incident, 6, 302
Incident management, 301, 307
India, 188, 200, 205
Information security, 3, 377
Information Security Management System (ISMS), 8, 377, 378, 394
Information Systems Audit and Control Association (ISACA), 379
Information Technology Infrastructure Library (ITIL), 10
Insider, 34, 286
Integrity (digital), 15
Intel AMT, 60
International Telecommunications Union (ITU), 123
The Internet Corporation for Assigned Names and Numbers (ICANN), 119
Internet Engineering Task Force (IETF), 107, 108, 210
Internet of Things (IoT), 32, 38, 47, 93, 152, 318
Internet Protocol (IP), 121
Internet Protocol Security (IPSec), 130
Intrusion Detection System (IDS), 137, 139
IP address, 118
IP rotation, 119
Ireland, 217, 226
ISO, 4, 9, 17, 189, 382, 392, 410
IT security, 1

J
JavaScript, 41
Jurisdiction, 119, 123, 234, 355

K
Kerckhoffs's principle, 83, 157
Kevin Mitnick, 290
Keylogger, 37, 167

L
Laws, 12, 23, 32, 216, 224, 234, 319, 327, 355
Lightweight cryptography, 94
Logic bomb, 41

M
MAC
 address, 125
 mandatory access control, 61
 message authentication code, 70
Machine learning, 321
Macro, 39
Malware, 35
Managed Detection and Response (MDR), 314
Managed Security Service Provider (MSSP), 306, 346, 383
Managed Security Services (MSS), 314
Management, 240, 289, 377, 381, 395, 400
Man-in-the-middle (MitM), 18, 27, 73, 130, 149, 158, 179
Maturity level, 12, 341, 380, 387, 400
Measurement, 380
Microprocessor, 44, 48, 55
MISP, 313, 346
MITRE ATT&CK, 343
Mobile network, 32, 144, 155
Multi-Factor Authentication (MFA), 18, 184

N
National Cyber Security Center (NCSC), 306, 363, 366
National Institute of Standards and Technology (NIST), 10, 46, 89, 100, 361, 407
National Security Agency (NSA), 104, 278, 351, 359
NATO, 84, 278, 348, 350, 359
Natural Language Processing (NLP), 325, 330
Nigeria, 204
No eXecute (NX), 53
Non-repudiation, 19, 21, 79
Norway, 204
Norwegian Health Network (NHN), 212, 213

O
OAuth, 162, 199, 210
Online Certificate Status Protocol (OCSP), 113
OpenID Connect (OIDC), 199

Open-source intelligence (OSINT), 33, 222, 317, 339, 347
Open Systems Interconnection (OSI), 123
Open Web Application Security Project (OWASP), 259
Operating system, 43, 48, 58
OTP
  one-time pad, 84
  one-time password, 174

P

Passkeys, 172, 175
Password, 18
Password manager, 171
Password spraying, 29, 119
Patching, 249
Pegasus, 37
Pentesting, 248, 316, 319
People-process-technology, 11
People-product-partner-process, 10
Phishing, 18, 27, 290
Phishing resistant, 175
Physical security, 2, 9, 29, 271, 293
Post-quantum cryptography (PQC), 91
Pretty Good Privacy (PGP), 112
Privacy by design, 243
Privilege level, 48
Public Key Infrastructure (PKI), 99, 105, 141

Q

Quishing, 27

R

RACI chart, 306
Random-Access Memory (RAM), 44, 316
Ransomware, 37
Readiness, 7, 301
Reconnaissance, 33, 317, 339, 343
Red-teaming, 317, 413
Relying party, 197
Responsibility, 12, 225, 227, 356
Return On Investment (ROI), 408, 413
Risk, 6, 13, 30, 161, 235, 259, 302, 405
Risk analysis, 412, 421
Risk management, 361, 380, 405
Rootkit, 40
RSA, 74, 88, 90, 103
Russia, 350, 352, 353, 366

S

Sabotage, 349
Safety, 2, 273, 286
Schrems, 234
Secure by design, 243
Secure Hash Algorithm (SHA), 90
Secure Sockets Layer (SSL), 93
Security, 2
Security Assertion Markup Language (SAML), 198
Security by obscurity, 83, 157
Security culture, 283
Security Development Lifecycle (SDL), 243, 253
Security Information and Event Management (SIEM), 330, 346
Security Operations Centre (SOC), 306, 314, 383, 390
Service Level Agreements (SLA), 31, 265, 319
Service Provider (SP), 195, 196, 202, 205
Shannon, 86, 321
Side channel, 45, 60
SIM swap, 163
Singapore, 204
Single Sign-On (SSO), 197
Social engineering, 25, 290
Software Assurance Maturity Model (SAMM), 244
Software Bill Of Materials (SBOM), 47
Spoofing, 18, 27, 257
Spyware, 37
SQL injection, 29, 261
Statement of Applicability (SoA), 395
Steganography, 133
STRIDE, 257
Structured Threat Information eXpression (STIX), 313, 346
Supply chain risk, 30, 47, 332

T

Tamper resistant and tamper proof, 57
Tar pit, 139
TEMPEST, 278
Threat, 5
Threat Intelligence-Based Ethical Red-Teaming (TIBER), 318
Threat modelling, 26, 255, 418
TLD, 119
TLS inspection, 140
Tracking, 35, 159, 218, 223, 247
Traffic-Light Protocol (TLP), 347

Index   437

Transmission Control Protocol (TCP), 121, 127, 132
Transport Layer Security (TLS), 77, 93, 110, 127, 139
Trojan, 39, 331, 332
Trust, 58, 112, 293
Trust anchor, 44
Trusted computing, 44, 57
Trusted Execution Environment (TEE), 57
Trusted Platform Module (TPM), 44, 57
Trust model, 111
TTP (in CTI), 342
Two-factor authentication (2FA), 18, 184
2G, 3G, 4G, 5G, 6G, 155

U

UK, 32, 141, 205, 288, 306, 327, 366, 392
Unified Extensible Firmware Interface (UEFI), 57
USA, 205, 234, 349, 354, 358, 383
User Datagram Protocol (UDP), 121

V

Virtual machine, 49, 54, 266
Virus, 36
Vulnerability, 5, 39, 46, 406

W

Weaponization, 321, 339, 350, 353
Web Application Firewall (WAF), 137
WebAuthn, 175
Wi-Fi, 44, 93, 125, 146
Worm, 29, 40

X

X.509, 107, 112, 123
XDR (Extended Detection & Response), 314
XSS (Cross Site Scripting), 29

Z

Zero-day vulnerability, 47
Zero trust, 135, 139

SPRINGER NATURE

GPSR Compliance

The European Union's (EU) General Product Safety Regulation (GPSR) is a set of rules that requires consumer products to be safe and our obligations to ensure this.

If you have any concerns about our products, you can contact us on ProductSafety@springernature.com

In case Publisher is established outside the EU, the EU authorized representative is:

Springer Nature Customer Service Center GmbH
Europaplatz 3
69115 Heidelberg, Germany

The manufacturer's authorised representative in the EU is Springer Nature Customer Service Centre GmbH, Europaplatz 3, 69115 Heidelberg, Germany. If you have any concerns regarding our products, please contact ProductSafety@springernature.com

Printed and bound by CPI Group (UK) Ltd, Croydon, CR0 4YY

25/03/2026

02078193-0015